Polymer Structure Characterization
From Nano to Macro Organization

Polymer Structure Characterization
From Nano to Macro Organization

Richard A Pethrick
Department of Pure and Applied Chemistry, University of Strathclyde, Glasgow, UK

RSCPublishing

ISBN-13: 978-0-85404-466-5

A catalogue record for this book is available from the British Library

© The Royal Society of Chemistry 2007

All rights reserved

Apart from fair dealing for the purposes of research for non-commercial purposes or for private study, criticism or review, as permitted under the Copyright, Designs and Patents Act 1988 and the Copyright and Related Rights Regulations 2003, this publication may not be reproduced, stored or transmitted, in any form or by any means, without the prior permission in writing of The Royal Society of Chemistry, or in the case of reproduction in accordance with the terms of licences issued by the Copyright Licensing Agency in the UK, or in accordance with the terms of the licences issued by the appropriate Reproduction Rights Organization outside the UK. Enquiries concerning reproduction outside the terms stated here should be sent to The Royal Society of Chemistry at the address printed on this page.

Published by The Royal Society of Chemistry,
Thomas Graham House, Science Park, Milton Road,
Cambridge CB4 0WF, UK

Registered Charity Number 207890

For further information see our web site at www.rsc.org

Preface

Behind the apparently innocuous smooth structure of hair or a polymer fibre lies a complex structure which dictates the physical properties of that material. This book attempts to give the reader the necessary background to understand the factors that influence molecular organization and control the way in which these structures are formed. The book is written to be useful as support material for undergraduate and postgraduate courses on molecular organization and structure. As the subtitle implies, in order to truly appreciate the factors that influence the properties of many molecular materials it is necessary to be able to observe the materials over length scales which range form nanometres to millimetres. Within this scale range many materials exhibit different levels of organization, and it is to understand the factors which control this structure building that is the aim of this book. The coverage of the book has been limited to consideration of the 'solid' state. Organization in the liquid state—colloids and lyotropic liquid crystals—has been included as it helps understand the way in which many biological systems are able to undertake self-assembly in solution prior to forming an ordered solid.

It is hoped that this book will aid the teaching of crystal growth in small molecules as well as polymers, development of an understanding of both the chemical and physical characteristics of liquid crystalline materials and provide the tools to attempt to rationalize the varied structures which nature creates. These topics are often covered as part of undergraduate courses in chemistry, physics and materials science. The more detailed discussion of the topics on polymer crystallization and morphology form part of postgraduate or advanced masters courses in materials science. This monograph does not attempt to produce a comprehensive review of the literature on these topics, but rather tries to illustrate some of the basic principles with selected examples. Large textbooks have been written on topics such as polymer crystallization, morphology, *etc.*, and it would be an impossible task to cover all aspects of the subject in detail in a small monograph. It is hoped, however, that this selected digest presents the topics at an understandable level and provides a good foundation upon which more detailed exploration of the literature can be based.

Similarly a number of the techniques used in the study of morphology and various related aspects have been summarized. Each technique is worth a volume in its own right and the reader is encouraged to consult more specialist texts to gain a greater insight into their use and applications. It is hoped that the material presented will provide the reader with a sufficient appreciation of the methods to be able see how the information they provide can be used to gain greater insight into the way molecules are organized within solids.

Morphology and structure in solids are the results of a delicate interplay of forces which act at atomic, molecular and macroscopic levels. Liquid crystalline materials have become of importance through their use in displays; however, the principles underlying their organization and self-assembly are very important in understanding how simple molecules behave as well as biomolecular systems.

The general structure of the monograph follows the format that has been used for a number of years in teaching these subjects at undergraduate and postgraduate level. Each chapter should builds on the previous chapters to help the reader gain an appreciation of the factors that are critical in determining that nature of the organization which is developed in a particular system. Whilst the thrust of the monograph is consideration of order; disordered systems play an important part in materials technology and the area of amorphous materials logically results from a combination of a number of factors influencing the 'structure' or rather the lack of it being developed in the solid.

To understand many of the topics covered in this book it is necessary to appreciate the way in which information has been obtained. Scattered through the book are sections on various experimental techniques. They have been introduced at appropriate points in the volume rather than, as is often done, being collected into a single chapter. It is hoped that this method of organization will be helpful. More detailed discussions of the methods are covered in specialist texts; however, it is hoped that the summaries presented here should give the reader sufficient understanding of the methods to be able to appreciate their use in the context of morphological investigations.

In preparing this monograph, a number of textbooks have been consulted and the arguments presented by certain authors have been adopted, where they present a clear and logical development of the topic. In particular the discussion of polymer crystal growth follows clearly that presented by Gedde in his textbook, *Polymer Physics*. For the interested reader a number of these excellent texts have been listed at the end of each chapter. Where appropriate in the text, specific examples of the research work at the University of Strathclyde have been included to assist with the discussion. This is primarily a teaching monograph and no attempt is made to be comprehensive in coverage of the literature or presenting all possible views on any particular topic. The author is very aware of the vast volume of information that is available on this general topic and only hopes that this simplified introduction will help students and researchers to make some progress in understanding this fascinating subject, the principles outlined in the monograph are generally applicable to all molecular systems; the principal differences arise as a result of the detailed balance between inter- and intramolecular contributions to the mean forces field. It is hoped that armed with the general introduction to the subject, the reader may feel better equipped to approach the more specialist texts and the vast quantity of literature that exists on this subject. If this objective has been accomplished then the book will have succeeded.

I would like to acknowledge the contribution which my colleagues, former colleagues and collaborators have made to educating me in various aspects of the topics covered: Stanley Affrossman, Frank Leslie, John Sherwood, Kevin Roberts, Randell Richards and Christopher Viney. The content of this book is purely my responsibility, but they introduced me to some of the topics and helped me develop my understanding of these areas.

R. A. Pethrick
Department of Pure and Applied Chemistry

Contents

Chapter 1 Concept of Structure–Property Relationships in Molecular Solids and Polymers

1.1	Introduction	1
1.2	Construction of a Physical Basis for Structure–Property Relationships	2
	1.2.1 Ionic Solids	2
	1.2.2 The Crystal Surface	5
	1.2.3 Molecular Solid	5
	1.2.4 Low Molar Mass Hydrocarbons	8
	1.2.5 Poly(methylene) Chains	10
1.3	Conformational States of Real Polymer Molecules in the Solid State	11
	1.3.1 Crystalline Polymers	11
	1.3.2 Disordered or Amorphous Polymers	12
1.4	Classification of Polymers	12
	1.4.1 What Factors Determine Whether a Polymer Will Form a Crystalline Solid or Not?	13
	Recommended Reading	14
	References	14

Chapter 2 Crystal Growth in Small Molecular Systems

2.1	Introduction	16
	2.1.1 Crystal Types	17
2.2	Crystallization	21
	2.2.1 Supersaturation and Crystallization	22
2.3	Nature of Crystal Structures: Morphology and Habit	23
	2.3.1 Morphology Prediction	24
2.4	Homogeneous Crystal Growth	25
	2.4.1 Empirical Description of Nucleation	27
	2.4.2 Stages of Crystal Growth	29
	2.4.3 Heterogeneous Crystal Growth	30

	2.4.4	Nucleation and Growth Rates	31
	2.4.5	Methods of Attachment to the Growth Surface	32
	2.4.6	Bravais–Friedel–Donnay–Harker Approach	32
	2.4.7	Periodic Bond Chains	33
	2.4.8	Attachment Energy	34
	2.4.9	Ising Model Surface Roughening	35
2.5	Sources of Nucleation Sites on Surfaces, Steps and Dislocations	36
	2.5.1	Two-Dimensional Nucleation	36
	2.5.2	Dislocations and Related Defects	37
	2.5.3	Screw Dislocation (BCF) Mechanism	39
	2.5.4	Rough Interface Growth (RIG) Mechanism	40
	2.5.5	Relative Rates of Crystal Growth	40
	2.5.6	Computer Prediction of Morphology	40
2.6	Macrosteps	42
	2.6.1	Impurities	42
2.7	Analysis of the Data from Step Growth	43
2.8	Refinements of the Theory	45
2.9	Methods of Microstructural Examination	46
Recommended Reading	49
References	49

Chapter 3 Liquid Crystalline State of Matter

3.1	Introduction	52
	3.1.1 The Liquid Crystalline State	52
	3.1.2 Historical Perspective	52
	3.1.3 Mesophase Order	53
	3.1.4 Nematic Liquid Crystals (N)	53
	3.1.5 Smectic Liquid Crystals	54
	3.1.6 Cholesteric Liquid Crystal (C)	55
3.2	Influence of Molecular Structure on the Formation of Liquid Crystalline Phases	55
	3.2.1 Influence of Chain Rigidity	55
	3.2.2 Influence of Size of Rigid Block	57
	3.2.3 Influence of Sequence Structure in Chain	58
	3.2.4 Variations Within a Homologous Series of Molecules	58
	3.2.5 Changes in Substituents	59
3.3	Common Features of Many Liquid Crystal Forming Molecules	59
	3.3.1 Nematic Liquid Crystals	60
	3.3.2 Influence of the Linking Group on the Thermal Stability of the Nematic Phase	62
	3.3.3 Terminal Group Effects	64

		3.3.4 Pendant Group Effects	65
		3.3.5 Terminal Substitution Effects	66
	3.4	Cholesteric Liquid Crystals	67
	3.5	Smectic Liquid Crystals	67
	3.6	Theoretical Models for Liquid Crystals	70
		3.6.1 Statistical Models	71
		3.6.2 Development of Statistical Mechanical Models	72
		3.6.3 Distributions and Order Parameters	72
	3.7	Elastic Behaviour of Nematic Liquid Crystals	75
	3.8	Computer Simulations	78
	3.9	Defects, Dislocations and Disclinations	79
	3.10	Applications	82
	3.11	Polymeric Liquid Crystals	82
	3.12	Polymeric Liquid Crystalline Materials	82
		3.12.1 General Factors Influencing Polymeric Liquid Crystalline Materials	83
		3.12.2 Main Chain Crystalline Polymers	84
		3.12.3 Side Chain Liquid Crystalline Polymers	89
		3.12.4 Nature of Flexible Spacer and its Length	89
		3.12.5 Nature of the Backbone	89
		3.12.6 Polymer Network Stabilized Liquid Crystal Phase	91
	3.13	Structure Visualization	92
	3.14	Conclusions	92
	Recommended Reading		92
	References		92

Chapter 4 Plastic Crystals

	4.1	Introduction	99
	4.2	Plastic Crystalline Materials	99
	4.3	Alkanes and Related Systems	103
	4.4	Conclusions	105
	Recommended Reading		105
	References		105

Chapter 5 Morphology of Crystalline Polymers and Methods for its Investigation

	5.1	Introduction	107
	5.2	Crystallography and Crystallization	107
	5.3	Single Crystal Growth	112
		5.3.1 Habit of Single Crystals	113
	5.4	Crystal Lamellae and Other Morphological Features	115
		5.4.1 Solution-Grown Crystals	115
		5.4.2 Chain Folding	115

Contents   xi

		5.4.3	Crystal Habit	116
		5.4.4	Sectorization	116
		5.4.5	Non-planar Geometries	116
	5.5	Melt Grown Crystals	117
		5.5.1	Melt-Crystallized Lamellae	117
		5.5.2	Polymer Spherulites	118
	5.6	Annealing Phenomena	123
	5.7	Experimental Techniques for the Study of Polymer Crystals	124
		5.7.1	Optical Microscopy	125
		5.7.2	Microtomes	126
		5.7.3	Basic Light Microscopy	127
		5.7.4	Light versus Electron Microscopy	128
		5.7.5	Phase Contrast Microscopy	129
		5.7.6	Polarized Light Microscopy	129
		5.7.7	Origins of Birefringence	130
		5.7.8	Orientation Birefringence	130
		5.7.9	Strain Birefringence	130
		5.7.10	Form Birefringence	131
		5.7.11	Polarization Colours	131
		5.7.12	Modulation Contrast Techniques	132
		5.7.13	Interference Microscopy	133
	5.8	Electron Microscopy	133
		5.8.1	Sample Preparation: Etching and Staining	133
	5.9	X-Ray Diffraction	134
	5.10	Raman Scattering and Phonon Spectra	135
	5.11	Degree of Crystallinity	135
		5.11.1	Density and Calorimetric Methods	136
		5.11.2	X-Ray Scattering	137
		5.11.3	General Observations	138
	5.12	Conclusions	138
	Recommended Reading	139
	References	139

Chapter 6 Polymer Crystal Growth

6.1	Introduction	141
	6.1.1	Thermodynamics of Polymer Molecule in the Melt	141
	6.1.2	Nucleation	142
6.2	Minimum Energy Conditions and Simple Theory of Growth	143
6.3	Nature of Chain Folding	146
6.4	Crystals Grown from the Melt and Lamellae Stacks	148
	6.4.1	Location of Chain Ends	150

	6.5 Crystallization Kinetics	150
	6.6 Equilibrium Melting Temperature	153
	6.7 General Avrami Equation	156
	6.8 Comparison of Experiment with Theory	159
	6.9 Growth Theories	159
	6.9.1 Lauritzen–Hoffman Theory	160
	6.9.2 Sadler–Gilmer Theory	169
	6.10 Crystallization via Metastable Phases	171
	6.11 Molecular Fractionation	172
	6.12 Orientation-Induced Crystallization	175
	Recommended Reading	176
	References	176

Chapter 7 Glasses and Amorphous Material

	7.1 Introduction	179
	7.2 Phenomenology of the Glass Transition	180
	7.2.1 Dynamic Mechanical Thermal Analysis	181
	7.2.2 Dielectric Relaxation Spectroscopy (DRS)	183
	7.2.3 Positron Annihilation Lifetime Spectroscopy (PALS)	189
	7.3 Free Volume and the Williams–Landel–Ferry Equation	190
	7.4 How Big is the Element That Moves in the T_g Process?	192
	7.5 Physical Characteristics of T_g	194
	7.5.1 Factors Influencing the Value of T_g	194
	7.5.2 Molar Mass Effects	194
	7.5.3 Plasticization Effect	195
	7.5.4 Incorporation of Comonomer and Blends	195
	7.5.5 Effects of Chemical Structure	196
	7.6 Kauzmann Paradox	198
	7.6.1 Pressure Dependence of the Glass Transition	198
	7.6.2 Physical Ageing	198
	7.7 Distribution of Free Volume in a Glass	200
	7.8 Fragility	202
	7.9 Theories of T_g	204
	Recommended Reading	205
	References	205

Chapter 8 Polymer Blends and Phase Separation

	8.1 Introduction	207
	8.2 Thermodynamics of Phase Separation	208

Contents xiii

		8.2.1	Thermodynamics of Polymer–Polymer Miscibility	209
		8.2.2	Enthalpy and Entropy Changes on Mixing	211
	8.3	Phase Separation Phenomena		213
		8.3.1	The Phase Diagram for Nearly Miscible Blends	213
	8.4	Parameters Influencing Miscibility		215
		8.4.1	Molar Mass Dependence of Phase Diagrams	215
		8.4.2	Effect of Pressure on Miscibility	217
		8.4.3	Addition of Block Copolymers	217
		8.4.4	Refinements of Theory	217
	8.5	Kinetics of Phase Separation: The Spinodal Decomposition		218
	8.6	Specific Examples of Phase-Separated Systems		219
		8.6.1	High-Impact Polystyrene	219
		8.6.2	Rubber Toughened Epoxy Resins	220
		8.6.3	Thermoplastic Toughened Epoxy Resins	222
		8.6.4	Epoxy Resins	223
	8.7	Block Copolymers: Polystyrene-*block*-Polybutadiene-*block*-Polystyrene (SBS) Block Copolymer		224
		8.7.1	General Characteristics	225
		8.7.2	Thickness of the Domain Interface	228
	8.8	Polyurethanes		229
	Recommended Reading			231
	References			231

Chapter 9 Molecular Surfaces

	9.1	Introduction		233
	9.2	Gibbs Approach to Surface Energy		233
		9.2.1	Contact Between a Liquid and a Surface	234
		9.2.2	Derivation of Young's Equation and Definition of Contact Angle	234
	9.3	Surface Characterization		238
		9.3.1	Classical Surface Assessment Methods, Contact Angle Measurements	238
		9.3.2	Visualization of the Polymer Surface	240
		9.3.3	Atomic Force Microscopy	244
	9.4	Spectroscopic Assessment of the Surface: Attenuated Total Reflection Infrared, Fluorescene and Visible Spectroscopy		247
	9.5	X-Ray and Neutron Diffraction Analysis		247
		9.5.1	Neutron and X-ray Reflectivity	247
	9.6	Ion Beam Analysis: Electron Recoil and Rutherford Backscattering		252

	9.7 Vacuum Techniques: X-ray Photoelectron Spectroscopy (XPS), Secondary Ion Mass Spectroscopy (SIMS), Auger Electron Spectroscopy (AES)	253
	9.7.1 X-Ray Photoelectron Spectroscopy	253
	9.7.2 Electron Mean Free Path, Attenuation and Escape Depth	258
	9.7.3 XPS Depth Profiling	262
	9.7.4 Secondary Ion Mass Spectrometry (SIMS)	264
	9.8 Fourier Transform Infrared (FTIR) Imaging	268
	Recommended Reading	269
	References	269

Chapter 10 Polymer Surfaces and Interfaces

10.1 Introduction	271
10.1.1 Crystalline Polymers	271
10.1.2 Amorphous Polymers	272
10.1.3 Polymer Blends	273
10.2 Theoretical Description of the Surface of a Polymer	273
10.2.1 Surface Tension of Homopolymers	273
10.2.2 Theories of Homopolymer Surface Tension	274
10.3 Surface Segregation	275
10.4 Binary Polymer Blends	276
10.5 End Functionalized Polymers	278
10.6 Phase Segregation and Enrichment at Surfaces	278
10.7 Electrohydrodynamic (EDH) Instabilities in Polymer Films	280
Recommended Reading	282
References	282

Chapter 11 Colloids and Molecular Organization in Liquids

11.1 Introduction	284
11.2 Ideal Non-mixing Liquids	285
11.3 Minimum Surface Energy Conditions	287
11.4 Langmuir Trough	289
11.5 Langmuir–Blodgett Films	290
11.6 Micelle Formation	291
11.7 Stability Energy and Surface Area Considerations in Colloids	293
11.8 Stability of Charged Colloids	293
11.9 Electrical Effects in Colloids	294
11.10 Electrical Double Layer	294
11.11 Particle (Micelle) Stabilization	295

		11.11.1	Charge Stabilization: Derjaguin–Landau–Verivey–Overbeek (DLVO) Theory	296
		11.11.2	Steric or Entropic Stabilization?	398
		11.11.3	Entropic Theory	399
	11.12	Phase Behaviour of Micelle Systems		300
	11.13	Phase Structures in Polymer Systems		302
		11.13.1	Block Copolymers and Associated Phase Diagrams	302
		11.13.2	Pluronics	304
	Recommended Reading			305
	References			306

Chapter 12 Molecular Organization and Higher Order Structures

	12.1	Introduction		308
	12.2	Hair		308
	12.3	Structure in Cellulose Fibres		310
	12.4	Natural Silks		317
		12.4.1	Silk Fibre Chemistry	317
		12.4.2	Silk Fibre Processing	317
		12.4.3	Silk Fibre Microstructures	319
	References			320

Subject Index 321

Figure Acknowledgements

The following figures have been obtained from:

Figure 2.4 K. G. Libbrecht, *Rep. Prog. Phys.*, 2005, **68**(4), 855-895.
Figure 2.18 R. E. Hillner, S. Manne, P. K. Hansma, A. S. J. Gratz., *Faraday Discuss.*, (1993) 95, 191.
Figure 4.4 N. E. Hill, W. E. Vaughan, A. H. Price and M. Davis (1969), *Dielectric Properties and Molecular Behavior*, Van Nostrand Reinhold, Hoboken, N. J.
Figure 5.3 L. Mandlekern, *Physical Properties of Polymers*, American Chemical Society, Washington DC, (1984) Chapter 4.
Figure 5.6 A. Keller, *Kolloid Z. Z. Polym.*, (1967), **219**, 118.
Figure 5.7 and 5.8 A. S. Vaughan, D. C. Bassett, *Comprehensive Polymer Science*, Ed G. Allan, Pergamon, Oxford (1989) Vol 2, 432.
Figure 5.10 D. A. Hemsley, *Applied Polymer Light Microscopy* ed D. A. Hemsley, Elsevier Applied Science, London, (1989) 67.
Figure 5.11 P. K. Datta, R. A. Pethrick, *Polymer*, 1978, **19**, 145.
Figure 5.12 A. J. Pennings, *J. Polym. Sci. Polym. Symp* 1977, **59**, 55.
Figure 5.13 and 5.14 B. P. Saville, *Applied Polymer Light Microscopy* ed D. A. Hemsley, Elsevier Applied Science, London, (1989) 112.
Figure 6.14 J. H. Magill, *J. Mater. Sci.* (2001), **36**, 3143.
Figure 7.6 S. Havriliak S. Negami, *Dielectric and Mechanical Relaxation in Materials* Hanser, Munich, 1997.
Figure 7.7 G. P. Mikhailov, T. I. Borisova, *Polym Sci USSR*, 1961, **2**, 387.
Figure 7.8 R. A. Pethrick, F. M. Jacobsen, O. A. Mogensen, M. Eldrup, *J. Chem Soc. Faraday Trans 2*, 1980, **76**, 225.
Figure 7.9 R. A. Pethrick, B. D. Malholtra, *Phys. Rev. B*, 1983, **22**, 1256.
Figure 7.12 J. E. Kinney, M. Goldstein, *J. Res. Natl. Bur. Stand. Sect. A*, 1978, **78A**, 331.
Figure 7.13 and 7.14 L. C. E. Stuick, *Physical Ageing in Amorphous Polymers and Other Materials*, Elsevier, Amsterdam, 1978.

Figure 8.2 R. Koningveld, W. H. Stockmayer, E. Neis, *Polymer Phase Diagrams, A Textbook*, Oxford University Press, 2001.
Figure 8.14 M. P. Stoykovich, P. F. Nealey, *Mater. Today*, 2006, **9**(9), 20.
Figure 9.2 B. Cherry, *Polymer Surfaces*, Cambridge University Press, Cambridge (1981).
Figure 9.4 I. H. Hall, *Structure of Crystalline Polymers*, Elsevier Applied Science, London 1984.
Figure 9.5 F. S. Baker, J. P. Craven, A. M. Donald, *Techniques for Polymer Organization and Morphology Characterization*, ed R. A. Pethrick, C. Viney, Wiley, 2003.
Figure 9.7 D. C. Bassett, R. H. Olley, A. S. Vaughan, *Techniques for Polymer Organization and Morphology Characterization*, ed R. A. Pethrick, C. Viney, Wiley 2003.
Figure 9.13 G. Beamson, D. Briggs, *High Resolution XPS of Organic Polymers*, Wiley Chichester, UK, 1992, 119.
Figure 9.14 S. G. Gholap, M. V. Badiogar, C. S D. Gopinath, *J. Phys. Chem. B*, 2005, **109**, 1391.
Figure 9.15 J. Hegg, C. Kramer, M. Wolter, S. Michaelis, W. Plieth, W. J. Fischer, *Appl. Surf Sci.*, 2001, **180**, 36.
Figure 9.18 G. Beamson, D. Briggs, *Mol. Phys.*, 1992, **76**, 919.
Figure 9.20 and 9.21 D. Briggs, *Encyclopedia of Polymer Science*, Wiley, New York 1988, **6**, 399.
Figure 11.16 P. Alexandridis, U. Olsson, B. Lindmann, *Langmuir*, 1998, **14**(10), 2627.
Figure 12.5 S. Y. Ding, M. E. Himmel, *J. Agric. Food Chem.*, 2006, **54**, 597.
Figure 12.6 A. K. Bledzki, J. Gassan, *Prog. Polym. Sci.*, 1999, **24**, 221.
Figure 12.8, 12.9 and 12.10 A. M. Emons, B. M. Mulder, *Trends Polym. Sci.*, 2000, **5**, 35.

CHAPTER 1
Concept of Structure–Property Relationships in Molecular Solids and Polymers

1.1 Introduction

Low molar mass organic molecules and polymeric materials are often found as solids and their physical properties are a consequence of the way in which the molecules are organized: their *morphology*. The morphology is a result of specific molecular interactions which control the processes involved in the individual molecules packing together to form a solid phase. Depending on the extent of the molecular organization, a crystalline solid, liquid crystals or amorphous solid may be formed. As we shall see later, the organization that is created at a molecular level sometimes also tells us about the macroscopic form of the material, but in other cases it does not, hence the subtitle of the book: 'from nano to macro organization'.

Synthetic polymers, often referred to as *plastics*, are familiar in the home as furniture, the frames for double glazed windows, shopping bags, furnishings (carpets, curtains and covering for chairs), cabinets for televisions and paper and paint on the walls. Outside the house plastics are used for rainwater pipes, septic, water and fuel storage tanks, garden furniture, water hoses, traffic cones and sundry other items which we see around us. Removal of all articles containing polymers from a room would leave it bare. Synthetic plastics form the basis for many forms of food packaging, containers for cosmetics, soft drink containers and the trays used in microwave cooking of food. Natural polymers such as wood, cotton and wool all exhibit a high degree of order and many biopolymers play a critical role in the human body.

Whilst a focus of this monograph is structure in polymeric materials, many of the factors that control the organization of these big molecules are best studied with lower molecular weight analogues. It is therefore appropriate to spend some time understanding small molecular systems before the consideration of the complexity of polymers is undertaken.

The physical properties of a material are dictated by its ability to self-assemble into a crystalline form. Polymer chemists have for many years sought to establish structure–property relationships that predict various physical properties from of knowledge the chemical structure of the polymer. Staudinger[1,2] recognized that polymers or macromolecules are constructed by the covalent linking of simple molecular repeat units. This structure is implied in the phrase *poly* meaning many and *mer* designating the nature of the repeat unit.[3,4] Thus poly(ethylene) is the linkage of many ethylene units:

$$H_2C=C_2H \rightarrow -(CH_2-CH_2)_n-$$

Recognition of the nature of this process of *polymerization* made it possible to produce materials with interesting and useful properties, and brought about the discipline of polymer science. The value of 'n' indicates the number of monomers in the polymer chain.

In the last forty years, a very significant effort has been directed towards understanding the relation between the chemical structure of the polymer repeat unit and its physical properties.[5,6] In the ideal situation, knowing the nature of the repeat unit it should be possible to be able to determine all the physical properties of the bulk solid. Whilst such correlations exist, they also require an understanding of the way in which the chemical structure will influence the chain–chain packing in forming the solid. Similar correlations can be created for the understanding of other forms of order in lower molecular weight materials.

1.2 Construction of a Physical Basis for Structure–Property Relationships

Why do structure–property relations apparently work?

1.2.1 Ionic Solids

In order to understand the basis for structure–property correlations, it is appropriate to consider the structure of simple ionic solids. A solid sodium chloride single crystal (Figure 1.1) can be constructed starting from the atomic species of sodium [Na^+] and chlorine [Cl^-].[7] Since each ion carries a single charge, pairing the ions to form an NaCl *pair* would create a lower energy state. This *pair* will have a minimum separation and would be charge neutral. The line formed between the two atoms can be considered to lie on the x-axis. If another *pair* of atoms is brought close to the first pair, then once more a minimum energy situation would be created if the two pairs of atoms are aligned but their orientations are in the opposite sense. This arrangement will have a lower energy than the isolated pair and the distance between the atoms will be slightly reduced compared with the isolated pair. A further reduction in energy will accompany bringing to the cluster a further pair of atoms. This latter pair will

Concept of Structure–Property Relationships in Molecular Solids and Polymers

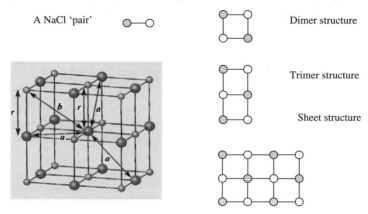

Figure 1.1 NaCl-type of crystal structure.

align in an opposite sense to the pair to which it attaches itself. Once more there will be a small change of separations to reflect the formation of a lower energy state. It is relatively easy to see that this process can be repeated and a sheet of atoms would be formed. As we will see later this same principle is used in considering attachment and growth of molecular and polymeric crystals.

The sheet of atoms formed by the process described above is not the lowest energy structure that can be formed. If this original sheet is sandwiched between similar states such that each of the Na and Cl atoms becomes surrounded by atoms of the opposite sign then a true minimum will be observed. If this order structure cannot be formed, because the entropy (disorder) is high, then the ensemble of atoms will be in the melt or gaseous state.

In the case of the NaCl crystal, this lowest energy structure is a cubic close-packed structure and results from each atom having six neighbouring atoms of opposite sign. Changes in the size of the ions and their charges lead to different types of packing being favoured. However, it is relatively easy to see that an *average energy* can be ascribed to a basic unit of the structure and this will reflect the physical properties of the bulk. Whilst the energy of the first pair can be calculated explicitly, adding additional elements means that the force field has to be averaged and will give rise to the problem of how one calculates the interaction of many bodies all interacting. The energy is the result of electrostatic (Coulombic) interactions between unit charges and in principle can be calculated by averaging all the interactions that will act on an atom chosen as the reference. The above example illustrates not only the lowering of the energy by surrounding an atom by other atoms but it illustrates that atoms in the surface will have a higher energy and we will meet this concept again when we consider polymers organizing at interfaces.

The total number and relative magnitudes of the Coulombic interactions and whether they are attractive or repulsive are taken into account by using a factor known as the *Madelung* constant, A. The lowest energy for the lattice $\Delta U(0K)$ (Coulombic) can for an ionic lattice be expressed by

$$\Delta U(0K) \text{ (Coulombic)} = \left(\frac{LA|z_+||z_-|e^2}{4\pi\varepsilon_0 r}\right) \quad (1.1)$$

where $|z_+|$ and $|z_-|$ are, respectively, the modulus of the positive and negative charges, e is the charge on the electron ($=1.602 \times 10^{-19}$ C), ε_0 is the permittivity of a vacuum ($=8.854 \times 10^{-12}$ F m^{-1}), r is the intermolecular distance between the ions (in m), L is Avogadro's number (6.022×10^{-23} mol^{-1}) and A is the Madelung constant.

The Madelung constant takes into account the different Coulombic forces, both attractive and repulsive, that act on a particular ion in a lattice. In the NaCl lattice, six Cl$^-$ atoms surround each Na$^+$ atom. The coordination number -6 describes the number of atoms which surround the selected reference atom. X-ray analysis indicates that each atom is a distance of 281 pm from its nearest neighbour. To calculate the Madelung constant we consider the four unit cells that surround the selected reference atom. Firstly there are twelve Cl$^-$ ions each at a distance a from the central ion, and the Cl$^-$ ions repel one another. The distance a is related to r by the equation

$$a^2 = r^2 + r^2 = 2r^2 \text{ and thus } a = r\sqrt{2} \quad (1.2)$$

Next there are eight Na$^+$ ions each at a distance b from the central Cl$^-$ ion, giving rise to attractive forces. Distance b is related to r by the equation

$$b^2 = r^2 + a^2; \quad b^2 = r^2 + (r\sqrt{2})^2; \quad b^2 = 3r^2; \quad b = r\sqrt{3} \quad (1.3)$$

Further attractive and repulsive interactions occur, but as the distance involved increases, the Coulombic interactions decrease.

The Madelung constant, A, contains terms for all the attractive and repulsive interactions experienced by a given ion, and so for the NaCl lattice the Madelung constant is given by

$$A = 6 - \left(12 \times \frac{1}{\sqrt{2}}\right) + \left(8 \times \frac{1}{\sqrt{3}}\right) - \ldots \quad (1.4)$$

where the series will continue with additional terms for interactions at greater distances. In general, the larger the distances involved the smaller the contribution to the energy and the magnitude of A is dominated by the first and second neighbour interactions. Note that r is not included in the equation and the value of A calculated is for all sodium chloride types of lattices. Thus the Madelung constant is a single parameter that describes with other constants the energy of the lattice. In this system, the dominant forces are electrostatic and hence the picture of the atoms as spheres is a reasonable approximation to reality. The physical properties of sodium chloride can be calculated on the basis of a knowledge of the interaction between the atoms. This simple principle can be extended to molecular species and to polymers. Obviously as the molecular structure becomes more complex the problem of the calculation increases dramatically; however, the additivity principle often applies and reflects the appropriateness of *mean field approximations* in many cases. The *A* parameter is associated with a specific atom pair and changing the atoms will give another characteristic value. Examination of a number of pairs of such systems allows specific interactions to be identified which can be used additively to predict the properties of an unknown system. In the case of an atomic solid the dominant forces are electrostatic. In most organic materials, short-range van der Waals repulsive and attractive interactions are dominant and longer range electrostatic and dipolar interactions play a very important role in defining the final structure.

1.2.2 The Crystal Surface

A further important feature of physical predictions can be obtained from this simple model. If we consider the surface of the solid, it is relatively easy to see that the atoms in this sheet will have a slightly different energy from those in the bulk of the material. This excess energy was recognized by Gibbs and discussed in terms of *surface tension* for a liquid. Bringing a further layer of atoms to this surface—crystal growth—can lower the energy of the atoms in the surface or if the atoms are different this process is usually considered as absorption. NaCl is a very simple model and the question we will next address is whether this concept can be applied to covalently bonded systems.

1.2.3 Molecular Solid

The next step in the development of an understanding of the physical properties of polymers is to consider how a molecule such as dodecane forms single crystals. Crystals of dodecane are usually grown from a solution or from the melt by slow cooling. The dodecane molecule, $CH_3-(CH_2)_{10}-CH_3$ (Figure 1.2), has an all-*trans* conformation as a consequence of nonbonding repulsive interactions between hydrogen atoms on neighbouring carbon atoms. A higher energy *gauche* state exists in which the interaction between neighbouring atoms is greater than in the *trans* conformation. The distribution between the *gauche* and *trans* conformations is predictable in terms of statistical mechanics.

Following the process used to create the NaCl crystal, a low-energy state can be achieved if two of these all-*trans* dodecane molecules are brought close together and aligned. Following the logic presented above, a lowering of the energy will occur and a further reduction will be observed when a third and fourth molecule are brought up to the first two. This process would produce a layer of molecules extending in the y–z plane. A further reduction in energy would be achieved by the addition of another sheet of molecules on top of the first and so forth. The forces that govern the interaction between the molecules

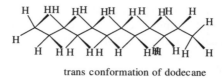

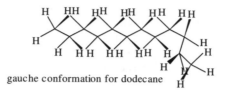

Figure 1.2 *Trans* and *gauche* conformations in dodecane.

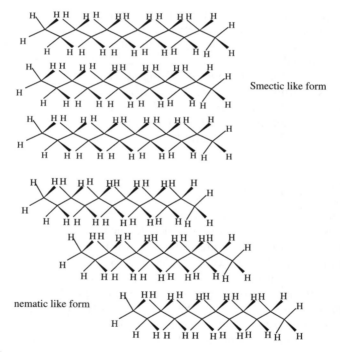

Figure 1.3 *Smectic* and *nematic* forms of packing in dodecane.

are now van der Waals interactions rather than the stronger electrostatic Coulombic interactions.

A question that could be asked is whether an even lower energy would arise if the molecules were to pack in a staggered array rather than being perfectly aligned. The proposed difference in order gives rise to *nematic* and *smectic*

Concept of Structure–Property Relationships in Molecular Solids and Polymers 7

phases in liquid crystalline phases (Figure 1.3). The staggered form is analogous to the *nematic* phase, with the molecules aligned in one direction but disordered in at least one other direction. In the *smectic* phase, the molecules are aligned in a plane but may be misaligned between the layers and are closer to the lowest energy crystalline ordered structure than the *nematic* phase. The *nematic* will have a number of methyl–ethylene bond interactions and these will be less favourable than the ethylene–ethylene bond interactions. The topic of liquid crystals is discussed more fully in Chapter 3.

Dodecane can exist in a number of higher energy forms in which one or more *gauche* structures are incorporated into the chain backbone. The process of conformational change will involve the hydrogen atoms on neighbouring carbon atoms being brought into an eclipsed conformation. This eclipsed conformation is a higher energy state and inhibits the free exchange between the *trans* and the *gauche* conformation in which the energy has once more been minimized.

At any temperature above absolute zero there will be a finite population of the higher energy *gauche* state dictated by the Boltzmann distribution:[8]

$$\frac{n_g}{n_t} = \frac{g_1}{g_2}\exp\left(\frac{-\Delta E}{RT}\right) \tag{1.5}$$

where ΔE is the energy difference between the *gauche* and *trans* states, g_1 and g_2 are, respectively, the degeneracy of the *trans* and *gauche* states at the temperature T and R is the gas constant. Since there are two gauche states, which are energetically degenerate, then the statistical factor is $\frac{1}{2}$ and ΔE is the energy difference between the *trans* and *gauche* states. This temperature dependence of the conformation of many molecular species plays a critical role in determining their behaviour when cooled to form a solid. At the temperature of the melt phase, there will be expected to be a significant population of *gauche* conformations. The *trans* conformation is the lowest energy state and is able to nucleate crystal growth.

The molecules which are in the surface will also have the lowest energy state and whilst there will be defects in the crystal there is no reason to believe that the surface structure should not be different from that of the bulk. The energy of interaction between the dodecane chains can be seen to be the average of the interaction of one methylene chain with another. As with the case of NaCl, a single interaction parameter should describe the physical properties of the solid. In creating this average energy the interactions of the end chains needs to be included. Studies of the melting points for the paraffin homologous series indicate that the lower members lie on different curves depending whether they have odd or even numbers of carbon atoms.[9–11] The equilibrium crystal structure is also different and can in part be explained by the way in which the methyl groups at the end of the molecules interact. As the chains become longer this odd–even effect disappears and is nonexistent for n greater than about 20 (Figure 1.4).

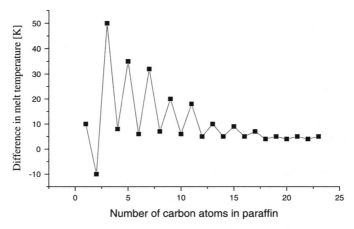

Figure 1.4 Difference between freezing points of successive paraffins.

1.2.4 Low Molar Mass Hydrocarbons

The dodecane molecule in the liquid state will be expected to have on average one *gauche* state per molecule at room temperature[12] but in the solid, however, it will have a structure that is predominantly made up of the all-*trans* form. The enthalpy of interaction compensates for the required loss of entropy in the crystallisation process. In the case of the *n*-alkanes, the bond lengths for the C–C and C–H bonds are, respectively, 1.5 and 1.10 Å and the C–C–C bond angle is 112° from studies of the solid.[13,14] The H–C–H bond angle has been found to be 109°. The conformational changes can be described by a potential energy diagram (Figure 1.5).

Abe *et al.*[15] have shown that the potential energy profile can be reproduced by selecting a barrier to the interchange between the *trans* and *gauche* forms of 12 540 J mol^{-1} and the energy difference between the two conformations has a value of 2090 J mol^{-1}. The energy and barrier to rotation are a result of nonbonding repulsive and attractive interactions between the hydrogen–hydrogen and hydrogen–carbon atoms on neighbouring carbon atoms.

The original calculations carried out by Scott *et al.*[16] used simple pairwise interactions. More sophisticated quantum mechanical calculations have confirmed the correctness of these original predictions. The conformation of *n*-butane, the simplest *n*-alkane, is described by three rotation angles (Figure 1.6).

Conformation energies computed for this molecule[15] in the neighbourhood of the *trans* and *gauche* minima for bond 2 indicate that the *gauche* minimum occurs at 112.5° rather than the expected 120° and has a value of 2215 J mol^{-1}.

Similar calculations for the next member of the series, *n*-pentane, give a potential surface, the contours being drawn in values of 1 kcal mol^{-1} (4.18 kJ mol^{-1}) above the minimum conformation, the *trans–trans* [tt] conformation

Concept of Structure–Property Relationships in Molecular Solids and Polymers

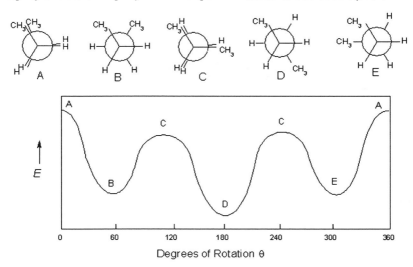

Figure 1.5 Potential energy curve obtained for *n*-butane for rotation about the 2,3 central bond of the molecule.

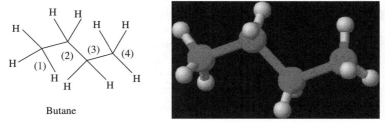

Figure 1.6 Labelled structure for *n*-butane. The angle ϕ_1 is the torsional angle for the bond joining C1 and C2, ϕ_2 is the torsional angle for the bond joining C2 and C3, ϕ_3 is the torsional angle for the bond joining C3 and C4.

with $\phi_2 = \phi_3 = 0$ (Figure 1.7). For the purpose of this calculation, the terminal methyl groups were fixed at $\phi_1 = \phi_4 = 0$, *i.e.* the terminal methyl groups are fixed in their staggered conformations. The portion of the energy surface for $\phi_2 < 0$ is produced by inversion through the origin in the ϕ_2, ϕ_3 plane.

The tg$^\pm$ and g$^\pm$t minima are equivalent to those for *n*-butane. The g$^+$g$^+$ and g$^-$g$^-$ minima (not shown) occur in the vicinity of $\phi_2 = \phi_3 = \pm 110°$. Thus the *gauche* minima for two adjoining bonds in *gauche* conformations of the same sign are mutually displaced a few degrees from the values ($\sim 112.5°$) which would be assumed by each if both of its neighbours were *trans*.

The *trans* minima for neighbouring bonds 1 and 4 are also perturbed a few degrees. These effects arising from subtle interactions between pairs of H atoms on third neighbour carbons are small. The calculated energy for the g$^+$g$^+$ pair is 4932 kJ mol^{-1}, which is close to twice the value for one *gauche* bond alone.

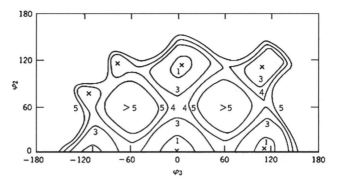

Figure 1.7 Energy contour map for the internal rotation in n-pentane with $\phi_1 = \phi_4 = 0$. The contours are shown at intervals of 1 kcal mol^{-1}. Minima are indicated by ×.

Hence the energy for neighbouring bonds in *gauche* states of the same sign may be treated as being additive. It is also interesting to note that the breadth of the *gauche* well is little affected by its neighbour.

Surprisingly, calculations of methylheptane[17] and methylpentane[18] have shown that the predictions are very close to the experimental values. These molecules, despite the fact that they can exhibit complex potential surfaces, are exhibiting simple additive interactions in the same way as that observed in n-butane. Once more this establishes the possibility of using a molecular description to be able to predict the physical properties of these molecules.

1.2.5 Poly(methylene) Chains

In principle, the poly(methylene) chain would represent a more complex potential surface than that of pentane. The number of different conformations following this simple scheme is 3^{n-3}. A typical polymer molecule may have 10000 carbons and thus $3^{9997} \approx 10^{4770}$ conformations, *i.e.* an enormously large number of states. Surprisingly the energy surface obtained is essentially identical to that of n-pentane (Figure 1.7). The energy surface obtained suggests a five-fold scheme, with rotational states at angles $\phi_i \approx 0, 77, 115, -115, -77°$. These could be labelled t, g^{*+}, g^+, g^-, g^{*-}, with combinations $g^{\pm}g^{\pm}$ forbidden by their large energies. A further simplification that is found to be practically justified is the deletion of the g^{*+} and g^{*-} states since they are assumed to have energies which are close to the $g^{*\pm}g^{\pm}$ and $g^{\pm}g^{*\pm}$ states. The result of this approximation is that the five-fold scheme is replaced by a three-fold representation that resembles that shown in Figure 1.5. However, the smaller numbers of minima have properties that represent the effects of a larger number of nonbonding interactions. The values of the energies that are usually used to describe the curve are 2.1 kJ mol^{-1} for the energy difference and 8.4 kJ mol^{-1} for the eclipsed state. Flory and others have shown that this simple approach can be applied successfully to many other chains.[19,20] The basis of the so-called Rotational Isomeric States Model (RISM) used extensively for the prediction of the physical properties of polymers is thus based on simple additive effects of nonbonding interactions between the atoms attached to the backbone carbon atoms.[20-22]

1.3 Conformational States of Real Polymer Molecules in the Solid State

In a real polymer system, the chain will attempt to crystallize in the lowest energy state. The lowest energy state for a polymer such as poly(methylene) will be an extended all-*trans* structure. Studies of single crystals of poly(methylene) formed from dilute solution resemble the predicted structure of a single crystal; however, there are a number of other factors which will influence the nature of the crystal structure or morphology observed. It is appropriate to divide polymers into various types depending on their chemical repeat unit.

1.3.1 Crystalline Polymers

Polymers such as poly(methylene) which have a high degree of symmetry associated with the polymer backbone have a simple potential energy surface. There will exist a finite possibility of finding chains having long sequences of *trans* elements that are favourable for the formation of nucleating sites for crystal growth. The *gauche* elements are less readily packed and hence will be accommodated at the limits of these aligned *trans* orientated regions. If the crystals are grown at low temperature the *gauche* content will be predicted to be low and hence the 'defects' can be accommodated at the interfaces. However, as we shall see later, the real situation is rather different and more detailed considerations of the way in which the chains can pack as well as the structure of the backbone are required to be able to interpret the structure of the solid state for a particular polymer. In general, however, if the polymer has a simple backbone structure it is likely to form a crystalline polymer solid. Thus polytetrafluoroethylene, like polyethylene, forms a crystalline phase. Other polymers with simple structures are poly(ethylene oxide), poly(methylene oxide), poly(propylene oxide), poly(isotactic propylene), *etc*. Interestingly the isotactic polypropylene polymer forms a hard rigid crystalline solid, whereas the syndiotactic or atactic polymers are soft disordered materials with low levels of crystallinity. The tactic forms of polypropylene are shown in Figure 1.8.

In the case of isotactic polypropylene, the methyl groups dictate a helical conformation to the backbone and this results in a regular structure which is able to crystallize readily.

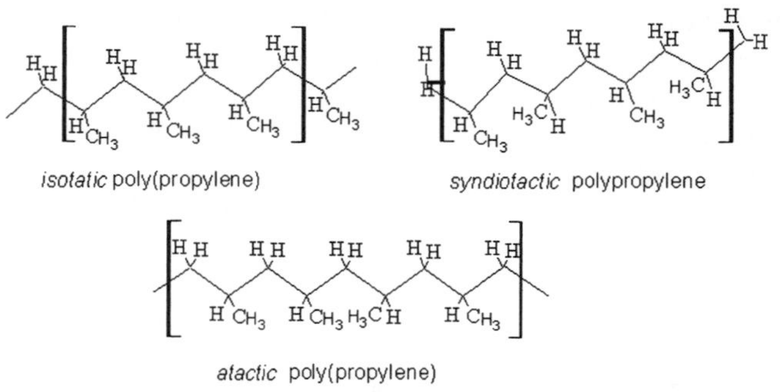

Figure 1.8 The tactic forms of polypropylene. Both the isotactic and syndiotactic forms have elements of symmetry and hence can crystallize, whereas the atactic form does not have a symmetry element and is amorphous.

1.3.2 Disordered or Amorphous Polymers

A polymer such as atactic polystyrene has a very bulky phenyl group pendant to the chain backbone and the groups are irregularly distributed in space. The phenyl groups are not favourably disposed to interact and crowding leads to a situation where the *trans* structure ceases to be the lowest energy state and energy profile simplifies to two conformations: an accessible *gauche* state and a distorted *trans* state. Using this approach, accurate simulations of the size of the isolated polymer molecule in dilute solution have been made.[23,24] In the solid state, polystyrene retains this disordered state and its morphology is that of an amorphous material. In general, if the polymer backbone contains a very bulky pendant group then it is highly probable that unless there are some very strong interactions the polymer will exhibit an amorphous structure in the solid state. The isotactic form of polystyrene allows the phenyl chains to interact in a favourable manner and crystalline forms are obtained from this polymer. Detailed studies of the kinetics of formation of the crystalline structure in this polymer system indicate that conformational dynamics are important in determining the observed behaviour.

1.4 Classification of Polymers

For flexible polymers, *i.e.* those that are able to undergo internal rotation about their backbones, the following classification can be made according to various constraints that we can place on the solidification process (Figure 1.9). There are two other classes of polymer worthy of mention: these are rigid rod like molecules where the shape of the polymer has a major effect on its ability to pack, and crosslinked polymers. In the latter system, the morphology of the polymers will be influenced by the size of the chain elements that exist between the network crosslinks. In a polymer system such as a silicone rubber, the

Concept of Structure–Property Relationships in Molecular Solids and Polymers 13

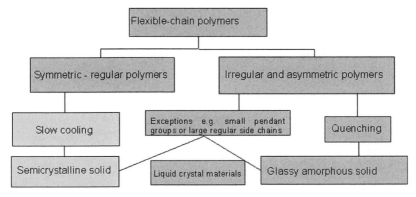

Figure 1.9 Classification of morphology by the chemical structural polymer type.

network dimensions may involve significant lengths of polymer chain and the resultant material is elastomeric. In contrast, an epoxy resin may be formed from relatively small chain elements and the high density of crosslinks results in the material being very rigid. The modulus of the material is dictated by two facts: the size of the polymer chain between the crosslinks and the conformational dynamics of that chain. If the barrier to internal rotation is high, as in the case of an aromatic-containing epoxy resin, then that material will be hard and glassy. If, as in the case of a typical silicone rubber created by the crosslinking of a polydimethylsiloxane (PDMS), the chain is flexible and an elastomeric material develops. Decreasing the molar mass of the PDMS results in a progressive increase in the chain–chain interactions and a subsequent increase in the modulus and hardness of the material until ultimately when the link is $-Si(Me)_2O-$ the material takes on glassy characteristics.

1.4.1 What Factors Determine Whether a Polymer will Form a Crystalline Solid or Not?

The regularity of the polymer backbone is the key factor; *isotactic* polypropylene crystallizes forming a rigid stable solid, whereas *atactic* polypropylene does not and forms a rubbery elastic solid. For flexible polymers, the structure of the solid is dictated by the symmetry of the polymer backbone. For the formation of a semi-crystalline solid it is necessary for there to be either an element of symmetry in the repeat unit chemical structure or strong interactions to aid the packing of the molecule and initiate the alignment that is required for the crystal growth process.

Atactic polymers do not crystallize, with two exceptions:

(i) When the X group in $(-CH_2CHX-)_n$ is very small, allowing regular packing of the chains regardless of whether the different pendant groups are randomly placed. Poly(vinyl alcohol) with its small hydroxyl X

group and strong polymer–polymer interactions is the kind of exception which demonstrates crystallization.
(ii) The X group forms a longer regular side chain. Side chain crystallization may occur provided that the pendant groups are of sufficient length, usually greater than about six repeat units.

Random copolymers, where the repeat units do not have a regular sequence structure, are incapable of crystallizing except when one of the constituents is at a significantly higher concentration than the other constituent. Linear low-density polyethylene with a crystallinity of about 50% contains 98.5 mol% of methylene units and about 1.5 mol% of CHX units, where X is $-CH_2CH_3$ or a longer homologue. Polymers that are potentially crystallizable may be quenched to a glassy amorphous state. Polymers with large side chains, or having an inflexible backbone chain are more readily quenched to an amorphous glassy state. The structure of the solid is therefore a combination of the intrinsic effects of the chemical structure and thermal factors involved in the creation of the solid.

Block copolymers, as we shall see later, are able to phase separate and it is possible for one of these phases to have semi-crystalline structure. Whether or not a crystalline phase is observed depends on the relative molar masses of the elements that form the polymer chain and their ability to pack into the required structure.

Polymers with rigid backbones, if they are sufficiently straight, will align like matchsticks and will either form 'liquid crystal' like structures or semi-crystalline mesomorphic phases.

Recommended Reading

P.J. Flory, *Statistical Mechanics of Chain Molecules*, Wiley Interscience, New York, 1969.
W.L. Mattice and U.W. Suter, *Conformational Theory of Large Molecules*, Wiley Interscience, New York, 1996.
M. Rubinstein and R.H. Colby, *Polymer Physics*, Oxford University Press, Oxford, 2003.

References

1. H. Staudinger, *From Organic Chemistry to Macromolecules*, Wiley-Interscience, New York, 1970.
2. H. Staudinger, *Chem. Ber.*, 1924, **57**, 1203.
3. H. Mark and G.S. Whitby (eds), *Collected Papers of Wallace Hume Carothers on High Polymeric Substances*, Wiley-Interscience, New York, 1940.

4. P.J. Flory, *Principles of Polymer Chemistry*, Cornell University Press, Ithaca, NY, 1953.
5. H. Morawitz, *Polymer: The Origins and Growth of a Science*, Wiley-Interscience, New York, 1985.
6. P.J.T. Morris, *Polymer Pioneers*, Centre for the History of Chemistry, Philadelphia, 1986.
7. C.E. Housecroft and E.C. Constable, *Chemistry*, Pearson Education, Essex, UK, 2nd edn, 2002, p. 242.
8. E. Wyn Jones and R.A. Pethrick, in *Topics in Stereochemistry*, ed. E.L. Eliel and N.L. Allinger, Wiley Interscience, New York, 1970, vol. 5, p. 205.
9. A.R. Ubbeholde, *The Molten State of Matter*, John Wiley, New York, 1978, p. 160.
10. K. Larsson, *J. Am. Oil Chem. Soc.*, 1966, **43**, 559.
11. L. Reinisch, *J. Chim. Phys. Physiochem. Biol.*, 1968, **65**, 1903.
12. R.A. Pethrick, M.A. Cochran, P.B. Jones and A.M. North, *J. Chem. Soc., Faraday Trans.*, 1972, **68**, 1719–1728.
13. H.J.M. Bowen and L.E. Sutton, *Tables of Interatomic Distances and Conformations in Molecules and Ions*, Chemical Society, London, 1958(Supplement 1965).
14. H.M.M. Shearer and V. Vand, *Acta Crystallogr.*, 1956, **9**, 379.
15. A. Abe, R.L. Jernigan and P.J. Flory, *J. Am. Chem. Soc.*, 1966, **88**, 631.
16. D.W. Scott, J.P. McCullough, K.D. Williamson and G. Waddington, *J. Am. Chem. Soc.*, 1951, **73**, 1707.
17. R.A. Pethrick, A.M. Awwad and A.M. North, *J. Chem. Soc., Faraday Trans. 2*, 1983, **79**, 731–743.
18. R.A. Pethrick, A.M. Awwad and A.M. North, *J. Chem. Soc., Faraday Trans.*, 1982, **78**, 1687–1698.
19. P.J. Flory, *Statistical Mechanics of Chain Molecules*, Wiley Interscience, New York, 1969.
20. W.L. Mattice and U.W. Suter, *Conformational Theory of Large Molecules*, Wiley Interscience, New York, 1996.
21. U.W. Gedde, *Polymer Physics*, Chapman & Hall, London, 1995.
22. M. Rubinstein and R.H. Colby, *Polymer Physics*, Oxford University Press, Oxford, 2003.
23. D.Y. Yoon, P.R. Sunderararajun and P.J. Flory, *Macromolecules*, 1975, **8**, 776.
24. R. Rapold and U.W. Suter, *Macromol. Theory Simul.*, 1994, **3**, 1.

CHAPTER 2
Crystal Growth in Small Molecular Systems

2.1 Introduction

Before considering the complexities of crystal growth in polymeric systems, we shall briefly look at the way in which small molecules form crystals. Slow cooling of the melt of an organic molecular system, such as benzophenone,[1,2] will lead to the growth of large single crystals, which can be several centimetres in size and have been a speciality of the research of Professor J. N. Sherwood at Strathclyde for many years (Figure 2.1).

In such a crystal, the molecules are ordered and adopt a minimum energy structure. Each of the faces will correspond to a particular orientation of the molecules within the crystal lattice. Because of the molecular orientations within the unit cell, the crystal faces may or may not have a different surface

Figure 2.1 A single crystal of benzophenone. The scale bar indicates 1 cm.

energy. In general, the single-crystal form of the material is a purer material than the *amorphous* disordered form.

Crystallization and crystal growth are very important parts of the manufacture of pharmaceutical chemicals and pigments. Paracetamol exhibits different crystal structures that have different solubilities.[3] The crystal growth process can be influenced by the presence of impurities that in general will be excluded from the growing crystal, but can in some cases be adsorbed on specific surfaces and influence growth. These molecules that modify the energy of specific faces are called *habit* modifiers. Impurities that have chemical structures that are similar to those of the bulk crystalline material can often be every effective *habit* modifiers. The exclusion of impurities from growing crystals forms an important method of purification for pharmaceutical and other fine chemicals. In certain cases impurities can be toxic and it is essential that they be removed from the pharmaceutical product. In some cases the impurities form the seeds from which the crystal growth is nucleated. To be able to predict the *morphology*—the bulk structure of the crystals—it is necessary to understand the factors that control the kinetics of the growth step and the influence of impurities on the growth process.

2.1.1 Crystal Types

The single crystals demonstrate elegant shapes as a result of the three-dimensional ordering of the molecular entities. The primary repeating pattern, the unit cell, replicates into the macro structures. Each entity sits at or near the intersection (lattice points) of an imaginary grid, the lattice (Figure 2.2). The smallest repeating unit is the asymmetric unit.

In Chapter 1 the unit cell for sodium chloride was presented (Figure 1.1). The periodic arrangement of any motif (e.g. group of atoms) of points located such that each has identical surroundings, produces an infinite arrangement that is called a *lattice*. The repeating period of the space lattice is called the *unit cell*. The *unit cell* contains a number (Z) of asymmetric units and it can be described by the dimensions a, b and c and the angles α, β and γ, with α being the angle between b and c, β between a and c and γ between a and b, i.e. $\alpha = \angle b.c$, $\beta = \angle a.c$ and $\gamma = \angle a.b$. These lengths and angles are the lattice constants or lattice parameters of the unit cell. The lattice parameters can for a crystalline

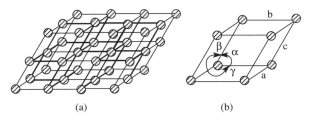

Figure 2.2 Crystal lattice (a) and the corresponding unit cell (b).

Table 2.1 The seven crystal systems and associated symmetry.[6,7]

Crystal system	Cell parameters	Minimum symmetry
Cubic	$a=b=c;\ \alpha=\beta=\gamma=90°$	Four 3-fold rotation axes
Tetragonal	$a=b\neq c;\ \alpha=\beta=\gamma=90°$	One 4-fold rotation or rotation–inversion axis
Orthorhombic	$a\neq b\neq c;\ \alpha=\beta=\gamma=90°$	Three perpendicular 2-fold rotation or rotation–inversion axes
Trigonal	$a=b=c;\ \alpha=\beta=\gamma\neq 90°$	One 3-fold rotation or rotation–inversion axis
Hexagonal	$a=b\neq c;\ \alpha=\beta=90°,\ \gamma=120°$	One 6-fold rotation or rotation–inversion axis
Monoclinic	$a\neq b\neq c;\ \alpha=\gamma=90°\neq\beta$	One 2-fold rotation or rotation–inversion axis
Triclinic	$a\neq b\neq c;\ \alpha\neq\beta\neq\gamma\neq 90°$	None

material be readily determined by X-ray or electron diffraction. Cells with only one unique motif are referred to as *primitive*. It is possible to generate a primitive cell from a given lattice, but in many cases end-, face- or body-centre representations are preferred because they may show greater symmetry than the primitive cells.

Bravais[4] postulated that there were fourteen different ways of arranging the lattice points in three-dimensional space. These are consistent with seven crystal systems that are listed in Table 2.1. The primitive lattice cell (P) has a lattice point only at the corner of the cell. Face centred (F) involves a lattice point at the centre of the opposite pairs of faces, while base centred (C) has a lattice point at the centres of the basal planes of the cell. Finally, body centred (I) involves a lattice point at the centre of the cell. The idealized Bravais structures are shown in Figure 2.3.

Crystals exhibit a high degree of symmetry. A number of different symmetry operations are possible on the lattice structures:

- Rotation around an *n*-fold axis, where the motifs are generated using cylindrical coordinates (r, ϕ), $(r, \phi + 360°/n)$, $(r, \phi + 2\times 360°/n)$, *etc.*; n can take values 1, 2, 3, 4 or 6.
- Inversion centre located at (90, 0, 0) where the motifs are located at (x, y, z) and $(-x, -y, -z)$, x, y and z being Cartesian coordinates.
- Rotary-inversion axes, which involve a combination of rotation ($\alpha = 360°/n$) and inversion and are indicated by $\bar{n}$, which can take values $\bar{1}$, $\bar{2}$, $\bar{3}$, $\bar{4}$ and $\bar{6}$.
- Mirror planes.
- Screw axes, which involve a combination of translation along the screw axis and a rotation about the same. It is designated n_δ, where n is the rotation by an angle $\alpha = 360°/n$ and δ is an integer related to the translational component t, where $t = (\delta/n)c$, in which c is the length of the unit cell along the screw axis.
- Glide plane, which combines a translation in the plane and a reflection across a plane.

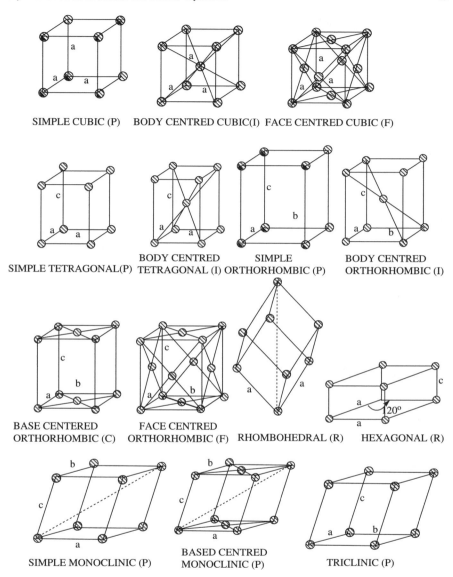

Figure 2.3 The fourteen Bravais lattices.[4]

The entire group of symmetry operators that completely describe the symmetry of the atomic arrangements within a crystal is called the *space group*. There are 250 space groups distributed among the 14 Bravais lattice groups. Another group of symmetry elements is the *point group*, which operates on the points that are usually groups or atoms. The allowed point group operators are rotation axes, axes of rotary inversion, inversion centres and mirror planes, and these altogether add up to 32 possible point groups.

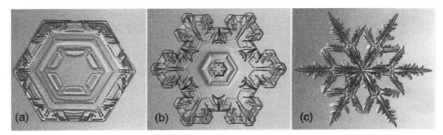

Figure 2.4 Examples of several different morphological types of snow crystals found in nature. Reproduced from reference 5.

The crystal is a *fractal* structure and the organization of the primitive unit cells can often be seen in the shape of the macroscopic crystal. A classic example of the *fractal* repetition of the unit cell is the crystalline structure of a snowflake.[5] The unit cells have the ability to build into a variety of complex shapes, yet each unit cell retains its perfect structure. The primary unit cell structure in the case of a snowflake is hexagonal and undergoes dendritic growth to produce an array of different macro crystals (Figure 2.4). The final shape of the snow crystal will depend on the conditions used in the growth process (temperature, humidity, *etc.*), which leads to a wide variety of observed *morphologies*.

In crystals, it is necessary to denote the plane directions and is done conventionally either by Miller's indices or by lattice planes. Directions are given as the lowest vector in referring to the coordinate system, $x(a)$, $y(b)$ and $z(c)$. A vector parallel to the chain axis is denoted [001]. The first plane intersects the origin of the coordinate system. The next plane intersects the three axes at $x = a/h$, $y = b/k$ and $z = c/l$. The task is to find an integral combination of h, k and l that is finally presented in parentheses (hkl). All planes containing the chain axis, *i.e.* those parallel to the chain, have the general formula $(h, k, 0)$. The lattice index system indicates not only the orientation of the planes but also the shortest distance between planes. The set of planes denoted (010) is a subset of (020). The orientation of the two sets of planes is the same but the interplane distances (d_{hkl}) are different: $d_{010} = b$ and $d_{020} = b/2$.

Negative values of hkl are indicated by bars $(0\bar{1}0)$. Several sets of planes appearing in highly symmetrical crystal structures may be denoted together with brackets of the type $\{hkl\}$, e.g. the planes in a cubic structure (100), (010), (001), $(0\bar{1}0)$, $(\bar{1}00)$, $(0\bar{1}0)$ and $(00\bar{1})$ are denoted simply $\{001\}$.

Miller's index system is similar to the lattice plane index system but with the difference that the hkl values are the lowest possible integer values. The Miller's index notation for both the sets of planes with the lattice plane indices (010) and (020) is simply (010). Miller's indices thus provide information only about the orientation of the planes and disregard the interplanar distances involved.

The most densely packed diffraction planes along the chain axis for polyethylene are denoted (002) in the lattice plane notation. The distance between the lattice planes is thus $c/2 \approx 0.127$ nm. In the Miller's index notation they are

Crystal Growth in Small Molecular Systems 21

(001). Whereas in small molecule systems crystal planes are not necessarily obviously related to the molecular structure, in polymers packing of chains or helices will naturally generate layered structures and the relevance of the interplanar distance to the nature of the polymer–polymer interaction potential becomes more obvious.

The scattering data have to be analysed in terms of the reciprocal lattice. The reciprocal lattice is defined in terms of the translation vectors of the unit cell: $\bar{a}$, $\bar{b}$ and $\bar{c}$. A of set of vectors of the reciprocal cell, $\bar{a}^*$, $\bar{b}^*$ and $\bar{c}^*$, exists fulfilling the following conditions: $\bar{a}.\bar{a}^* = 1$, $\bar{a}.\bar{b}^* = 0$, $\bar{a}.\bar{c}^* = 0$, $\bar{b}.\bar{a}^* = 0$, $\bar{b}.\bar{b}^* = 1$, $\bar{b}.\bar{c}^* = 0$, $\bar{c}.\bar{a}^* = 0$, $\bar{c}.\bar{b}^* = 0$, $\bar{c}.\bar{c}^* = 1$. It can also be shown that: $\bar{a}^* = (\bar{b} \times \bar{c})/(\bar{a}.\bar{b} \times \bar{c})$; $\bar{b}^* = (\bar{c} \times \bar{a})/(\bar{a}.\bar{b} \times \bar{c})$; $\bar{c}^* = (\bar{a} \times \bar{b})/(\bar{a}.\bar{b} \times \bar{c})$.

The scalar product $\bar{a}.\bar{b} \times \bar{c}$ is equal to the volume of the unit cell, $\bar{a}^*$ is perpendicular to plane bc, $\bar{b}^*$ to plane ac and $\bar{c}^*$ to plane ab. In an orthorhombic cell, the reciprocal cell vectors are parallel to the original cell vectors: $|\bar{a}^*|=1/|\bar{a}|$; $|\bar{b}^*|=1/|\bar{b}|$; $|\bar{c}^*|=1/|\bar{c}|$. The reciprocal of the reciprocal vectors (cell) is the original cell. Thus: $\bar{a}=(\bar{b}^* \times \bar{c}^*)/(\bar{a}^*.\bar{b}^* \times \bar{c}^*)$; $\bar{b}=(\bar{c}^* \times \bar{a}^*)/(\bar{a}^*.\bar{b}^* \times \bar{c}^*)$; $\bar{c}=(\bar{a}^* \times \bar{b}^*)/(\bar{a}^*.\bar{b}^* \times \bar{c}^*)$. In real space hkl is equal to a point ($\bar{r}^*$) in the reciprocal space: $\bar{r}^* = h\bar{a}^* + k\bar{b}^* + l\bar{c}^*$; thus $\bar{r}^*$ is perpendicular to (hkl) and the interplanar spacing (d_{hkl}) can be calculated from $d_{hkl} = 1/|\bar{r}^*|$. It is helpful to think of the reciprocal lattice representation of the crystal lattice in which the planes of the crystal are each represented by a lattice point of the reciprocal lattice. This point in reciprocal space is located in a direction from the origin that is perpendicular to the (hkl) planes in real space.

2.2 Crystallization

The processes of crystallization and crystal growth, like many other processes in chemistry, are controlled by thermodynamic and kinetic factors. Thermodynamics will dictate the preferred, lowest energy form, but the rate at which this is achieved will depend on the processes involved in the molecular attachment: kinetic factors. In the simplest model, the molecules are placed at the points of lowest energy on the ideal lattice structure. It is usually assumed that the entity that is being attached is a single molecule; however, it could also be a dimer or a cluster of molecules. In certain situations, for instance growth of benzoic acid from a non-polar solvent, the entity which may be involved is a dimer or higher order cluster:

In the crystal lattice, the forces experienced by the molecule may be different from those that control the formation of the dimer in solution, and small but important conformational changes can occur that will influence the nature of the morphology generated. In general, the crystallizing entity will be solvated in solution and its energy will reflect its interaction with the solvent molecules. Rearrangement of an entity that is initially attached to the surface may lead to a lower energy structure and this process is called Ostwald ripening.[8]

Crystal growth will usually be carried out from either the melt phase or from a saturated solution of the compound in a suitable solvent. The most perfect single crystals are grown from solution, and this process is the easiest to understand. Crystallization involves dissolving the pure compound in a solvent at high temperature and then lowering the temperature to a point at which *nucleation* occurs: stable clusters of molecules are formed. At the point of nucleation, the solubility of the material in the solvent has become critical and controls precipitation. The crystallization solution at this point has become *supersaturated*.[9,10] A solution that contains an excess of the solute at a given temperature is described as being supersaturated.

2.2.1 Supersaturation and Crystallization

Supersaturation is the essential driving force for all crystallization processes that occur from solution. Ostwald[8] first classified supersaturation in terms of 'labile', 'metastable' and 'supersaturated' depending on whether spontaneous nucleation did or did not occur (Figure 2.5). Crystallization can be promoted from solution either by cooling the solution, which leads to a decrease in solubility, or by evaporation. Crystal growth requires that nucleation should first occur and this can be achieved by one of a number of processes depending on the nature of the solution being examined.[9,10] Nucleation can be divided into two main processes: primary and secondary. The former is associated with growth from the melt in the absence (homogeneous) or presence (heterogeneous) of impurities. Secondary nucleation occurs if seed crystals are present in the mother liquid phase during nucleation.

At *supersaturation* the chemical potential of the solution and that of the solid that is formed on crystallization are equal:

$$\mu_{\text{solution}} = \mu_{\text{solid}} \tag{2.1}$$

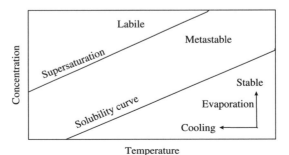

Figure 2.5 A typical solubility curve for a compound showing the three regions.[9,10]

where μ_{solution} is the chemical potential of the solute in solution and μ_{solid} is the chemical potential of the solute in the solid (crystalline) phase. If μ_{solid} is less than μ_{solution}, *i.e.* $\Delta\mu = \mu_{\text{solution}} - \mu_{\text{solid}}$ is positive, the solution is said to be supersaturated and when $\Delta\mu$ is negative the solution is undersaturated. For crystal growth to occur, the solution has to be *supersaturated*. Because the chemical potentials of the solution and solid are often not easily determined, it is more common to express the supersaturation of a solution in terms of the concentration (*c*) of the solute in the solvent. The supersaturation is therefore the condition where the solute exists at a concentration in excess of its equilibrium solubility value and the free energy imbalance provides the driving force for crystal growth. The driving force from crystal growth is therefore determined by the magnitude of the concentration relative to the equilibrium solubility value. The relative supersaturation (σ) usually defined as a percentage value:

$$\sigma = \frac{c - c^*}{c^*} = S^{-1} \qquad (2.2)$$

where c is the actual solute concentration, c^* is the equilibrium solute concentration and S is the supersaturation ratio. The value of c^* will depend on temperature and hence the value of σ will also be temperature dependent.

Depending on the initial concentration of the solute in the solvent, the point at which supersaturation is achieved will vary with temperature or through the process of solvent evaporation. The first process that is usually undertaken in order to successfully grow crystalline materials is the determination of the solubility curve in the growth solvent (Figure 2.5).

For a typical compound, the solubility diagram will exhibit three regions: the stable solution (low concentration–high temperature), metastable region and a labile region (high concentration–low temperature) where crystal growth occurs very rapidly. The stable region lies beneath the solubility curve and the solution is undersaturated and crystallization impossible. Cooling or evaporation increases the relative concentration to a point where the solution is now metastable and the solution is supersaturated and crystallization can be spontaneous. Within the metastable region, a solution can undergo controlled crystallization. Beyond this concentration region, the solution becomes increasingly unstable and a point is reached, supersaturation, at which instantaneous precipitation of the material is likely to occur.

2.3 Nature of Crystal Structures: Morphology and Habit

A rigorous terminology for the bulk shape of a crystal has been developed and leads to discussion of *morphology* and *habit*.[11] The morphology will depend upon the relative orientation and growth rates of the crystal faces. The growth velocity of a crystal face is measured by the outward rate of movement in a direction perpendicular to that face as shown in Figure 2.6. In most cases, for a

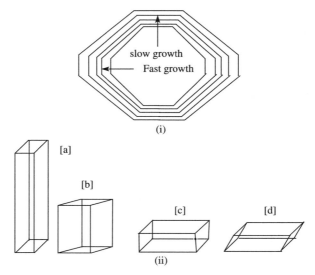

Figure 2.6 A schematic showing the effects of the different rates of growth on the crystal habits (a, b and c) and morphologies (c and d).

pure compound the velocity of growth will vary from face to face and reflects the nature of the orientation of the molecule within the unit cell of the crystal. If we examine the alignment of the molecule within the crystal, it is possible to identify which polar or functional groups will be orientated towards a particular face, with the result that the energy of each face is slightly different. The differences in energy of each face will in turn influence the growth rates for those faces. The shape of the crystal, or *habit*, is determined to a significant extent by the slowest growing faces.

The general shape that a crystal exhibits is defined as its *habit* (prismatic, needle-shaped or plate-like). The environment in which the crystal growth takes place will often influence the *habit* obtained. It is possible by changing the growth conditions to change the *habit* whilst retaining the same crystal morphology. In Figure 2.6 are presented various crystal *habits*: (a) prismatic, (b) needle-like and (c) plate-like. It is possible for a crystal to change its morphology but retain the same *habit*, this is illustrated in Figure 2.6c and 2.6d. Furthermore a crystal can change its *habit* and morphology, as shown in Figure 2.5a and 2.5d, where the *habit* has changed from being needle-like to plate-like and the crystal angles have changed.

2.3.1 Morphology Prediction

The morphology and *habit* exhibited by a crystal are controlled by a combination of the internal crystal structure of the unit cell and the external growth parameters. The morphology that is observed is in general a consequence of two factors: the method of nucleation and the dominant factors influencing the growth of the crystal. The nucleation is controlled by the ability to form a stable cluster of molecules and crystal growth rate is determined by the ability of further molecules to become attached to the nucleus.[12] For simplicity we will consider initially the case of homogeneous growth from solution. As we will see later, the growth process can be subdivided into a series of sequential steps that reflect the way various factors influence the nucleation and growth processes.

2.4 Homogeneous Crystal Growth

The classical approach to crystal growth[13-17] considers the thermodynamic changes that occur on crystallization. The overall free energy difference, ΔG, between a small solid particle, the nucleus, and the solute in solution is the sum of the excess free energy between the surface of the particle and bulk of the particle, ΔG_s, and the excess free energy between a very large particle and solute in solution, ΔG_v:

$$\Delta G = \Delta G_s + \Delta G_v \qquad (2.3)$$

Assuming that the clusters are spherical and have a radius r, then ΔG_s is a positive quantity proportional to r^2, while ΔG is a negative quantity proportional to r^3. Thus:

$$\Delta G = 4\pi r^2 \gamma + \frac{4}{3}\pi r^3 \Delta G_v \qquad (2.4)$$

where γ is the interfacial tension between the cluster and surrounding solution and ΔG_v is the free energy change associated with the transformation per unit volume (Figure 2.7). Since ΔG_s and ΔG_v are of opposite sign and depend differently on the size of the growing crystal, the free energy of formation will pass through a maximum. Below the critical value, the nuclei that are formed are disrupted and dispersed before crystal growth can occur. Once this critical value has been reached the nuclei become stable and crystal growth can take place.

The maximum value, ΔG_{crit}, corresponds to the critical nucleus size, r_c. The value of the critical nucleus can be found by differentiating ΔG with respect to the radius of the nucleus in eqn (2.4) and setting $d\Delta G/dr=0$:

$$\frac{\partial \Delta G}{\partial r} = 8\pi r \gamma + 4\pi r^2 \Delta G_v = 0 \qquad (2.5)$$

Thus

$$r_c = -\frac{2\gamma}{\Delta G_v} \qquad (2.6)$$

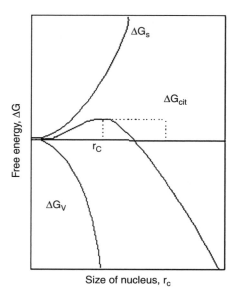

Figure 2.7 Free energy diagram for cluster formation.

Substituting eqn (2.6) into eqn (2.4) leads to

$$\Delta G_{\text{crit}} = \frac{16\pi\gamma^3}{3(\Delta G)^2} = \frac{4\pi\gamma r_c^2}{3} \qquad (2.7)$$

This relation can be used with the free energy relationship:

$$\Delta G_v = \Delta H_v - T_c \Delta S_v \qquad (2.8)$$

where ΔH_v and ΔS_v are the volume enthalpy and entropy of crystallization, respectively. At equilibrium, T_c is the equilibrium saturation temperature and $\Delta G_v = 0$. Equation (2.8) becomes

$$\Delta S_v = \frac{\Delta H_v}{T_c} = \frac{\Delta H_c}{V_m T_c} \qquad (2.9)$$

where V_m is the molar volume. Given $\Delta T = (T_c - T)$ and using eqn (2.8) and (2.9) the following equation is obtained:

$$\Delta G_v = \frac{\Delta H_c}{V_m} - \frac{T \Delta H_c}{V_m T_c} = \left(\frac{\Delta H_c}{V_m}\right)\left(\frac{\Delta T}{T_c}\right) \qquad (2.10)$$

For small supersaturations within the metastable zone (Figure 2.5), the degree of supersaturation can be related to the extent of under cooling (ΔT) through the Gibbs–Thomson relationship:[15]

$$\ln(S) = \frac{\Delta H_c \Delta T}{k T_c^2} \qquad (2.11)$$

where k is the Boltzmann constant. By eliminating ΔG_v and ΔH_c from eqn (2.6), (2.10) and (2.11) then

$$r^* = \frac{2\gamma V_m}{k T_c \ln(S)} \qquad (2.12)$$

Equation (2.12) shows that the critical cluster size is inversely proportional to the degree of supersaturation. As the supersaturation increases so the critical size decreases and consequently the solutions become less and less stable. Crystal growth becomes a stable process because as the nuclei are formed the concentration of solute in solution is reduced and the system becomes thermodynamically stable. Substituting eqn (2.12) into eqn (2.7) gives

$$\Delta G_v = \frac{16\pi \gamma^3 V_m^2}{3 k^2 T_c^2 [\ln(S)]^2} \qquad (2.13)$$

The rate of nucleation, J, is the number of nuclei formed per unit time per unit volume. According to Becker and Döring,[14] J can be expressed by an Arrhenius type of relationship:

$$J = A \exp\left(-\frac{\Delta G^*}{kT}\right) \qquad (2.14)$$

where A is the pre-exponential factor. Combining eqn (2.13) and (2.14), J can be expressed as

$$J = A \exp\left[-\frac{16\pi \gamma^3 V_m^2}{3 k^3 T^3 [\ln(S)]^2}\right] \qquad (2.15)$$

This equation indicates the complexity of the factors that influence the nucleation process: the temperature (T), degree of supersaturation (S), interfacial tension (γ) and molecular volume (V_m).

2.4.1 Empirical Description of Nucleation

An alternative method of modelling nucleation behaviour has been developed based on a kinetic approach.[17] In this model the mass nucleation rate, J_n, may be expressed by the following relationship:

$$J_n = K_n \Delta C_{max}^m \qquad (2.16)$$

where K_n and m are the nucleation rate constant and nucleation reaction order, respectively. In eqn (2.16), ΔC_{max} is the maximum allowable supersaturation.

The nucleation rate can be expressed as

$$J_n = \left(\frac{\varepsilon}{1-(\varepsilon-1)}\right)\left[\left(\frac{\partial C^*}{\partial T}\right)\left(\frac{\partial T}{\partial t}\right)\right] \qquad (2.17)$$

where T is the temperature, t is the time, dC^*/dT is the rate of concentration change with temperature and ε is the ratio of the molecular weights of solvated and unsolvated species. When solvation does not occur, the first term in eqn (2.17) is equal to unity. For a linear range of solubility, the metastable limit, ΔC_{max}, may be written as a function of the maximum allowable undercooling temperature ΔT_{max} before nucleation commences:

$$\Delta C_{max} = \left(\frac{\partial C^*}{\partial T}\right)\Delta T_{max} \qquad (2.18)$$

At the metastable limit, a shower of nuclei produce sufficient surface area to prevent a significant further increase in supersaturation with respect to time. Due to cooling then:

$$\frac{\partial \Delta C}{\partial t} = 0 \text{ at } \Delta C_{max} \qquad (2.19)$$

that is:

$$\frac{\partial C}{\partial t} = \frac{\partial C^*}{\partial t} = J_n \qquad (2.20)$$

Hence at the metastable limit, the mass nucleation rate (or rate of change of concentration) is equal to the super saturation rate due to cooling. Equation (2.20) can be expressed in terms of the cooling rate, b (dT/dt):

$$J_n = b\left(\frac{\partial C^*}{\partial T}\right) \qquad (2.21)$$

Equating eqn (2.16) and (2.21) for J_n gives

$$K_n \Delta C_{max}^m = b\left(\frac{\partial C^*}{\partial T}\right) \qquad (2.22)$$

and substituting for the supersaturation in terms of eqn (2.21) gives

$$b\left(\frac{\partial C^*}{\partial T}\right) = K_n\left[\left(\frac{\partial C^*}{\partial T}\right)\Delta T_{max}\right]^m \qquad (2.23)$$

Taking logarithms of eqn (2.23) gives

$$\log b = (m-1)\log\left(\frac{\partial C^*}{\partial T}\right) + \log K_n + m\log(\Delta T_{max}) \qquad (2.24)$$

Crystal Growth in Small Molecular Systems

This equation indicates that the dependence of the logarithm of the maximum undercooling, ΔT_{max}, on the logarithm of the cooling rate, b, should be linear and the slope is equal to the nucleation reaction order, m.

The above theories do not consider in detail the growth process once the nucleation has occurred. Growth takes place via a number of steps and in principle is the same for both homogeneous and heterogeneous processes; however, in the latter we have to consider the possible influence of impurities on the relative rates of growth.

2.4.2 Stages of Crystal Growth

The incorporation of the growth units, single molecule or molecular cluster, onto a crystal face can be divided into several key stages. The processes can be visualized schematically as in Figure 2.8, and are designated as follows:

(i) *Bulk transport of the growth entity to the surface.* The solvated molecular entity or cluster of molecules first has to diffuse through a concentration gradient to the surface and may in the case of an ionic system have to pass through a boundary layer.
(ii) *Attachment of the growth entity to the surface.* The growth entity may in the simplest case be a single molecule, but may be a dimer, a cluster or a solvated species.
(iii) *Desorption.* Return of a growth entity or solvated species to the solution.
(iv) *Surface diffusion of the growth units to the step edge.* The absorbed entity, which may be a partially desolvated species, will move across the surface until it finds a suitable low-energy site. This might be a vacancy but is more probably a growing edge or dislocation. The edge or

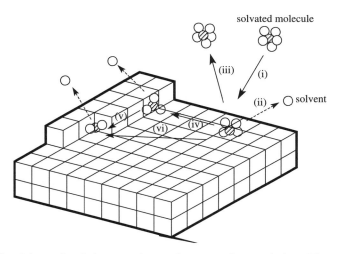

Figure 2.8 Schematic of the crystal growth process from solution. The crystallizing molecule is shown as the shaded circle and the solvent as the open circle.

dislocation will represent a lower energy state and hence the entity will become more strongly absorbed at these sites.

(v) *Migration of the growth unit along the step edge to a kink site.* The absorbed entity although in a low-energy site will still retain the ability to move as long as it is within a site of the same energy. Hence it is possible for an absorbed molecule to move along an edge to find a kink that will be of lower energy. As in the case of an entity absorbed on the surface, desorption from a step is always possible and hence migration to a kink is a mechanism whereby the probability of desorption is decreased.

(vi) *Volume diffusion to the step and direct integration of the growth units.* When the crystal is in equilibrium the processes above will be reversible, the rate of adsorption depending on the bulk concentration and the rate of desorption a function of temperature. As the bulk concentration is increased above the equilibrium value the rate of adsorption becomes greater than that of desorption and so the step will proceed to grow.

Whether or not this latter process occurs will depend on the relative energies of the binding at a particular site at which first attachment occurs to the energy of binding at an adjacent site.

At each stage in the process, loss of solvent molecules from the growth entity occurs and the free energy of the system is progressively lowered. The approach of the solvated entity to the surface will occur through a concentration gradient due to the depletion of material close to the growing surface. Once adsorbed (iv) the growth unit can diffuse across the surface to an energetically more favourable position such as a step or kink (v). The growth unit can alternatively be desorbed from the surface (iii) and return to the bulk solution. The growth unit is incorporated into the growing face once it attaches to a kink site (v). On a growing surface, kinks will be relatively abundant on the steps of a crystal. A 'flat face' step can therefore be seen as a continually changing surface that is never exactly flat and is a dynamic entity, as illustrated in Figure 2.9.

Whilst this analysis is appropriate for solution growth, similar situations will be developed in the melt, except the growing entity will often be a cluster of molecules. In both situations the formation of the step is a critical step in determining the final morphology of the material.

2.4.3 Heterogeneous Crystal Growth

In the case of crystallization from a melt, it has been observed that the rate of nucleation initially follows an exponential growth curve as the degree of undercooling is increased, reaching a maximum and subsequently decreasing.[15] This effect is attributed to the increase in viscosity as the melt is cooled inhibiting the nucleation process by suppressing the mobility of the entities from the nucleus moving, as illustrated from studies of crystallization of citric acid.[16]

Crystal Growth in Small Molecular Systems

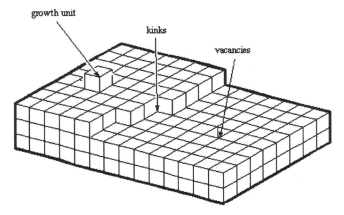

Figure 2.9 Schematic of a surface undergoing dynamic growth. In addition to kinks there will be isolated growth sites and also vacancies created by incomplete addition to steps.

2.4.4 Nucleation and Growth Rates

Whether it is homogeneous or heterogeneous growth, the nucleation rate J is difficult to measure, as the critical clusters formed correspond to a small number of molecules and hence are very small. In practice, the induction time (τ) is determined, which is the time between supersaturation and the first appearance of visible crystals. Assuming that the first appearance of the crystals is primarily controlled by the nucleation step, then τ is inversely proportional to the rate of nucleation:[18,19]

$$\tau = \frac{1}{\tau} \tag{2.25}$$

Combining eqn (2.15) and (2.25), the following is obtained:

$$\ln\left(\frac{1}{\tau}\right) = \ln(J) = \ln(A) - \left[-\frac{16\pi\gamma^3 V_m^2}{3k^3}\right]\left[\frac{1}{T^3[\ln(S)]^2}\right] \tag{2.26}$$

Using eqn (2.26), the interfacial energy can be evaluated from measurements of the induction time as a function of supersaturation. The radius of the critical nucleus can be derived from eqn (2.12).

Crystal growth is assumed to involve migration of the molecule or a cluster of molecules to a site on the surface that has a favourable energy of interaction. In order that a particular crystal face may grow, incorporation of growth units must occur in a sequential manner. The propagation of such successive growth layers will determine the overall rate of growth and hence the development of a particular morphology.

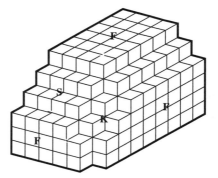

Figure 2.10 Designation of the flat surfaces (F), steps (S) and kinks (K).

2.4.5 Methods of Attachment to the Growth Surface

A growing crystal will not have perfectly flat surfaces and there are three types of face that can be identified: kinked (K), stepped (S) and flat (F) surfaces (Figure 2.10). This designation of the crystal surfaces is due to Kossel[20–22] and leads to the concept of surface roughening. A molecule adsorbed onto a kink site will be bounded on three sides whereas at stepped and flat surfaces the growth unit would only be bound by two sides and one side, respectively. Thus a face with a high density of kinks (K) would grow rapidly whilst flat faces (F) would grow the slowest. Any additional molecules adsorbed on the rough faces (the stepped and kinked faces) would never yield the even surfaces that are visible on most single crystals. It is therefore the flat faces that are observed to control the crystal growth.

2.4.6 Bravais–Friedel–Donnay–Harker Approach

Gibbs[12] attempted to predict the rate of crystal growth and indicated that the process required a minimization of the total free energy associated with the growing surfaces. It was subsequently shown that for a crystal at equilibrium, the faces generated are at a distance from the origin of the growth that is proportional to their individual surface energies. Developing this idea a simple three-dimensional plot of surface energy as a function of crystal orientation can be constructed. Bravais[4] and Friedel[23] used this concept to develop a simple method relating the internal structure to the morphology of a crystal.[24]

The Bravais law states that the morphologically most important forms would be those having the greatest interplanar spacings (d_{hkl}) and implies that the growth rates of the faces are inversely proportional to the interplanar crystal spacings. Donnay and Harker,[25] who developed rules relating the crystal symmetry to growth planes, refined the growth theory. These rules account for the effect of translational symmetry operators that cause the surface structure to be repeated more than once in the period d_{hkl}. For example, as we will explore later, the interplanar spacing of a face perpendicular to a 2_1

screw axis would be halved. The consequence of this consideration is that higher ordered planes can grow in preference to lower ordered ones.

The Bravais–Friedel–Donnay–Harker (BFDH) morphological simulations are based solely on lattice geometry and symmetry.[22,23] The BFDH simulations indirectly take into account the bond strengths present by considering slice thickness, but neglect mechanistic factors such as atom and bond types that also affect crystal growth. However, since these models are usually based on experimental observation they can help infer the dominant interactions that are controlling the crystal growth process.

The rate of crystal growth (R) predicted by the theory is expressed empirically as

$$R \alpha \frac{1}{d_{hkl}} \tag{2.27}$$

2.4.7 Periodic Bond Chains

Hartmann and others[24–27] have attempted to predict the morphology starting from consideration of the bonding within the crystal structure and the interactions between the crystallizing units. Their theory assumes that intermolecular forces govern the morphology of a crystal and identified the existence of uninterrupted chains of molecules 'strongly' bonded within crystal lattices, called periodic bond chains (PBCs). In PBC theory, the crystal growth mechanism is considered to involve the formation of consecutive bonds between crystallizing units during crystallization. A crystal network is made up of many different PBCs with differing energies that can be then classified into 'weak' and 'strong' PBCs. This concept is consistent with the idea that the energy of different faces will control the growth in the various observed directions. The theory leads to the idea of an 'attachment energy' (E_{att}) of a face, defined as the bond energy released per molecule when a new elementary growth layer, called a slice, of thickness d_{hkl} is attached to an existing crystal face. It is defined as

$$E_{cr}^{ss} = E_{att}^{ss} + E_{sl}^{ss} \tag{2.28}$$

where E_{cr}^{ss} is the energy of crystallization and E_{sl}^{ss} is the energy released on the formation of a growth slice of thickness d_{hkl}. The PBC theory assumes that rate of bond formation and hence the displacement velocity of a face will decrease with an increase in the attachment energy. Consequently the morphology that is developed is directly related to the attachment energy.[28,29] At low supersaturation the rate of crystal growth, R, is given by the following empirical relationship:

$$R \alpha E_{att}^{ss} \tag{2.29}$$

and provides a useful relationship for the prediction of the morphology of crystals grown below the roughening transition. To illustrate the way in which

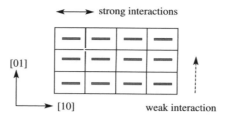

Figure 2.11 The way in which a simple two-dimensional morphology can be created using PBC theory.

growth occurs, we will consider the two-dimensional growth of a crystal which takes place predominantly in the horizontal direction (Figure 2.11).

In Figure 2.11, the strong interactions between the molecules lie along the horizontal direction and the weak interactions are perpendicular to the molecular axis. Therefore the growth will occur predominantly in the horizontal direction resulting in large (01) faces and small (10) faces. This simple situation will rarely occur and usually there will be interactions of varying strength between the molecules in various directions leading to more complex morphologies being observed.

In summary, for a crystal to grow in the direction of a strong bond the chain must be uninterrupted throughout the structure, and if a bond chain contains more than one type of bond it is the weakest bond present that determines the development of that orientation. Faces of a crystal can subsequently be classed into three types of faces according to periodic bond chains: F faces that have at least two PBC vectors parallel, S faces which are parallel to at least one PBC and K faces which have no PBC parallel.

2.4.8 Attachment Energy

For molecular materials, the strength of intermolecular interactions can be calculated by summing the interactions between a selected central molecule and each of the surrounding molecules within the crystal.[30] The simplest approach uses the atom–atom method in which the forces are assumed to be short range and involve the use of a van der Waals or similar potential modified to include dipolar terms. These calculations can be modified to include specific interactions such as hydrogen bonding, *etc.*, and allow calculation of the crystal lattice energy that can be compared with the enthalpy of melting of the crystal. Scaling such calculations against measured heat of melting allows reasonable estimates to be made of the significant intermolecular bonds interacting and hence predictions of the morphology. The prediction of the morphology involves the calculation of the slice energy that is calculated by summing the interactions between a central molecule and all the molecules within a slice thickness d_{hkl}. Calculation of the attachment energy involves summing the interactions between a central molecule and all the molecules outside the slice. The centre of the slice can be defined as the centre of gravity of a molecule or atom. The slice and attachment energies are then averaged over all these sites. As with PBC theory, the calculated attachment energies are directly related to the morphological importance of a crystal face. The rate of crystal growth, since it is related to the attachment energy, has the same proportionality as eqn (2.29).

2.4.9 Ising Model Surface Roughening

Growth at higher concentrations is assumed to involve addition of molecules to rough rather than to smooth surfaces. The Ising model is an attempt to describe the growth of a rough surface. A factor α has been introduced by Jackson[30] to describe the transition from smooth to rough growth. The factor α has the form

$$\alpha_{hkl} = \left(\frac{E_{sl}}{E_{cr}^{ss}}\right)\left(\frac{\Delta H(T)^{diss}}{RT}\right) \quad (2.30)$$

where $\Delta H(T)^{diss}$ is the enthalpy of dissociation (or melting for melt growth). If we consider eqn (2.28) and (2.29) we can see that if the temperature increases, or if the in-plane surface intermolecular bonds, defined by E_{sl}^{ss}, decrease, α will decrease. At a critical value α_c that is approximately 2, an order–disorder phase transition takes place at the growth interface that becomes macroscopically rough.

- If α is larger than α_c the crystal face is in essence flat on an atomic scale and it has to grow by spiral growth or a two-dimensional nucleation mechanism, since an edge separating a solid and a fluid domain has an edge free energy larger than zero.
- If α is smaller than α_c one mixed solid–fluid phase occurs, the edge free energies are zero and the crystal interface solid surface is in essence rough and the crystal can grow without a layer mechanism. At this stage the overall crystallographic orientation (*hkl*) will also be lost.

In the Ising surface model the crystal fluid phase is partitioned at the interface between the solid and fluid, with the crystal interface considered as a gradient in solid and fluids cells, going from complete solid to a complete fluid phase. Using the concepts introduced by Onsager,[31] the dimensionless Ising temperature (θ_{ijk}) that corresponds to the order–disorder phase transition can be calculated:

$$\theta_{ijk} = \left(\frac{RT}{2\phi}\right)_c \quad (2.31)$$

where ϕ is the strongest bond in the crystal lattice. By analogy with the α_{hkl} factors in eqn (2.30) we can define an experimental roughening factor

$$\theta = \frac{E_{cr}^{ss}}{\phi}\left(\frac{2RT}{\Delta H(T)^{diss}}\right) \quad (2.32)$$

and thus for $\theta > \theta_c$ the crystal surface is rough and hence crystal growth will proceed.

The essential feature of any growth model is the recognition of the importance of steps in the growth process. The surface is not perfectly flat but will be covered with steps arising from various processes. Attachment involves the interaction of the absorbed growth entity with these steps.[32] The growth rate is therefore given by

$$R \alpha \frac{1}{\theta_{hkl}^c} \qquad (2.33)$$

The principal difference between the above theories is the way in which attachment is treated to either a flat or rough surface. Aggregates of molecules on the surface form a nucleus for growth and will ultimately form a *step*; growth will also be possible at a dislocation.

2.5 Sources of Nucleation Sites on Surfaces, Steps and Dislocations

Any factors that can affect either the formation or movement of the growth entity will influence the type of morphology that is created, and therefore is a fundamental issue when considering crystal growth.

2.5.1 Two-Dimensional Nucleation

To gain some insight into the factors controlling step growth we will consider two-dimensional nucleation of growth. To create a layer on a flat surface, a stable cluster (Figure 2.8 and 2.9) has to form on which growth is initiated and step propagation will follow. In the two-dimensional nucleation mechanism, a growth unit will be adsorbed on the crystal surface and will require attachment of other units before stable growth can occur. During this stage the molecules attached to the surface will retain partially their solvation shells and will be able to move over the surface. As the growing units join up, the radius of the nucleus will increase and an increase in the free energy is observed (Figure 2.7) up to a maximum value. Further increase in the size of the nucleus leads to a decrease in the free energy. The size of the nucleus at which the free energy is a maximum is designated the critical radius, ρ_c, and is determined by eqn (2.6) and (2.7). If the nucleus radius is less than ρ_c it is unstable and dissolution is favoured; if the radius is greater than ρ_c it is stable and growth is favoured:

$$\rho_c = \frac{\omega \alpha}{kT \ln(1 + \sigma)} \qquad (2.34)$$

Crystal Growth in Small Molecular Systems

where ρ_c is the critical radius, ω is the unit volume of the growth unit, α is the free energy of the step edge, k is the Boltzmann constant, T is the absolute temperature and σ is the relative supersaturation.

Once a stable nucleus has been established, further growth can follow creating a new step on the flat surface and rapid growth will follow. For the process of step growth to be repeated another nucleus will have to be formed on this new surface and the nucleation step is the rate-determining process for the crystal growth. The theory predicts the probability of forming a nucleus is a very sensitive function of supersaturation and has been shown to be negligible at low supersaturations. Thus at low supersaturations growth would be expected to be zero if it were controlled solely by two-dimensional nucleation. However, experiment shows that contrary to prediction appreciable growth can occur at low degrees of supersaturation. The nucleation in these situations would appear to be easier than predicted by the theory and the ability to achieve growth at minimal supersaturation is attributed to the presence of imperfections in the crystals, called *growth defects*.

A further reason why growth becomes possible is the existence in solution of impurities that are able to nucleate crystal growth. Dust and other such material can nucleate crystallization and has to be carefully eliminated if perfect single crystals are to be grown.

2.5.2 Dislocations and Related Defects

2.5.2.1 Screw Dislocations

The occurrence of defects in the crystal lattice can act as nucleating sites for step growth and in particular dislocations with a screw component of the Burgers vector normal to that of the crystal face (Figure 2.12). The most important type of dislocation is a screw dislocation that causes a discontinuity in the crystal surface called a screw dislocation, which is a line defect in the crystal surface (Figure 2.12). The height displacement brought about by this slip creates a step

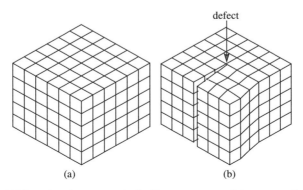

Figure 2.12 Dislocation in a crystal; the Burgers vector lying normal to the crystal.

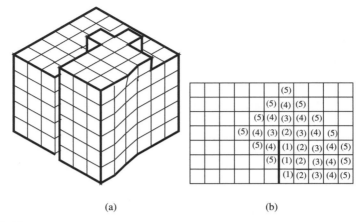

Figure 2.13 Schematic of growth at a dislocation that will lead to screw dislocation growth: (a) snapshot of the growth process; (b) sequence of adsorption of growth units around the defect leading to a screw climbing growth.

onto which the growth units can adsorb. Steps associated with the screw dislocation terminate at the dislocation. In order to propagate the step, growth will occur and wind around the dislocation (Figure 2.13). Away from the dislocation the steps advance relatively linearly and the number of the steps created per unit time by the dislocation determines the growth rate of the crystal. The critical radius of the two-dimensional nucleus restricts the curvature of the growth spiral: if the radius of the curve is less than ρ_c the advancing step will dissolve.

The initial attachment will occur at the dislocation, and the first layer, designated (1) in Figure 2.13, will follow closely the line of the defect. A second layer will easily be attached to the first line and will start to move the edge around the defect. Subsequent layers of attachment will move the edge around the defect and create the spiral growth indicated in Figure 2.13.

However, due to the upper surface being a spiral ramp, a completed growth layer is never obtained. Consequently the step will never disappear and growth can continue unperturbed. This mechanism eliminates the need for two-dimensional nucleation on which to construct new growth layers. A screw dislocation can be right handed, clockwise going from higher to lower level, or left handed, anticlockwise going from higher to lower level. Screw dislocations give rise to growth spirals and the distinction of 'handedness' becomes of importance when steps originating from distinct sources start to intersect. This mechanism has been observed in polymeric materials and will be illustrated later in the book.

2.5.2.2 Like Sign Dislocations

If two dislocation sources of the same sign (right or left handed) are very close together, less than half a step spacing, their spirals may join and form nonintersecting parallel spirals (Figure 2.14). These spirals effectively act as a

Crystal Growth in Small Molecular Systems 39

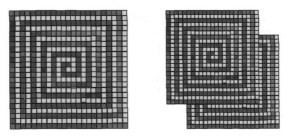

Figure 2.14 Two cooperating spirals (left), one depicted in white and the other in black; and two interlocking spirals (right).

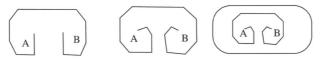

Figure 2.15 Progressive growth of two step dislocations of opposite sign (A and B).

single spiral but of double the height of a single spiral and are called cooperating spirals. The rate of step production per unit time would increase up to a factor of two depending on dislocation proximity and therefore increase the growth rate.

Alternatively if the same spirals were further apart, greater than half a step spacing, they would become a pair of interlocking spirals (Figure 2.14) and have the activity of one spiral. The less dominant spiral (the less developed spiral) would merely be contributory to the more dominant dislocation spiral. This case applies no matter how many screw dislocations are present; the most dominant spiral will still control the growth rate.

2.5.2.3 Opposing Sign Dislocations

If two dislocations of opposite sign emerge on a crystal face they will bridge to form a closed loop (Figure 2.15). This step will grow around both of the dislocations if the distance between the dislocations is greater than the critical diameter of the two-dimensional nucleus. The step grows and bows around the two dislocation points and eventually backs on to itself to form a completed loop.

Yet again as with other screw dislocations this ledge is a self-perpetuating step source and successive loops are formed without the need of nucleation. If the separation of the dislocations is less than the critical diameter, the step growth is prevented and the dislocation sources become inactive.

2.5.3 Screw Dislocation (BCF) Mechanism

Burton, Cabrera and Frank[33] have attempted to take into account the effect that occurs at low supersaturation and requires screw dislocations to provide a permanent source of surface steps and hence binding sites for growth to proceed. The surface growth rate R is proportional to the supersaturation (σ) squared:

$$R \propto \sigma^2 \tag{2.35}$$

2.5.3.1 Birth and Spread (B&S) Mechanism

This mechanism occurs at moderate supersaturation and involves two-dimensional surface nucleation on a developing crystal.[34] The theory describes the growth of a layer from a nucleus that exists on a flat surface. The growth rate is roughly proportional to the exponential of the supersaturation:

$$R \propto \sigma^{5/6} \exp(\sigma) \tag{2.36}$$

2.5.4 Rough Interface Growth (RIG) Mechanism

At high supersaturation the rough interface growth mechanism occurs and the crystal grows without the presence of well-defined surface layers at the interface.[30] Due to the rough surface, the approaching entities are provided with numerous binding positions (Figure 2.10). This method of growth shows a linear dependence on supersaturation.

2.5.5 Relative Rates of Crystal Growth

Combining the predictions of the above theories it is possible to map the way in which the crystal growth rates vary with concentration.[35] The relative rates of the different processes are shown in Figure 2.16.

2.5.6 Computer Prediction of Morphology

Using the theories outlined above and simple force fields based on modified Lennard-Jones potentials of the type discussed in Chapter 1 it has been possible to calculate the morphologies that are predicted by theory.[32] The crystal morphology can be simulated using a classical Wulff plot[36] that computes the smallest polyhedron enclosed by the various crystal faces (hkl) using relative centre to face distances which are derived from the various models. Such projection of crystal shapes can be accomplished with the aid of the Gnomonic projection[37] that is made with the aid of available computer programs.[38] A computer program created by Roberts et al.[39–41] designated Habit has been used to examine a range of crystal morphologies. The predictions for the case of naphthalene are shown in Figure 2.17. In this case the Hartman–Perdok model (Figure 2.17c) and attachment model (Figure 2.17d) compare favourably with experiment. In other systems the other models have been shown to be successful.

Crystal Growth in Small Molecular Systems

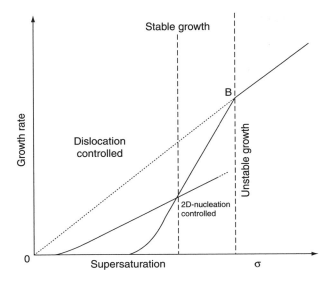

Figure 2.16 The relationship between growth rate and supersaturation showing the change from BCF to B&S to RIG mechanisms as the supersaturation increases.

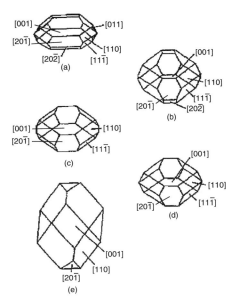

Figure 2.17 Results of the morphological simulation for naphthalene using various models in comparison with experiment: (a) experimentally observed morphology; (b) BFCH model; (c) Hartman–Perdok model; (d) attachment model; (e) Ising model.

2.6 Macrosteps

In the theories discussed above, it is assumed that the entity that is being absorbed is a single molecule or possibly a dimer. It is always possible that small clusters of molecules could be formed in the solution and diffuse to the surface. Thermodynamics would indicate that this is rather unlikely unless the entity has a radius that is greater than that required for nucleation. Therefore it is usually assumed that in conditions close to supersaturation such a mechanism is not very favourable. However, in a melt phase the possibility of larger entities being attached does now become possible and would lead to rough crystals.

The cluster that then attaches itself to the surface will undergo rearrangement, Oswald ripening, to achieve a lower energy structure. This structure, however, may not be the same as that which would be associated with the growth form dilute solution. It is assumed that the crystallizing entity will shed all the molecules of solvation in the region of the kink and it is the unsolvated species that is incorporated into the crystal. However, it is always possible that the strength of the solvation interactions can be sufficiently strong that the loss of the final solvent molecule is very slow. In such a situation the rate of step growth will be significantly slowed down. A further factor that can influence step growth is if it is possible to bunch the elementary steps. This is rather like cluster formation and leads to the formation of macrosteps and macro growth hillocks.

2.6.1 Impurities

Although the theory has been developed for homogeneous nucleation, in practice it is extremely difficult to produce materials that are completely free of impurities. In the presence of impurities the growth is heterogeneous. The addition of very small amounts of additives to the growth solution has been shown in many cases to have a marked effect on crystal growth. The impurities are often of two types: (i) impurities that occur in the compounds which reflect the method used in the synthesis of the compound and (ii) molecules that are added to the crystal growth solution to influence the morphology. The latter are called *habit* modifiers. They are usually molecules that have a structure that is similar to that of the molecule being crystallized but lack one of the functions. As such if they are incorporated into the lattice they will influence the balance of strong and weak forces (Figure 2.8). For instance, benzaldhyde will be a modifier for the growth of benzoic acid.

dimer - benzoic acid

benzaldehyde

The dimer of benzoic acid is a non-polar entity whereas the benzaldehyde monomer will have a significant dipole. It is therefore very important to purify organic materials to a very high level if one wishes to obtain the true crystal morphology. Generally, solutions may contain between 10^6 and 10^8 solid particles per cubic centimetre and these foreign particles can act to lower the interfacial energy necessary for cluster formation:[17]

$$\Delta G = 4\pi r^2(\gamma - \gamma_s) + \frac{4\pi r^3 \Delta G_v}{3} \qquad (2.37)$$

where γ_s is the interfacial tension between the hetero-nuclei and any molecular nuclei present within the solution. Heterogeneous nucleation occurs at a lower supersaturation than homogeneous nucleation ($\gamma > \gamma_s$) and hence the induction time and degree of undercooling necessary for bulk crystallization is lowered. This equation assumes that the hetero-nuclei appear homogeneous throughout the solution. It is difficult to model mathematically heterogeneous nucleation.

Impurities will in general add to the step or kink and retard growth of the crystal. When impurities are adsorbed onto the terrace the progressing step becomes pinned behind it and to advance has to bow between the impurities. If the distance between the impurities is less then $2\rho_c$ the step curvature will be too great and the step will dissolve thus stopping further growth. Additives can affect all of the crystal faces in an equal manner but usually, due to the different structure of the steps in different crystallographic directions, an additive will affect all the faces to varying degrees. It is for this latter reason that sometimes a particular habit modifier is added to a crystallization system.

The low molecular weight hydrocarbon fragments in diesel fuel readily crystallize at temperatures around 0 °C and form platelets.[42,43] These platelets can in cold weather block filters in the fuel supply pipes and lead to the vehicle becoming immobilized. The addition of a *habit* modifier suppresses the crystal growth in one direction and, instead of plates being formed, needles are created. The needles have the ability to pass through the fuel filters and hence allow the vehicle to continue operation in cold conditions. This illustrates a positive use of impurities in the control of crystal growth.[44]

2.7 Analysis of the Data from Step Growth

Observation of the nature of the crystal faces (Figure 2.10) in terms of the relative areas of the flat (F) or rough (S or K) surfaces provides an indication of the growth mechanism (two-dimensional nucleation or screw dislocation). Measurements of the step height give an indication of the growth unit size, and the step spacing (the distance between advancing steps), y, allows an estimation of the critical radius (ρ_c). From the critical radius an approximation of the step edge free energy can then be made:

$$y = 19\rho_c \tag{2.38}$$

and

$$\alpha = \frac{kT\sigma\rho_c}{\omega} \tag{2.39}$$

where ρ_c is the critical radius, ω is the volume of the growth unit, α is the step edge free energy, k is the Boltzmann constant, T is the temperature and σ is the supersaturation of the solution. Monitoring the proximity of two-dimensional nucleation (formation and dissolution) to step edges allows an approximation of the surface diffusion distance to be obtained. Moreover the measurement of step velocity gives an estimate of step kinetics. The effect of impurity addition can be classified from growth observation. Impurities can be adsorbed onto kinks, edges or terraces and affect none or all of the step directions. Examination of crystal growth with the addition of impurities can indicate the mechanism and extent of alteration that occurred. Recently the availability of atomic force microscopy (AFM) has allowed these previously inaccessible parameters to be determined experimentally[45–47] (Figure 2.18).

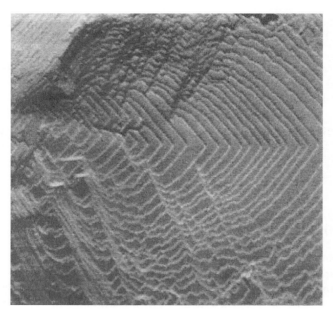

Figure 2.18 Atomic force microscopy images of an area 3.5 μm × 3.5 μm of poisoned growth of calcite.[46]

2.8 Refinements of the Theory

The principal challenge to the refinement of the theory is to describe accurately the processes associated with the migration of small clusters on the crystal surface and the consequent roughening processes that enhance crystal growth. As discussed above, one of the most difficult processes to model is the diffusion of the clusters across the surface and their entrapment at steps and defects. The physics of the diffusion of added atoms can be described by the Frank impurity mechanism using the BCF model of crystal growth.[48] The model is fundamentally two-dimensional and describes the quick formation of bunches of steps and their slower build-up in a lateral direction. In most regions, step bunches form quickly and equilibrium solutions to the equations can be developed. As pointed out in the discussion on modelling, in certain concentration ranges the Hartman–Perdok PBC theory has been found to be successful in its prediction of morphology.[49] Investigation of potassium dihydrogen phosphate (KDP), which is an ionic compound, presents particular challenges. The atomic structure at the outermost boundary is believed to control the growth mechanism and morphology of the crystal, while the presence or absence of cation–anion alternation at that boundary is supposed to determine the role of polarity in step kinetics. The observed pyramidal surface cell of KDP is considered to generate a strongly polarized growth front, but this may not always be the case. Surface electrostatics and the role of polarity on step kinetics cannot be inferred solely from the outermost ion array while ignoring the rest of the growth layer, including the step height. Cation–anion alternation at the surface termination is neither a necessary nor a sufficient condition for establishing surface polarity and, moreover, it does not necessarily imply alternation of positive and negative atomic charges on that plane. Consequently, and consistent with the symmetry point located in the middle of the experimentally determined d(101) layer cell of KDP, the observed pyramidal surface is unpolarized. The occurrence of the K^+-terminated, as opposed to the $H_2PO_4^-$-terminated, pyramidal surface of KDP grown out of an aqueous solution, as well as the role of ion impurities on the prismatic surface have been explained by this theory.

The use of computer modelling continues to advance and the integration of the Hartman–Perdok theory with the Ising theory for the roughening transition has been reported for the prediction of the crystal structures of venlafaxine, paracetamol and triacylglycerols.[50] An alternative approach to the description of the surface diffusion problem has been advanced also based on the BCF theory.[51] The flow of the solution over the surface influences the step bunching process. The flow within the solution boundary layer enhances step–step interactions and changes the resulting step pattern morphology on the growing surface.

The challenge to the theories is to be able to predict the various *polymorphs*— different morphologies that can arise when crystals are grown under different conditions.[51–53] The primary factors influencing polymorphism are solvent, temperature and impurities, *i.e.* habit modifiers. The recent use of computer predictions has allowed the development of a greater insight into the detailed factors that control the formation of particular morphologies.

2.9 Methods of Microstructural Examination

There are a number of methods now available for the examination of crystal morphology.[54] The shape can readily be determined by either optical or electron microscopy. Atomic force microscopy is now regularly used to study the surface of single crystals and considerable insight is being gained about the size of the steps and their relative location during the growth steps. The nature of the defect structure is somewhat more difficult to study. Imaging of defects that are features of the order of nanometres in dimension requires the use of X-ray diffraction methods. The usual method used is based on an approach put forward by Lang[55] and there are various experimental adaptations of these techniques:

(i) reflection topography[57–59]
(ii) transmission topography[55,60,61]
(iii) double crystal topography[56,62] and
(iv) white beam (synchrotron) topography.[63–68]

The Lang technique is shown in Figure 2.19. Incident X-rays enter a specimen crystal through a very narrow slit S_1 placed just before the crystal which is orientated to satisfy the Bragg diffraction condition. Since the X-rays are generated from scattering by an atom in the focal spot, they, even though collimated by the slit, have a divergency of the order of 100 seconds of arc. This degree of divergency is much larger than the intrinsic angular width of the diffraction peak for a perfect crystal that is of the order of 5 seconds of arc.[69] Therefore the incident wave must be regarded as a spherical wave that can be

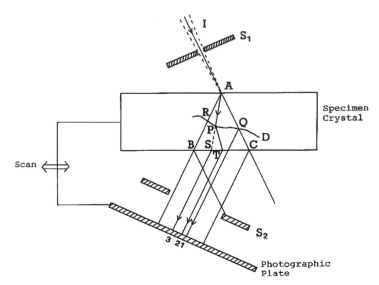

Figure 2.19 Schematic of the Laue X-ray diffraction experiment.

represented by a superposition of coherent waves. On entering the crystal, each reflecting plane aligned to satisfy the Bragg condition breaks every wave incident on it into two components, a transmitted and a diffracted wave. These waves interact with each other and set up two standing wave fields inside the triangular region ABC (Figure 2.19). This effect is know as anomalous transmission or the Borrmann effect[70,71] and the triangle ABC is called the Borrmann triangle. The Borrmann effect is one of the important consequences of the perfect periodicity of the lattice and provides a vivid demonstration of the dynamic theory of X-ray diffraction.

The two standing waves established in the Borrmann fan are Bloch waves of which the wave vectors differ. The difference in wave vectors leads to a difference in the propagation velocity and hence interference between the Bloch waves may occur. These interference effects, so-called Pendellösung effect, are observed in wedge-shaped crystals[72] and the resulting interference fringes are themselves a good indication of high crystal perfection.

The incident X-ray beam can be considered in two parts. A large part of the X-rays that do not satisfy the Bragg condition due to beam divergence pass straight through the crystal. This is called the direct beam. The remainder forms that part of the incident beam that satisfies the Bragg condition. Let us consider how these interact with defect D (Figure 2.19). Three types of defect image have been identified by Authier.[73] When the direct beam cuts the defect at Q, the misorientation due to lattice strain will cause a part of the beam to diffract. Thus the defect regions enhance the transmitted beam, increase its intensity and produce the so-called 'direct image' which appears black against a lighter background on the image plate. The limitation to this is the condition $\mu t > 10$, where μ is the X-ray absorption coefficient of the specimen and t is the thickness. Under this condition the direct beam suffers absorption within the sample and the direct image does not appear. This then defines the maximum usable thickness of the crystal. Within the Borrmann fan two types of image arise. On crossing the highly strained region at P the two waves decouple into their transmitted and diffracted components. When the waves re-enter the perfect part of the crystal they excite new wave fields. A part of the intensity is removed from the direction AP and the defect casts a shadow, producing the so-called 'dynamical image'. This is always less intense than the background and hence appears as a white image against a darker background. The newly created wave fields propagating along paths PT interfere with the original wave fields. The interference, which shows an oscillatory contrast, gives rise to the third type of image, the 'intermediary image', which has much poorer spatial resolution than the direct image.

In Figure 2.19, if the crystal and the photographic plate are stationary with respect to the incident beam, the recorded image is called a 'section topography' and represents a section through the crystal. If the crystal and plate are traversed together back and forth, the image is called a 'projection or transverse topography'. The projection topography integrates and records the full width of the crystal across the photographic plate. In projection topography dynamic and intermediary images become burred during the translation and are not well contrasted, whereas direct images form relatively clear projections of defects. These are the predominately observed images.

White beam X-ray topography (Laue technique) is probably the simplest available X-ray imaging technique. Although the theory of defect contrast in white beam X-ray topography is not fully established, the principles are similar to those described for the X-ray Laue method.

Unlike characteristic line sources, the beam divergence of synchrotron radiation is very small, $\sim 10^{-5}$ rad. The direct image results from the contribution of those wavelengths that do not participate in the diffraction from the perfect crystal, rather than from the contribution of the divergent beam as in the case of a monochromatic experiment. Despite this, there is little qualitative difference between the images obtained using nearly zero divergence white beam sources and the corresponding images obtained from characteristic line sources.[74] In synchrotron topography the superposition of several harmonic reflections may be inevitable on each Laue spot. This obviously can cause some increased image width and line degradation compared with the source images.

The principal limitation of both techniques is the resolution achievable. For X-rays, dislocation image widths are ~ 1 μm. Thus the maximum resolvable defect content is $\sim 10^5$ dislocations per square centimetre. Since specimens of high quality are readily achievable this limitation presents few problems. An example of a large single crystal is shown in Figure 2.20 together with a Laue image of a slice taken from the crystal.[1] The 'g' in Figure 2.20a indicates the dominant growth direction for this crystal. Although the defects are clearly visible in the Laue images, these crystals represent almost perfect crystals from the point of view of organic crystal growth. The apparent doubling of the edge in the crystal view (Figure 2.20a) is a consequence of interference effects, the so-called Pendellösung effect, which is observed in wedge-shaped crystals. The observation of these rings is a good indication of a high degree of perfection in

(a) (b)

Figure 2.20 Single crystal of benzophenone viewed as a Lang diffraction image (scale bar = 1 mm): (a) a crystal grown from ethanol with low defect density; (b) a melt-grown crystal with high defect density.

the crystals. The crystal shown in Figure 2.20b illustrates the higher defect density that is associated with melt-grown crystals. It can clearly be seen that there are a number of slip and spiral dislocations. The higher rate of crystal growth in the latter system allows incorporation of a higher number of defects in the final crystal structure.

Recommended Reading

J.W. Mullin, *Crystallization*, Butterworth-Heinemann, London, 3rd edn, 1993.
O. Sohnel and J. Garside, in *Precipitation: Basic Principles and Industrial Applications*, Butterworth-Heinemann, London, 1991.

References

1. C.S. Yoon, J.N. Sherwood and R.A. Pethrick, *Philos. Mag. A*, 1992, **65**(5), 1033–1047.
2. H.X. Cang, W.D. Huang and Y.H. Zhou, *Prog. Natural Sci.*, 1996, **6**(2), 235–242.
3. H.M. Cuppen, G.M. Day, P. Verwer and H. Meekes, *Cryst. Growth Design*, 2004, **4**, 1341–1349.
4. A. Bravais, *Etudes Crystallographiques*, Gauthie-Villars, Paris, 1866.
5. K.G. Libbrecht, *Rep. Prog. Phys.*, 2005, **68**(4), 855–895.
6. A.S. Myerson and R. Ginde, *Handbook on Industrial Crystallization*, ed. A.S. Myerson, Butterworth-Heinemann, London, 1993.
7. A.G. Walton, in *Nucleation*, ed. A.C. Zettlemoyer, Marcel Dekker, New York, 1969, pp. 225–307.
8. W. Ostwald, *Z. Phys. Chem.*, 1897, **22**, 289–330.
9. J. Garside and M.A. Larson, *J. Cryst. Growth*, 1986, **76**, 88–92.
10. L.N. Balykov, M. Kitamura and I.L. Maksimov, *Phys. Rev. B*, 2004, **69**(12), 125411.
11. J. Prywer, *Cryst. Growth Design.*, 2002, **2**(4), 281–286.
12. W.J. Gibbs, *Collected Works*, Longman Green, New York, 1928.
13. M. Volmer, *Kinetic der Phasenbildung*, Steinkoff, Dresden, 1939.
14. K. Becker and W.E. Döring, *Ann. Phys.*, 1935, **5**(24), 719–752.
15. G. Tamman, in *States of Aggregation* (translated by R.F. Mehl), Van Nostrand, New York, 1925.
16. J.W. Mullin and C.L. Leci, *J. Cryst. Growth*, 1969, **5**, 75–76.
17. J.W. Mullin, *Crystallization*, Butterworth-Heinemann, London, 3rd edn, 1993.
18. G.H. Nancollas and N.Q. Purdie, *Rev. Chem. Soc.*, 1964, **18**, 1–20.
19. O. Sohnel and J. Garside, *Precipitation: Basic Principles and Industrial Applications*, Butterworth-Heinemann, London, 1991.
20. V.V. Voronkov, *Sov. Phys. Crystallogr.*, 1968, **13**, 13.

21. D.E. Temkin, *Sov. Phys. Crystallogr.*, 1969, **14**, 179.
22. H.M. Cuppen, H. Meekes, E. Van Veenendaal, W.J.P. van Enckevort, P. Bennema, M.F. Reedijk, J. Arsic and E. Vlieg, *Surf. Sci.*, 2002 **506**, 183.
23. G. Freidel, *Lecon de Cristalographic*, Herman, Paris, 1911.
24. A.A. Chernov, *Acta Crystallogr.*, 1998, **54**(6), 859–872.
25. J.D.H. Donnay and D. Harker, *Ann Mineral.*, 1937, **22**, 446–467.
26. P. Hartmann and W.G. Perdok, *Acta Crystallogr.*, 1955, **8**, 49–52.
27. C.S. Strom, *J. Cryst. Growth*, 2001, **222**, 298–310.
28. P. Bennema, *J. Cryst. Growth*, 1996, **166**, 17–28.
29. F.F.A. Hollander, M. Plomp, J. van de Streek and W.J.P. van Enckevort, *Surf. Sci.*, 2001, **471**(1–3), 101–113.
30. K.A. Jackson, *Liquid Metals and Solidification*, American Society of Metals, Cleveland, OH, 1958.
31. L. Onsager, *Phys Rev.*, 1944, **65**, 1117.
32. K.J. Roberts, R. Docherty, P. Benema and L.A.M.J. Jetten, *J. Phys. D: Appl. Phys.*, 1993, **26**, B7–B21.
33. W.K. Burton, N. Cabrera and F.C. Frank, *Phil. Trans. R. Soc. London, Ser. A*, 1951, **243**, 299–358.
34. G.H. Gilmer and K.A. Jackson, in *Crystal Growth and Materials*, ed. E. Kaldis and H.J. Scheel, North Holland, Amsterdam, 1974, pp. 80–114.
35. R. Dochert, Modelling the morphology of molecular crystals, PhD thesis, University of Strathclyde, 1989.
36. G. Wulff, *Z. Krist. Min.*, 1901, **34**, 499.
37. F.C. Phillips, *An Introduction to Crystallography*, Longmans Green, London, 1963.
38. E. Dowty, *Am. Mineral.*, 1980, **65**, 465.
39. R. Docherty, G. Clydesdale, K.J. Roberts and P. Bennema, *J. Phys. D: Appl. Phys.*, 1991, **24**, 89.
40. R. Docherty and K.J. Roberts, *J. Cryst. Growth*, 1988, **88**, 159.
41. R. Docherty, K.J. Roberts and E. Dowty, *Comput. Phys. Commun.*, 1988, **51**, 423.
42. A.R. Gerson, K.J. Roberts and J.N. Sherwood, *J. Cryst. Growth*, 1993, **128**(1–4), 1176–1181.
43. A.R. Gerson, K.J. Roberts and J.N. Sherwood, *Acta Crystallogr. B*, 1991, **47**, 280–284.
44. A.R. Gerson, K.J. Roberts and J.N. Sherwood, *Powder Technol.*, 1991, **65**(1–3), 243–249.
45. F.F.A. Hollander, M. Plomp and J.P. Van De Streek, *Surf. Sci.*, 2001, **471**, 101–113.
46. R.E. Hillner, S. Manne, P.K. Hansma and A.S.J. Gratz, *Faraday Discuss.*, 1993, **95**, 191–197.
47. S. Hodgson, PhD thesis, University of Strathclyde, 2003.
48. C.R. Connell, *Physica D: Nonlinear Phenom.*, 2004, **189**(3–4), 287–316.
49. C.S. Strom, *J. Cryst. Growth*, 2001, **222**(1–2), 298–310.

50. P. Bennema, H. Meekes, S.X.M. Boerrigter, H.M. Cuppen, M.A. Deij, J. van Eupent, P. Verwer and E. Vlieg, *Cryst. Growth Design*, 2004, **4**(5), 905–913.
51. A.A. Chernov, *J. Cryst. Growth*, 2004, **264**(4), 499–518.
52. K. Sato, *J. Phys D*, 1993, **26**, B77–B84.
53. J. Bernstein, *J. Phys D*, 1993, **26**, B66–B76.
54. R.A. Pethrick, in *Techniques for Polymer Organisation and Morphology Characterisation*, ed. R.A. Pethrick and C. Viney, Wiley, London, 2003.
55. A.R. Lang, *J. Appl. Phys.*, 1958, **29**, 597.
56. U. Bonse and E. Kappler, *Z. Naturforsch.*, 1958, **13a**, 348.
57. W.F. Berg, *Naturwissenschaften*, 1931, **19**, 391.
58. C.S. Barrett, *Trans. AIME*, 1945, **161**, 15.
59. H. Barth and R. Hosemann, *Z. Naturforsch.*, 1958, **13a**, 792.
60. A.R. Lang, *Acta Crystallogr.*, 1957, **10**, 839.
61. A.R. Lang, *Acta Crystallogr.*, 1959, **12**, 249.
62. W.L. Bond and J. Andrus, *Am. Mineral.*, 1952, **37**, 622.
63. U. Bonse, in *Direct Observation of Imperfections in Crystals*, ed. J.B. Newkirk and J.K. Wernick, Wiley, New York, 1962, p. 431.
64. A. Gunier and J. Tennevin, *Acta Crystallogr.*, 1949, **2**, 133.
65. L.G. Schultz, *Trans. AIME*, 1954, **200**, 1082.
66. T. Tumoi, K. Naukkarinen and P. Rabe, *Phys. Status Solidi A*, 1974, **25**, 93.
67. J. Miltat, in *Characterisation of Crystal Growth Defects by X-ray Methods*, ed. B.K. Tanner and D.K. Bowen, Plenum Press, New York, 1980, pp. 401–420.
68. M. Hart, in *States of Aggregation* (translated by R.F. Mehl), Van Nostrand, New York, 1925, pp. 421–432.
69. M. Sauvage, in *States of Aggregation* (translated by R.F. Mehl), Van Nostrand, New York, 1925, pp. 433–455.
70. G. Borrmann, *Z. Phys.*, 1941, **42**, 157.
71. G. Borrmann, *Z. Phys.*, 1950, **127**, 297.
72. D.J. Fathers and B.K. Tanner, *Philos. Mag.*, 1973, **28**, 749.
73. A. Authier, *Adv. X-ray Anal.*, 1967, **10**, 9.
74. B.K. Tanner, O. Midgley and M. Safa, *J. Appl. Crystallogr.*, 1977, **10**, 281.

CHAPTER 3
Liquid Crystalline State of Matter

3.1 Introduction

Sometimes referred to as the fourth state of matter, the *liquid crystalline state* possesses the properties of both a liquid and a solid. The liquid crystalline state is usually associated with small molecules, but many polymeric systems exhibit similar types of order to those found in low molecular weight liquid crystals. It is appropriate to consider the factors that influence the formation of liquid crystalline phases in small molecules before considering polymer systems.

3.1.1 The Liquid Crystalline State

Whereas a perfect crystal exhibits a high degree of order, liquids possess total disorder. Between these extremes of crystalline order and liquid disorder, lie phases characterized by varying degrees of organization. The least ordered state is the *nematic* phase, the molecules in this phase being aligned in only one direction. The *smectic* phases correspond to molecules ordered in two directions and yet disordered in the third direction. The *smectic* phases correspond to various different types of partially ordered arrangements and resemble closely the crystalline phase of matter, discussed in Chapter 2. Many of the rules concerning molecular design and order observed for small molecules, and presented in this chapter, are also relevant to polymeric liquid crystalline materials.

3.1.2 Historical Perspective

The Austrian botanist Reinitzer,[1] when heating cholesteryl benzoate, observed a melting point at 145.5 °C, leading to a cloudy liquid which cleared at a temperature of 179.5 °C. He had discovered *cholesteric* liquid crystals. In 1922, Friedel[2] described a variety of different liquid crystal phases and proposed a classification scheme consisting of the three broad classes: *nematic, cholesteric* and *smectic* materials.

Liquid Crystalline State of Matter 53

3.1.3 Mesophase Order

Mesophases or *mesomorphic* phases are intermediate between solid and liquids and the molecules that exhibit this type of behaviour are termed *mesogens*. Most liquid crystals are organic molecules, and exhibit liquid crystalline behaviour over a defined temperature or concentration range or have order induced by the application of an external magnetic or electric ordering force. Liquid crystals that change their order with temperature are called *thermotropic*, whereas those that change with concentration of solvent are called *lyotropic*. A general feature of all liquid crystal materials is that the molecules have a rod-like characteristic, but, with a greater width than thickness, a better description would be lath-like in shape. However, the existence of a liquid crystalline phase is usually associated with a degree of rotational freedom around the molecular long axis, allowing the molecule to sweep out a cylinder by rotation. Most liquid crystals will possess some limited degree of flexibility.

3.1.4 Nematic Liquid Crystals (N)

The characteristic morphology of the liquid crystalline phase is usually observed using crossed polarized microscopy. The designation *nematic* comes from the Greek word νημα, which is the word for 'thread' which typifies the structures exhibited by this phase. These 'threads' correspond to lines of singularity in the *director* alignment called *disclinations* that are the disruption of the continuity of the *director*. The polarized light imposes on the system a reference alignment direction that probes the preferred direction of alignment of the molecules (Figure 3.1), which is indicated by the *director*. In *nematics*, this direction is referred to as the *anisotropic* axis and there are no long-range correlations between the centres of mass of the molecules. In the *nematic* phase, free translation is allowed in the direction of alignment whilst being constrained to remain approximately parallel to one another. Free rotation can occur around the anisotropic axis, with the result that the *nematic* phase is uniaxial and will have no polarity, although the constituent molecules may have polarity.

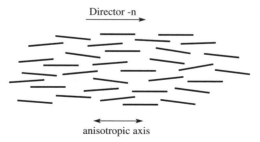

Figure 3.1 Schematic representation of a *nematic* liquid crystal phase. The lines represent the molecules and the unit vector *n*, called the *director*, describes the average direction of the molecular alignment along the uniaxial anisotopic axis.

3.1.5 Smectic Liquid Crystals

The word *smectic* comes form the Greek word σμηγμα, meaning 'soap' and reflects the mechanical properties displayed by many of these materials. All *smectics* are layered structures having well-defined interlayer distances. *Smectic* materials are therefore more ordered than *nematic* ones and occur at a temperature below that for the *nematic* phase. A number of smectic phases have been identified.[3]

Smectic A: S_A. Smectic A has a structure in which the molecules are arranged in layers, as shown in Figure 3.2a, but the molecules are disordered within the layers. In *thermotropic* liquid crystals, the *smectic* layer thickness way vary from a value that is closer to the length of the molecule to a value that is almost twice this value[4] and is typically in the range 20–80 Å. The lyotropic smectic A phase can have layer thicknesses up to several thousand angstroms. The director *n* represents the average molecular orientation of the molecules.

Smectic B: S_B. The *smectic* B phase is similar to the *smectic* A phase in that the molecules are aligned perpendicular to the layers but the molecules are in this case hexagonally close packed within the layers and are usually one molecule thick. The necessary disorder for these to be a liquid crystalline phase arises from rotation of the molecules about their long axis. It would be untrue to say that they are 'free' to rotate, but the molecules do not exhibit a well-defined orientation one to another (Figure 3.2c)

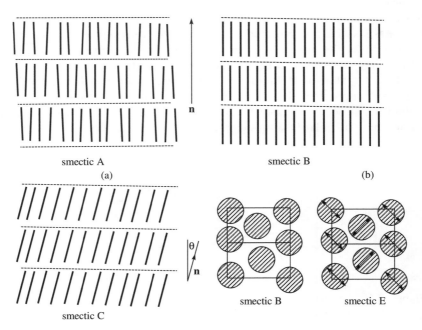

Figure 3.2 Schematic of the molecular orientation in smectic phases.

Smectic C: S_C. The alignment in the *smectic* C phase is similar to that in the *smectic* A phase except that the director is tilted at an angle θ relative to the layer normal (Figure 3.2b). The angle θ is usually temperature dependent and is called the *smectic* C tilt angle or *smectic* cone angle. The director ***n*** continues to be defined as the average direction of the molecular alignment. *Chiral smectic* C phases can occur when the constituent molecules are enantiomorphic. *Chiral smectic* C materials exhibit a helical structure in an analogous way to the 'twisting' that takes place across the *smectic* layers, the helical axis being in the direction of the layer.

Smectic E: S_E. Although the *smectic* E phase may occur on its own or in conjunction with a higher temperature *nematic* phase, it is most commonly encountered in the sequence of phases N → S_A → S_B → S_E, where the arrows indicate decreasing temperature.

Other smectic phases. Other smectic phases have been observed, where each phase relates to a particular type of crystal structure as defined in Chapter 2. For instance, smectic D is associated with a disordered cubic structure,[5] smectic H appears to be a tilted smectic B phase[6] and the smectic G phase has close similarities with the smectic E phase but is tilted.[7] A list of smectic phases is presented in Table 3.1.

3.1.6 Cholesteric Liquid Crystal (C)

Cholesteric liquid crystals are similar to nematic phases except that the molecular orientation between one layer and the next shows a progressive helical order. This helical structure arises from the chiral properties of the constituent molecules. Chiral molecules differ from their mirror image and have a left- or right-hand sense and are called enantiomorphic. The *director* is not fixed in space and rotates throughout the sample as shown in Figure 3.3.

3.2 Influence of Molecular Structure on the Formation of Liquid Crystalline Phases

A number of molecular factors influence whether or not liquid crystalline behaviour is observed within a specific molecular structure. These factors are also relevant when considering the behaviour of polymeric materials.

3.2.1 Influence of Chain Rigidity

The *n*-alkanoic acids (long-chain fatty acids) exhibit liquid crystalline characteristics. The alkane chain can be represented in an all-*trans* structure that gives the molecules the required high aspect ratio; however, being very flexible it will allow the dimer to readily form a crystalline structure. If, however, double bonds are now introduced into the structure in the form of alka-2,4-dienoic acids,[8,9] the structure is now sufficiently rigid, due to restricted rotation about

Table 3.1 Structural classification of smectic phases.[20,58]

Smectic	Comment
A (S_A)	Liquid-like layers with upright alignment of molecules
B (S_B)	Two distinct types of S_B have been identified: <ul><li>Three-dimensional (3D) crystal, hexagonal lattice, upright alignment</li><li>Stack of interacting 'hexatic' layers with in-plane short-range positional correlation and long-range 3D six-fold 'bond-orientational' order</li></ul>
C (S_C)	Tilted form of S_A
C* (S_{C*})	Chiral S_C with twist axis normal to the layers
D (S_D)	Cubic
E (S_E)	3D, orthorhombic upright alignment
F (S_F)	Monoclinic ($a > b$) with in-plane short-range positional correlation and weak or no interlayer positional correlation (tilted hexatic type of structure)
G (S_G)	3D, monoclinic ($a > b$)
G' ($S_{G'}$)	3D, monoclinic ($a > b$)
H (S_H)	3D, monoclinic ($b > a$)
H* (S_{H*})	Chiral S_H, with twist axis normal to the layers
H' ($S_{H'}$)	3D crystal, monoclinic ($b > a$)
I (S_I)	Monoclinic ($b > a$) possibly hexatic with slightly greater in-plane positional correlation than S_F
I* (S_{I*})	Chiral S_I

Many of the smectic phases are very similar and precise differentiation can be difficult between closely related phases.

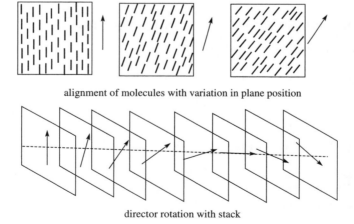

alignment of molecules with variation in plane position

director rotation with stack

Figure 3.3 Schematic of the change of molecular alignment with stack sequence of a *cholesteric* phase. Each layer has *nematic* order.

the double bonds of the sp^2 hybridized carbon atoms, for the compounds to exhibit *nematic* phase. Although the rigid sections would pack as for a crystalline structure, entropy will disorder the flexible saturated chains and inhibit crystallization.

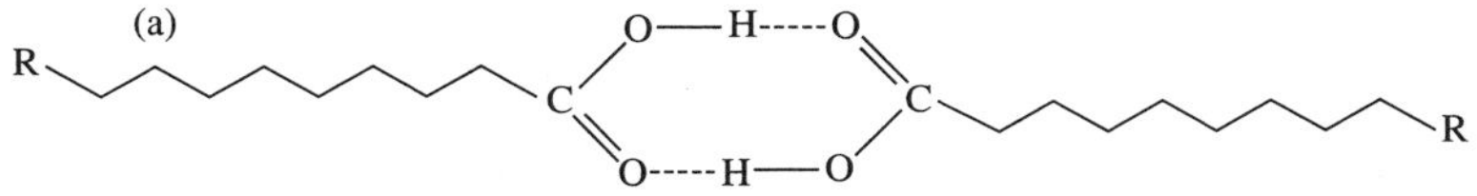

(a)

R = (CH$_2$)$_n$CH$_3$ n=integer   [1]

(b)

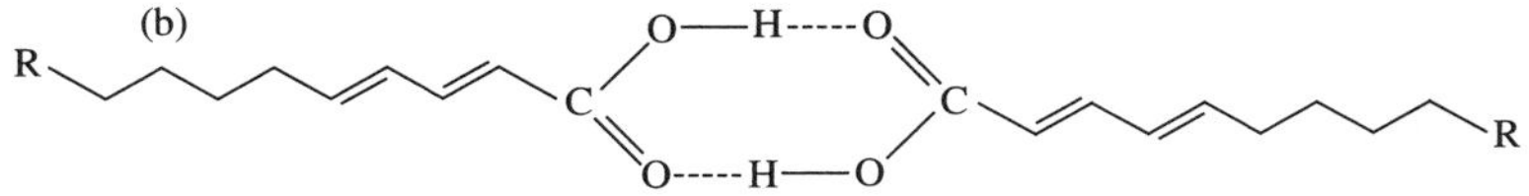

[2]

Looking for molecular structures that lead to liquid crystal behaviour, it is found that most molecules have a flexible chain in addition to a rigid element. Incorporation of a flexible chain generates the entropic disorder that inhibits the molecules forming crystalline structures. Theoreticians are able to consider the statistical mechanics of liquid crystals in terms of hard particles (rods, spheres, discs, *etc.*), but to understand the detailed behaviour we need to look at the molecular interactions in detail.

3.2.2 Influence of Size of Rigid Block

Whether a particular molecule will exhibit liquid crystalline characteristics depends upon the balance in inter- and intramolecular interactions. For instance, we can compare the following two diazomolecules:

(3)

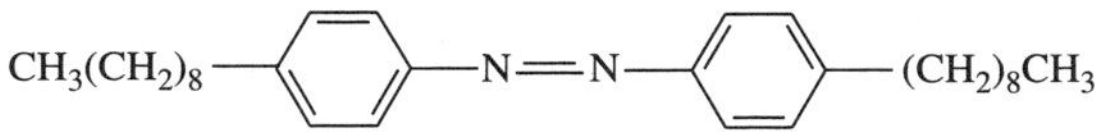

CH$_3$(CH$_2$)$_8$—⟨⟩—N=N—⟨⟩—(CH$_2$)$_8$CH$_3$

(4)

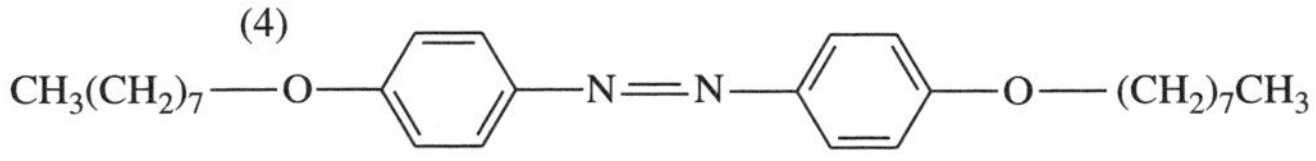

CH$_3$(CH$_2$)$_7$—O—⟨⟩—N=N—⟨⟩—O—(CH$_2$)$_7$CH$_3$

Both molecules (3) and (4) have similar rod-like structures with the same number of atoms in the terminal chains. Unless the small differences in the bond angles and bond lengths between the C–C and C–O bonds have significance, it is difficult to explain why (3) exhibits two *smectic* phases with crystalline (C) to *smectic* B (S$_B$) transition temperatures[10] at 37 °C, S$_B$ to S$_C$ at 40.5 °C and S$_A$ to isotopic (I) (liquid) at 53 °C, whereas (4) exhibits only a *nematic* phase with very different temperatures: C to N at 102 °C and N to I at 109 °C. An important difference between these apparently similar molecules is that the C–O bond will be conjugated to the aromatic ring and hence the length of the rigid element is greater in (4). In addition, the intermolecular repulsions for the systems would differ, since the –CH$_2$– and –O– functions behave differently electronically.

3.2.3 Influence of Sequence Structure in Chain

The cyano biphenyl molecules illustrate the effect of sequence structure on the ability to form various phases. The cyano biphenyl molecules form a very important group of materials being widely used in many liquid crystal display applications. Two examples are as follows:

(5) NC—⟨⟩—⟨⟩—O—∼∼

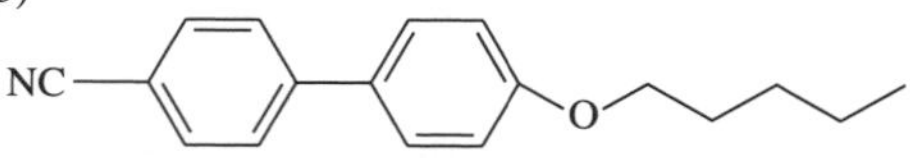

(6) NC—⟨⟩—⟨⟩—∼O∼

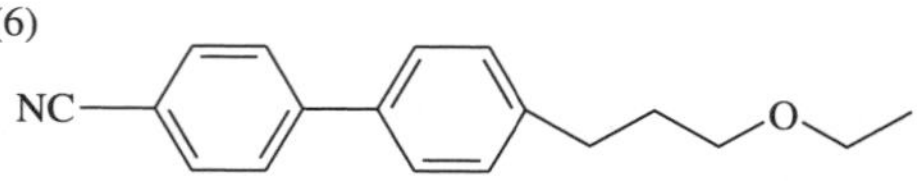

These molecules have the same chain lengths but differ only in the location of the oxygen. Compound (5) melts at 53 °C and gives a nematic phase and having a clearing point at 67.5 °C, whereas compound (6) melts at 36 °C and forms no *nematic* phase even on cooling to 0 °C, when re-crystallization occurs.[11] Once more the principal difference between these molecules is that (5) has the oxygen partially conjugated with the phenyl ring, whereas (6) has an aliphatic ether link and the oxygen increases the flexibility of the aliphatic chain and introduces greater disorder in the melt.

3.2.4 Variations Within a Homologous Series of Molecules

Cooling esters with the following formulae

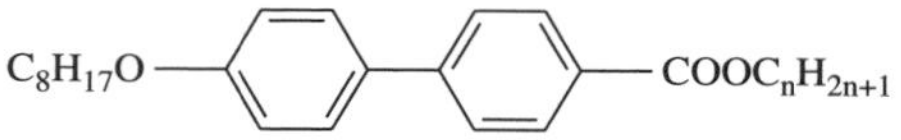

(7) C$_8$H$_{17}$O—⟨⟩—⟨⟩—COOC$_n$H$_{2n+1}$

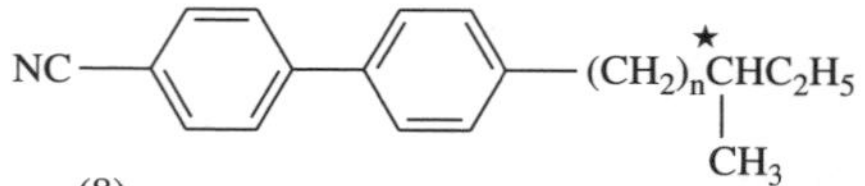

(8) NC—⟨⟩—⟨⟩—(CH$_2$)$_n$$\overset{\star}{\text{C}}HC_2H_5$
                                    |
                                    CH$_3$

gives rise to a variety of different liquid crystalline phases depending on the value of n in the terminal chain.[12] The asterisk indicates a *chiral* centre in the chain. The sequence of phases observed for the alkylbiphenyl esters are listed in Table 3.2.

There are a number of factors working within this series of molecules. Firstly, a S$_C$ phase appears at $n=4$, the S$_B$ and S$_E$ being also only observed for $n=3$. The alkyl cyano biphenyl molecules, structure (8), all have the same absolute

Liquid Crystalline State of Matter 59

Table 3.2 The phases observed for a series of alkylbiphenyl esters, structure (7).

Value of n in ester chain	Phase types observed on cooling
2	I → S_A → S_B → S_E → C
3	I → S_A → S_B → C
4	I → S_A → S_C → C
5	I → S_A → C

configuration at the asymmetric carbon atom, indicated with an asterisk. The molecules with values of *n* from 1 to 3 all exhibit *cholesteric* phases. However, if they are mixed together 1 with 2 or 2 with 3 *nematic* phases are observed. On moving the branching point from an even to an odd to an even point in the chain the sense of the *cholesteric* helix changes from right handed to left handed and back to right handed again.[13]

3.2.5 Changes in Substituents

Changes in substituents will influence the electronic polarizability of a molecule. Consider the following structure:

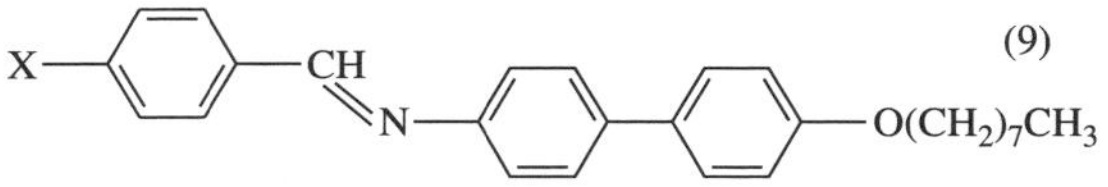

$$X--CH=N---O(CH_2)_7CH_3 \quad (9)$$

Methyl and chlorine groups have very similar sizes, so that molecules with X = Cl and CH_3 have the same rod shape.[14] However, the transitions observed for these molecules are very different: when X = Cl, C → N occurs at 193 °C and N → I occurs at 261.5 °C; when X = CH_3, C → S_A occurs at 178 °C, S_A → N at 178 °C and N → I at 222 °C. Such differences in behaviour cannot be considered to be subtle. To understand why these differences occur, the *mesogen* must be considered to be a dynamic entity consisting of a fairly rigid section, the core of the structure, and a flexible part. The entire molecule may be freely rotating about some axis that may not be along the most obvious line, and precise consideration of the stereochemistry of the molecule is essential to determine the preferred axis. Superimposed on the effects of geometry are the intermolecular interactions that define the potential in which the molecules rotate and move. The Cl group will add an additional dipolar contribution that will be reflected in the higher transition to the isotropic phase.

3.3 Common Features of Many Liquid Crystal Forming Molecules

Mesogenic behaviour is associated with molecules having a rigid core section, that will often contain an aromatic element, but this is not essential. Attached to the core will be a polarizable group, e.g. Cl, CN, NO_2, NMe_2, and also a long flexible chain. The latter allows the molecule to retain the entropic element that inhibits the formation of the crystalline phase. The flexible chain will usually be an alkyl or alkoxy group.

3.3.1 Nematic Liquid Crystals

The rigidity of the central core structure is often achieved using aromatic rings, as in the case of 4-cyano-4'-*n*-pentylbiphenyl[15] **(10)** or 4-cyano-4''-*n*-pentyl-*p*-terphenyl[16] **(11)**:

(10) C_5H_{11}—⟨⟩—⟨⟩—CN

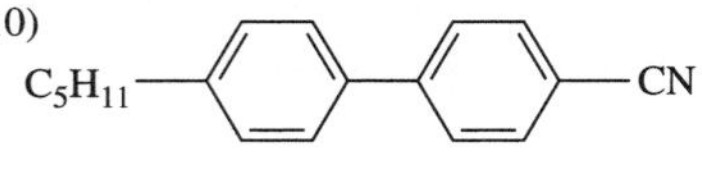

(11) C_5H_{11}—⟨⟩—⟨⟩—⟨⟩—CN

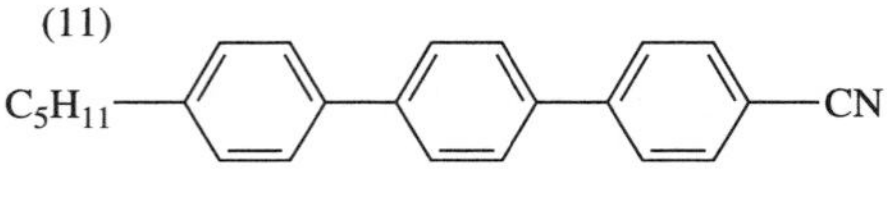

The biphenyl shows transitions C–N at 22.5 °C and N–I at 35 °C, whereas the terphenyl shows transitions C–N at 130 °C and N–I at 239 °C. In these molecules, planarity of the two aromatic rings is inhibited by strong interactions between the hydrogen atoms in the α-position to the C–C bond. Mesogenic behaviour is observed if the aromatic rings are joined by one of the following entities:[17]

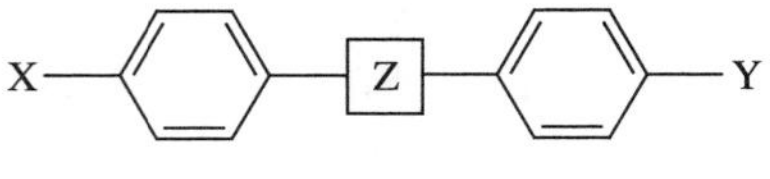

X—⟨⟩—[Z]—⟨⟩—Y

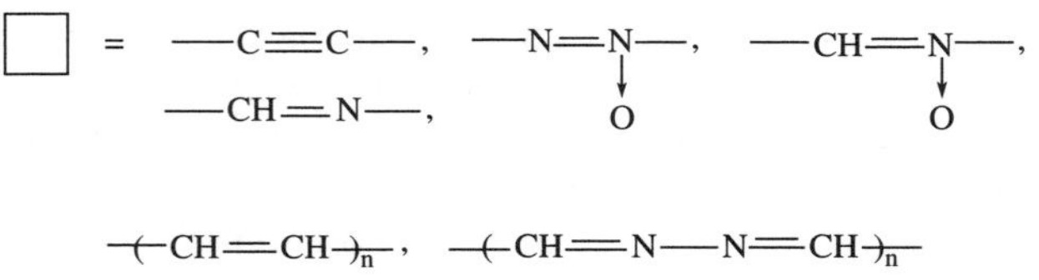

[Z] = —C≡C—,　—N=N—,　—CH=N—,
　　　　　　　　　　↓　　　　↓
　　　　　　　　　　O　　　　O
　—CH=N—,

—(CH=CH)$_n$—,　—(CH=N—N=CH)$_n$—

These linkages allow delocalization of electron density between the terminal aromatic rings, and have the effect if retaining the rigidity and planarity of the central core. The delocalized electron density can enhance the molecules anisotropic polarizability. Interestingly, the analogue of the above molecule in which the two aromatic rings are joined by a double bonded rather than a triple bond does not necessarily form a liquid crystal phase. The *trans* isomer retains the overall linear profile and can form a liquid crystalline phase whereas the *cis* does not.[18,19]

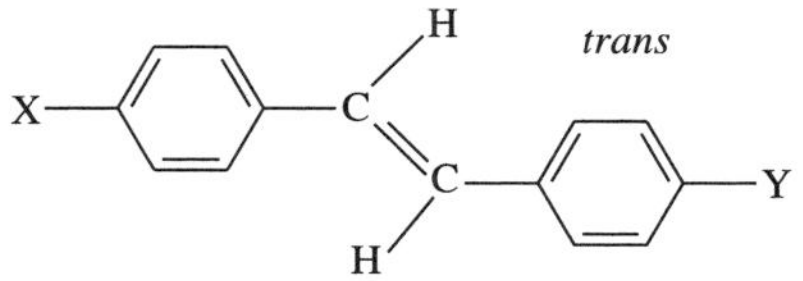

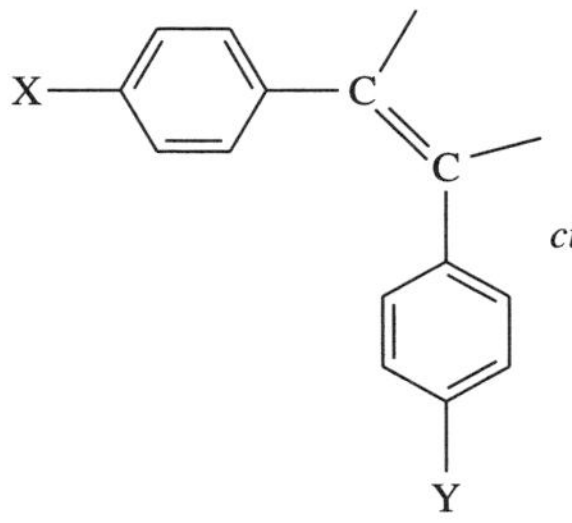

The importance of the effect of the length of the central rigid core of the structure on the thermal stability is illustrated by the following comparison:

(12) C_2H_5O—⌬—CH=N—⌬—OC_2H_5

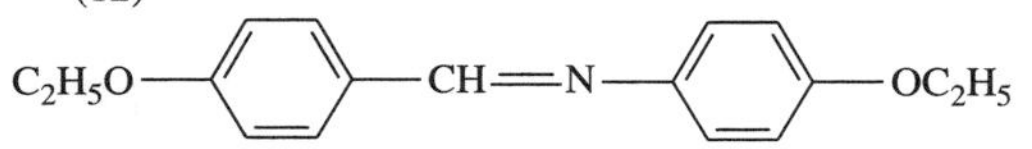

(13) C_2H_5O—⌬—CH=N—N=CH—⌬—OC_2H_5

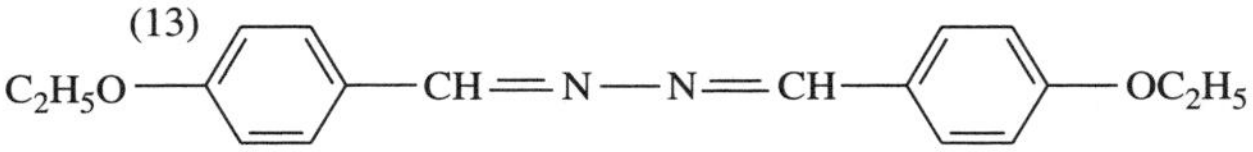

Molecule **(12)** exhibits transitions C–N at 148 °C and N–I at 143 °C, whereas **(13)** has transitions C–N at 172 °C and N–I at 199 °C. Replacement of a single *p*-phenylene ring by a 4,4'-biphenyl or 2,6-naphthalene ring system strongly increases the N–I temperature as illustrated in the following compounds:[16]

(14) C_6H_{13}—⌬—COO—⌬—CN

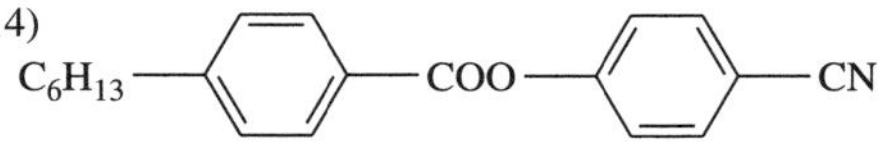

(15) C_6H_{13}—⌬—COO—[naphthalene]—CN

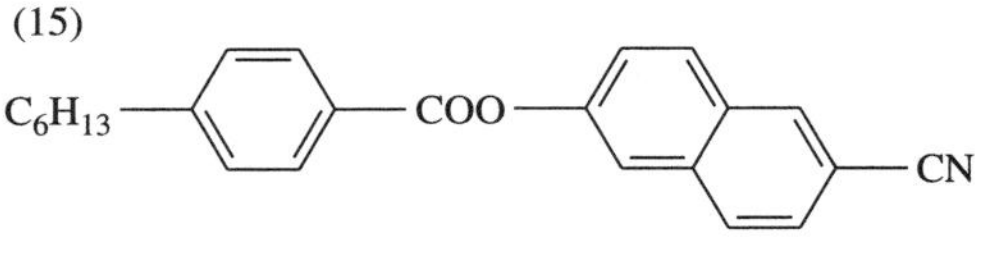

(16) C_6H_{13}—⌬—COO—⌬—⌬—CN

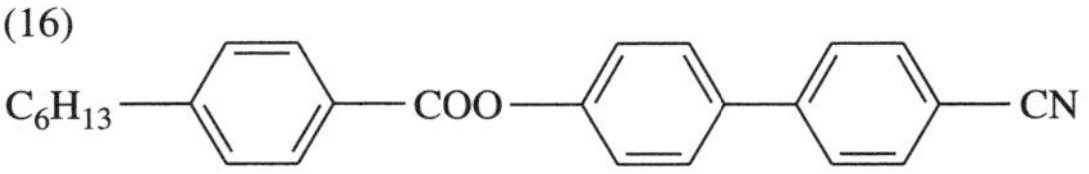

The respective transitions for **(14)** are C–N at 44.5 °C and N–I at 47 °C; for **(15)** are C–N at 72.8 °C and N–I at 138.4 °C; and for **(16)** are C–N at 91 °C and N–I at 229.6 °C. These examples emphasize the importance of extending the rigid core structure. The more elongated the molecule the greater the anisotopic polarizability and conversely changing the conjugated linking units to their flexible saturated equivalents, such as $-CH_2CH_2-$, $-OCH_2CH_2O-$ or $-CH_2O-$, usually leads to non-liquid crystalline materials or to *mesogens* with phases that have a lower thermal persistence. Even when three rings are involved and two are directly linked or are linked by a unit that preserves conjugation and molecular rigidity, a second linking unit can be more flexible and the resultant *nematic* phases are reasonably thermally persistent. If a considerable proportion of a lath-like molecule is rigid then it can pack parallel to neighbouring molecules and the more flexible groups become constrained to be in line with more rigid parts.[17]

(17)

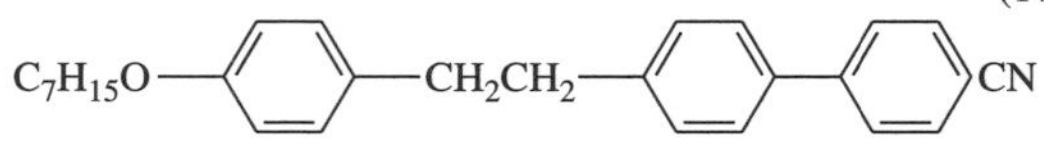

$$C_7H_{15}O--CH_2CH_2---CN$$

Molecule (17) despite containing a flexible linkage exhibits C → S_A at 61.3 °C, S_A → N at 125.8 °C and N → I at 147.8 °C. A plot of the N → I transition temperatures for mixtures of compound (10) with compound (17) is a curve rather than the normal linear plot found with binary mixtures. For compositions around the 50% region the N → I temperature is considerably higher than would be expected. This can be explained if we assume that in the pure compound (17) the breakdown of the *nematic* order is due to the onset of rotations about the CH_2–CH_2 link as the temperature is increased. At the lower N → I temperature of the mixtures, these rotations may be less important and compound (17) behaves as if it has a more rigid structure with a higher N → I temperature, probably around 180 °C. Such an N → I value would be more characteristic of a compound with three *p*-phenylene rings in a rigid molecule, e.g. 4″-substituted 4-cyano-*p*-terphenyl (11).[10]

3.3.2 Influence of the Linking Group on the Thermal Stability of the Nematic Phase

If we consider molecules with the general structure

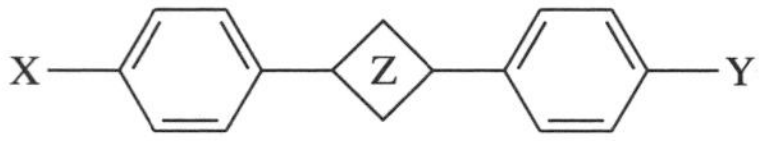

$$X--Z--Y$$

with minor exceptions dependent upon the nature of the end groups X and Y, it is found that the following order is observed in terms of stability of the N → I transition:

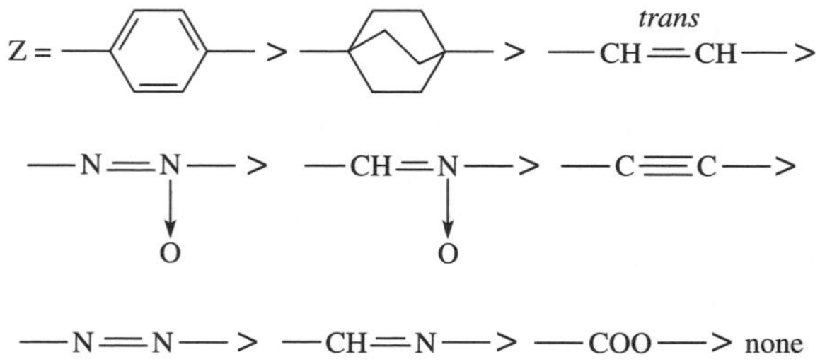

$$Z = -\text{phenyl}- > -\text{cyclohexyl}- > \overset{trans}{-CH=CH-} >$$

$$-N=N- > -CH=N- > -C\equiv C- >$$
$$\downarrow \downarrow$$
$$O O$$

$$-N=N- > -CH=N- > -COO- > \text{none}$$

Stable *mesophases* are produced by substituted *p*-terphenyls and *trans*-stilbenes, whereas *mesophases* of lower thermal stability are formed by Schiff bases,[20] esters and biphenyls. Stereochemical considerations are important and affect the position in the order, e.g. stibenes are planar, azoxybenzenes are slightly twisted and Schiff bases are considerably twisted. The relatively low position of the tolanes (Z=–C≡C–) in the order is surprising since crystalline diphenyl-acetylene[21] contains planar molecules and possibly in the mesophase the rings become non-coplanar. The consequences, however, could be less serious than those in Schiff bases, since the cylinder of electron density associated with the molecular orbital of the –C≡C– linkage may still allow conjugation to occur even when the rings are non-coplanar.

The high position of the bicyclic octane ring[22] reflects both its rigidity and also cylindrical characteristics. In a related set of molecules it has been found that

Z = [benzene] > [bicyclic octene] > [cyclohexene] >

N → I(°C) 285   276   273

[bicyclic octane] > [cyclohexa-1,3-diene] > [cyclohexane] >

269   251   245

trans
—CH=CH—
139

indicating that rings are much more effective than a stilbene linkage. The bicyclic octene ring is a little more effective than the bicyclic octane ring so that unsaturated linkages are of significance, although obviously not the only factor. This is also shown by the superiority of the two cyclohexadiene rings over the cyclohexane ring; one would anticipate that the cyclohexa-1,3-diene ring would have been superior to the cyclohexa-1,4-diene since the former should permit extended conjugation within the molecule. However, the butadiene system is quite seriously twisted in the ring structure and this unquestionably lowers the position of the cyclohexan-1,3-diene ring in the order. These results again emphasize the importance of a careful assessment of the effect of molecular shape geometry and conjugation on the anisotropic polarizability.[23,24]

In practice, the rules are complex and it is not correct simply to say that all cyclohexane rings will have the above effects. In the case where the cyclohexane ring is in a terminal function then the reverse effect can be observed. For example the N → I transitions in the benzoate esters are lower than those in the corresponding *trans*-cyclohexane carboxylic esters.[24,25]

C_nH_{2n+1}—[cyclohexane]—COO—[benzene]—R   (18)

Table 3.3 Transitions for compounds **(18)**, **(19)**, **(20)** and **(21)**.

	18				19			
n	r	C → N (°C)	N → I (°C)	n	R	C → N (°C)	N → I (°C)
5	OC$_4$H$_9$	49	58	5	OC$_4$H$_9$	49	81
4	CN	67.1	42.6	4	CN	54	67.5
6	C$_5$H$_{11}$	28	19	5	C$_5$H$_{11}$	36	48

	20				21			
n	R	C → N (°C)	N → I (°C)	n	R	C → N (°C)	N → I (°C)
5	CN	22.5	35	5	CN	31	55
6	CN	13.5	27	6	CN	42	47
7	CN	28.5	42	7	CN	30	59

C$_n$H$_{2n+1}$—⟨cyclohexane⟩—COO—⟨phenyl⟩—R   (19)

C$_n$H$_{2n+1}$—⟨phenyl⟩—⟨phenyl⟩—CN   (20)

C$_n$H$_{2n+1}$—⟨cyclohexane⟩—⟨phenyl⟩—CN   (21)

The transitions for the –OC$_4$H$_9$, –CN and –C$_5$H$_{11}$ forms of compounds **(18)** and **(19)** together with those of various alkyl compounds of **(20)** and **(21)** are summarized in Table 3.3.

There are obviously subtle effects that influence the precise N → I temperature, and a number of examples exist of the incorporation of a more flexible cyclohexane ring producing new *mesogenic* compounds, although sometimes it does not.[26,27]

3.3.3 Terminal Group Effects

As indicated above, it is possible to create new *mesogens* by changing the terminal group.[20] The effectiveness of various groups can change in relation to the core to which they are attached. For instance, consider the following core **(22)**:

R'O—⟨phenyl⟩—⟨phenyl⟩—N=CH—X   (22)

The effectiveness of the terminal groupings can be ranked as:

X = —⟨pyridyl (N para)⟩ > —⟨pyridyl (N meta)⟩ > —⟨phenyl⟩ > —⟨pyridyl (N ortho)⟩

Whereas in compounds with the general formula based on structure (23):

$$\text{R'O}-\underset{}{\bigcirc}-\text{COO}-\underset{}{\bigcirc}-\text{CH}=\underset{\underset{R''}{|}}{C}-X \quad (23)$$

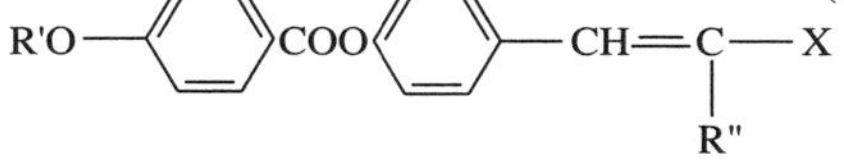

the effective ranking is

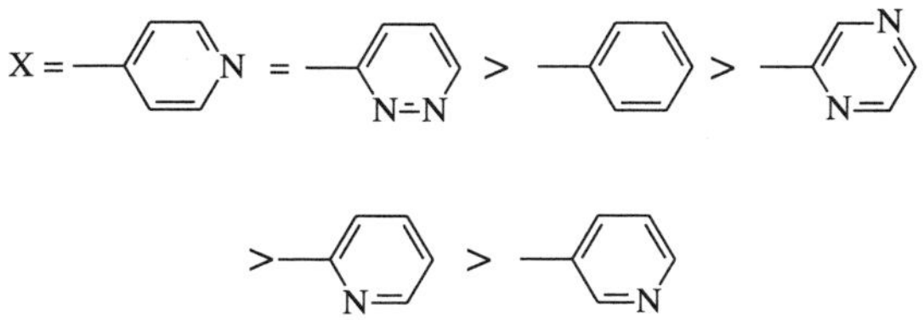

$$X = -\underset{N}{\bigcirc} = -\underset{N=N}{\bigcirc} > -\underset{}{\bigcirc} > -\underset{N}{\bigcirc}^{N}$$

$$> -\underset{N}{\bigcirc} > -\underset{N}{\bigcirc}$$

In the case of compounds with the general structure typified by (24):

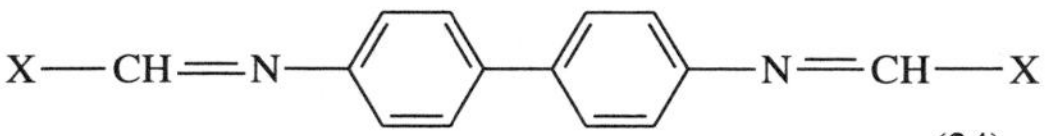

$$X-\text{CH}=N-\underset{}{\bigcirc}-\underset{}{\bigcirc}-N=\text{CH}-X \quad (24)$$

the order now becomes

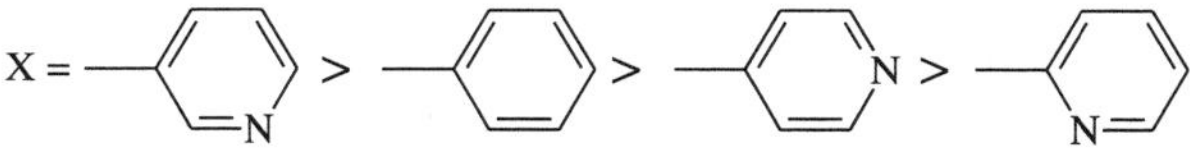

$$X = -\underset{N}{\bigcirc} > -\underset{}{\bigcirc} > -\underset{N}{\bigcirc} > -\underset{N}{\bigcirc}$$

It is clear that no consistent order exists for the effectiveness of the groupings and careful consideration has to be given to all the possible influences on the force field in predicting the nature of a particular system.[20]

3.3.4 Pendant Group Effects

Although it is generally accepted that the molecules should be long and rod-shaped, it is also possible to introduce pendant groupings that will disrupt the packing of the molecules and hence extend the *nematic* range. Consider structures of the type (25) and (26):

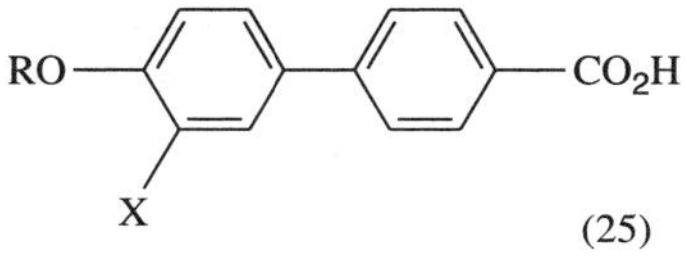

$$\text{RO}-\underset{\underset{X}{|}}{\bigcirc}-\underset{}{\bigcirc}-\text{CO}_2\text{H} \quad (25)$$

$$\text{RO}-\underset{}{\bigcirc}-\text{COO}-\underset{\overset{X}{|}}{\bigcirc}-\text{COO}-\underset{}{\bigcirc}-\text{OR} \quad (26)$$

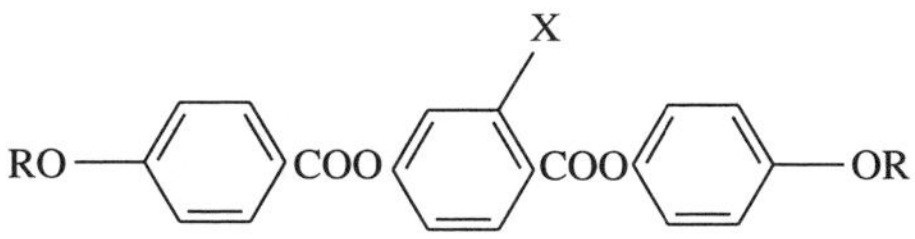

The N → I transition falls in proportion to the size of the substituent in going from X = F, Cl, Br, I, NO_2, CH_3.[26-28] A change in the size of the pendant group leads to a change in the anisotropic polarizability and an increased axial separation of the molecules gives a reduction in the intermolecular attractive forces.[28]

3.3.5 Terminal Substitution Effects

Whilst it is accepted that the shape of a molecule is critical in determining whether the molecule will exhibit *mesogenic* properties, it is the terminal group that often dictates the detailed phase behaviour. The order of terminal group effectiveness in terms of the N → I transition is: Ph > $NHCOCH_3$ > CN > OCH_3 > NO_2 > Cl > Br > $N(CH_3)_2$ > CH_3 > F > H. Replacement of the terminal ring hydrogen in a *mesogen* molecule by any of the commonly encountered substitutents will stabilize the *nematic* order. If a *para* substituent can be embedded in the conjugated system, a larger effect is usually observed as a result of its effect on the axial polarizability. Groups that are readily polarizable in general will have a larger effect than those that are not.[29]

Perhaps the terminal functions which have been of most interest are the alkyl $CH_3(CH_2)_n$– and alkoxy $CH_3(CH_2)_nO$– chains. These flexible chains will introduce the disorder that is necessary for the formation of the *nematic* phase.[30-32] It is generally found that:

- The N → I transition temperature alternates in a regular manner, the degree of alternation diminishing as n is increased. This is observed with the melting temperature of the normal alkanes and is attributed to the packing effects of the terminal methyl groups.
- The N → I temperature for ethers are in general 30–40 °C higher than those for the corresponding alkyl molecules.
- The N → I temperatures for a given series lie on two smooth curves which may either fall or rise as the series is ascended. If the N → I temperatures are high, ~200 °C, the curves will generally fall; however, if they are low, 20 °C or so, then the curves will rise. The crossover temperature between falling and rising curves would appear to be about 100 °C, in which region the curves are often rather flat.
- For a series of ethers, the even carbon chain members give the upper curve and the odd carbon members the lower curve. This situation is reversed for alkyl chains, where the oxygen in an ether effectively replaces a CH_2 group. This effect is also seen in the crystallization behaviour of polymer chains.
- If the alkyl chain adopts an extended *trans* conformation the axial polarizability increases about twice as much as that at right angles on passing from an even to an odd member of an alkyl substituted series. On passing from an odd to an even number the two polarizabilities increase about equally. At comparable molecular weights, the anisotopic molecular polarizability is therefore greater for odd than for even members and their N → I temperatures are higher; the situation is opposite for ethers. This observation implies that there is some conformational preference in the flexible chain.

3.4 Cholesteric Liquid Crystals

Whilst cholesteric liquid crystals, as the name implies, were originally based on cholesterol structures, it is useful to consider the structural features of other molecules that exhibit these phase characteristics.[33-35] In general, cholestric *mesogens* are twisted *nematic* phases, but the feature that separates them from other similar molecules is that they possess an asymmetric *chiral* centre. The structural and geometric factors which influence the temperature of the N → I transition are exactly those which influence the other *nematic* liquid crystals. It is usually found that the twisting power of the system diminishes as the chiral centre is moved away from the core structure, *i.e.* as n increases in the structure shown below:

$$\boxed{}-\langle\rangle-(CH_2)_n\overset{*}{C}H(CH_3)C_2H_5$$

If an oxygen atom replaces a CH_2 group in the flexible chain, the twisting power always decreases. More important, however, is the consistent rule that if the asymmetric centre is S in absolute configuration and the branching point is at an even number of atoms (E) from the core structure, the helix sense will be right handed (D) not left handed (L). We therefore have a simple rule of letters that combine as SED, SOL, REL or ROD, where O refers to a branching point at an odd number of atoms from the core and R refers to the other absolute configuration of the asymmetric centre.[20,36]

3.5 Smectic Liquid Crystals

There are eight recognized *smectic* types donated by S_A, S_B, S_C...S_H. Which phase is formed and the extent of its stability depends very critically upon the nature of the forces which exist between the molecules.[37] These forces are very subtle and it is difficult to generalize; however, certain points can be identified which are worth noting.

(i) If a *smectic* phase is to be formed then the rod-like molecules must usually form a layer crystal lattice that will generate the smectic phase on heating. However, even though a suitable molecular arrangement exists in the crystal lattice if the crystal forces are strong and the melting point of the compound is very high, a smectic phase may not be observed if the parallel stratified arrangement of the molecules is broken down completely or partially by translation in the direction of the major axis of the molecule. If this occurs then a nematic phase can be formed.

(ii) Replacement of a terminal ring hydrogen in a mesogen by any substitutent that does not destroy the linearity of the molecule or broaden it enhances the temperature of the nematic to isotropic transition.

However, substitutents such as –CN, –NO$_2$ and –OCH$_3$ which are high in the nematic terminal group efficiency order are low in the smectic order and can suppress smectic properties relative to nematic phase formation. However, in some instances, e.g. the long-chain 4-n-alkyl- and 4-n-alkoxy-4′-cyanobiphenyls, the system seems to adapt by forming a S$_A$ phase having an interdigitated bilayer structure and the smectic properties are then enhanced relative to S$_H$.[38-40] Terminal groups which contribute to the resultant dipole across the long axis, e.g. –COO–alkyl, –CH=CHCO–alkyl, –CONH$_2$ or –OCF$_3$, strongly promote smectic properties as do ionized functions, e.g. –COO$^-$M$^+$ or –NH$_3^+$X$^-$. Groups such as –Ph, –NHCOCH$_3$ and –OCOCH$_3$ strongly promote both smectic and nematic properties, but affect the smectic properties more markedly.

(iii) Extending the length of a terminal n-alkyl chain increases the smectic tendencies relative to the nematic tendencies of a system. Eventually a stage is reached when nematic properties are extinguished and the compounds are purely smectic. This behaviour is very general for S$_A$ and S$_C$ phases.

(iv) Alkyl chain branching on the C$_1$ of an ester alkyl chain (–COO–alkyl) has a much smaller effect on the thermal stability of S$_A$ phases than on *nematic* phases. Moving a branching CH$_3$ to C$_2$, C$_3$, *etc.*, gives progressively smaller effects. S$_B$ thermal stability is not greatly affected by a methyl branch, but in S$_E$ a 1-methyl group enhances thermal stability. The results in Table 3.4 illustrate the effects observed[41] for compound **(27)**.

Ph–Ph–CH=N–Ph–CH=CH–COO alkyl

(27)

(v) The effects of pendant groups have been studied extensively in system **(28)** that gives S$_C$ phases when the alkyl chain is sufficiently long. Whereas the *nematic* thermal stabilities decrease in proportion to the size of X, irrespective of its polarity, the S$_C \to$ N and the S$_C \to$ I transition temperatures again decrease but do so less in relation to the size of X if X is dipolar. Thus dipole moments do contribute to S$_C$ thermal stabilities. For example, X = CH$_3$ or Cl has a similar effect in decreasing $T[N \to I]$, but a larger effect on $T[S_C \to N]$ is given by

Table 3.4 Transition temperatures for compounds based on compound **(27)**.

Alkyl group	$C \to S_E$ (°C)	$S_E \to S_B$ (°C)	$S_B \to S_A$ (°C)	$S_A \to I$ (°C)
CH$_2$CH$_2$CH$_2$CH$_2$CH$_3$	92	101.5	168	204
CH(CH$_3$)CH$_2$CH$_2$CH$_2$CH$_3$	70	128	168	180
CH$_2$CH(CH$_3$)CH$_2$CH$_2$CH$_3$	68	113.5	157	190.5
CH$_2$CH$_2$CH(CH$_3$)CH$_2$CH$_3$	77.5	109	167	196
CH$_2$CH$_2$CH$_2$CH(CH$_3$)$_2$	93	108	168	199

$X = CH_3$. This illustrates the combination of both the effect of size and polarity on the S_C phases.

As for *nematic* phases, if the pendant group occupies a recess in the structure such that its full broadening effect is not operative, an increase in S_C thermal stability can arise, e.g. in 5-subsitututed 6-*n*-alkoxy-2-naphtholic acids. The increase in $T[S_C \rightarrow N]$ is larger than that for $T[N \rightarrow I]$, showing again that dipole interactions play a role in enhancing the thermal stability of the S_C transition.

Schiff bases have been studied in **(28)** and **(29)** which exhibit S_A and/or *nematic* phases.[42]

(28)

(29)

The 2- and 2'-substituents cause large reductions in $T[N \rightarrow I]$ because of the additional twisting about the inter-ring bond. In the 2-fluoro-substituted compound of **(28)**, $\Delta T[N \rightarrow I]$ is 49 °C. Even larger effects on the S_A phases are observed and $\Delta T[S_A \rightarrow N]$ is 71 °C for the 2-fluoro-substituted compound of **(28)**. Decreases in N and S_A thermal stability of only 0.5 °C and 9 °C, respectively, have been observed for fluoro substituents that exert no steric effect.[42] Unlike the trends in $T[N \rightarrow I]$ which follow the substitutent size, the trends in $T[S_A \rightarrow N]$ are irregular, suggesting that dipolar effects could again be playing a role.

(vi) Finally, in assessing a situation with regard to *smectic* or *nematic* tendencies, attention has to be paid to the location in the molecule of particular functions. For example, whereas a terminal COO–alkyl group favours *smectic* properties, the–COO group may be used successfully to link ring systems and produce strongly *nematic* materials. An illustration of this behaviour is benzoate esters such as **(30)** which are purely *nematic* (C → N occurs at 54 °C; N → I occurs at 89 °C despite the long alkyl chains at each end).

(30)

It is only possible in this chapter to outline briefly some of the factors that influence the formation of liquid crystalline phases and influence their stability, but these illustrations indicate how subtle effects of steric and dipolar interactions combine to give a large range of compounds which exhibit liquid crystalline behaviour. A more extensive discussion can be found elsewhere.[20]

3.6 Theoretical Models for Liquid Crystals

The general rules that emerge from analysis of the factors that influence the formation of a liquid crystalline phase are that:

- the molecules should contain a significant rigid block plus a flexible element which introduces the required level of disorder to avoid crystallization of the molecules,
- the length to the breadth of the molecules is usually of the order of 6:1, and
- the molecules in general will contain an element of the structure which is capable of significant long-range dipolar interactions.

Theoretical interest over the last thirty years in the modelling of the properties of liquid crystals has been driven by their application in displays. A variety of theoretical approaches have been applied to the description of the physical properties of liquid crystals. The statistical models developed are intuitively closer to our molecular understanding of these materials and will be considered initially. The more continuum approach, which allows visualization of macroscopic effects, will be considered later.

3.6.1 Statistical Models

If the liquid crystalline molecule is considered as a rod, then as the liquid is cooled and the density increases, the molecules will attempt to align and crystallize. To stop the molecules crystallizing it is necessary for the alignment of individual pairs of molecules to be inhibited. A molecule in the isotropic liquid phase has three degrees of freedom: two degrees of freedom in terms of rotation about the major and minor axes and one in terms of translation. Loss of these elements of freedom describes the liquid crystal phase transitions discussed above.

Starting from the isotropic phase, where the molecules have all three degrees of freedom, cooling will increase the density and rotation about the long axis becomes restricted. Series of models have been developed that consider the density of liquid in terms of the restriction of the order.[43-47] These theories identify a critical density at which the isotropic to *nematic* transition would be predicted. Constraint of the molecule in terms of its rotation about the long axis defines the *nematic* phase. If now the translational freedom is restricted and layered alignment is imposed on the molecules, then *smectic* order is created. The *smectic* phase can still retain disorder in rotational freedom about the short axis. Loss of this final degree of freedom will lead to the creation of a crystalline ordered structure. This simple approach provides a description for the isotropic → *nematic* → *smectic* → crystalline transitions.

Many molecules exhibiting liquid crystal properties have either a terminal alkyl or alkyl ether group. The terminal chains exhibit a number of different conformations, as in **(31)**:

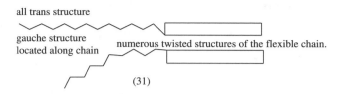

(31)

Within the liquid crystalline phase at a defined temperature there will be a statistical distribution of molecules with different chain conformations. As the liquid is cooled, so the lower energy all-*trans* conformation becomes preferred. There is a tendency for molecules with a similar structure to cluster and form nuclei for crystallization. Disordered clusters, however, form the basis of liquid crystalline phases, alignment being restricted to a preferred direction. The alignment of the molecules in space will not necessarily be continuous throughout the bulk of the material as would be found in a crystal. The variation of the direction of the molecular alignment can change in space in addition to the local order varying in the way that is broadly defined by the classification of *nematic* or *smectic* structures. The possible complexities of such a structure have led to the requirement to consider the nature of the organization in terms of an 'order' parameter.

The problem with the accurate modelling of the strength of the potential field that acts on a molecule depends on the type of model used. Most models use the van der Waals potential to describe the strong interactions between neighbouring molecules. In the calculation of the force field, it is usual to consider an average of six nearest neighbours. This approach immediately introduces the concept that the potential energy function should be a *mean field approximation* being an average of the interaction between specific elements of the molecular structure. A molecule in the *nematic* phase may be assured to have a higher degree of rotational freedom about its long axis than in a crystal and therefore it is reasonable to assume that a *mean field potential* is appropriate for the modelling of this type of situation. As will be appreciated from the above discussion of the molecular structure, most liquid crystalline materials will contain a significant dipole or anisotropic electronic polarizability. The dipolar and higher order anisotropic electronic polarizability effects are longer-range effects. The asymmetry of the force field will ultimately remove the rotational degree of freedom about the long axis, assist the alignment of molecules into layered structures and ultimately create the organization found in crystals. A refined model would consider the rod as being made up of a rigid element that contains dipole/electronic interaction and the flexible tail element that introduces the entropic element that helps create a variety of twisted and bent morphologies which are found in liquid crystals and helps suppress the formation of ordered crystalline material. This simple approach alters the description of the system in terms of a series of partition functions and allows a statistical mechanical approach to be developed for the description of the liquid crystalline phase.

3.6.2 Development of Statistical Mechanical Models

The first statistical theory was developed considering the critical density for the formation of the *nematic* phase and was based on a simple mean field approximation of the description of the force field.[43–45] This study leads to predictions that there should be a critical density ρ^* at which the phase transition will occur. Since on cooling the material the density will increase, there will also exist a critical temperature T^* at which such a transition occurs.

3.6.3 Distributions and Order Parameters

Using birefringence to examine the molecular alignment in a typical *nematic* indicates that there is a spread of orientations about the general direction of alignment. As a result there is an angle θ_1 between the applied field and the average direction of alignment of the molecules θ_2. The molecules will be distributed around this latter angle and expressed by the second Legrande polynomial [P_2], which automatically ignores higher order moments and hence assumes rotational averaging.

In an isotropic medium, the value of a physical property can be represented by a scalar quantity; however, for an anistropic medium the value depends on the measuring angle relative to the direction of alignment designated as the *director*. The properties of a *nematic* must be represented by a tensor which in the case of a uniaxial *nematic* for a property X can be conventionally expressed by $X_1=X_2=X_\perp$ and $X_3=X_\parallel$ which are, respectively, the values of the property X perpendicular and parallel to the director:

$$X = \begin{bmatrix} X_1 & 0 & 0 \\ 0 & X_2 & 0 \\ 0 & 0 & X_3 \end{bmatrix} \qquad (3.1)$$

we consider the molecule to resemble a rigid rod and select the optical axis by defining a unit vector **a** in the laboratory frame and give it an angle υ to the *director*, where **n** lies along the Z-axis and φ is the angle formed between the projection of **a** onto the X–Y plane and the X axis:

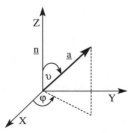

The components of the vector **a** are

$$a_z = \cos \upsilon; \ a_x = \sin \upsilon \cos \varphi; \ a_y = \sin \upsilon \sin \varphi \qquad (3.2)$$

A distribution function $f(\upsilon, \varphi)d\Omega$ can be defined which gives the probability of finding rods in a small solid angle $d\Omega=\sin \upsilon\, d\upsilon d\varphi$ around the direction (υ, φ). As **n** is defined along the Z-axis, $f(\upsilon, \varphi)$ is independent of φ, due to the uniaxial nature of the *nematic*. Consequently **n** has complete symmetry around the Z-axis and $\mathbf{n} \cong -\mathbf{n}$; therefore $f(\upsilon)=f(-\upsilon)$. A straightforward summation would be

$$S_1 = \int \cos \upsilon f(\upsilon, \varphi)\partial\upsilon\partial\varphi = \langle \cos \upsilon \rangle = 0 \qquad (3.3)$$

This is the second Legrande polynomial. Going to higher order terms gives

$$S_2 = \int \frac{1}{2}(3\cos^2\upsilon - 1)f(\upsilon, \varphi)\partial\upsilon\partial\varphi; \quad S_2 = \frac{1}{2}\langle 3\cos^2\upsilon - 1\rangle \qquad (3.4)$$

The factor of $\frac{1}{2}$ is introduced to avoid double counting of the interactions. This expression is useful, as for a perfectly aligned *nematic* $\upsilon=0 \Rightarrow S=1$, whereas in the random or isotropic liquid $\cos^2\upsilon=1/3 \Rightarrow S=0$. Thus *nematic* liquid crystals have values of S that will range form 0 to 1.

A model for the temperature dependence of the order parameter can be developed based on either assuming that short-range forces are dominant (van der Waals interactions are controlling) or long-range forces are dominant (electrostatic interactions are controlling). The long-range approach has been developed by Maier and Saupe.[48,49]

This *mean field theory* uses a weak anisotropic potential $F(\upsilon)$ which is given the form

$$F(\upsilon) = \left(\frac{A}{V^2}\right)S\left(\frac{3}{2}\cos^2\upsilon - 1\right) \qquad (3.5)$$

where A is the strength of the potential and V is the molar volume. The effective orientation potential for a single molecule then becomes

$$U(\cos\upsilon) = u_2 S_2 P_2(\cos\upsilon) \qquad (3.6)$$

Thus the average single particle approximation is proportional to the order parameter squared, where u_2 is related to γ and to T_c, the clearing temperature that marks the nematic to isotropic transition:

$$\gamma = 4.5415 T_c \qquad (3.7)$$

This assumes that the van der Waals repulsive forces dominate U_2, but are perturbed by longer range electrostatic and dipolar interactions and are approximately temperature independent. The Maier–Saupe theory[48,49] predicts S_2 being a universal function of reduced temperature:

$$\tau = \left(\frac{TV^2}{T_c V_c^2}\right) \qquad (3.8)$$

and leads to S_2 being equal to 0.43 at T_c for all materials. At this temperature $(A/VkT) = 4.55$. A reasonably accurate order parameter–temperature relationship is given by the expression

$$S_2 = (1 - 0.98\tau)^{0.22} \tag{3.9}$$

Further refinements of the theory have been made and these lead to more accurate predictions of the behaviour.

The alternative approach of using short-range interactions to create the potential leads to problems with the prediction of the *clearing point*—the *nematic* to isotropic transition. This theory models close packed clustering and hence a pseudo melt transition.

The order parameter is a theoretical tool that can be measured experimentally. It is possible to determine the anisotropy of a particular property for a *nematic* relative to the anisotropy of a perfectly aligned material. Consider a molecule lying at an angle v to n as indicated below:

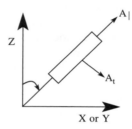

The value of a particular property along the molecular axis is $A_|$ and the value along the ordinary axis is A_t and correspond to the values which would be obtained from studies of a perfect single crystal. Thus the property A can be expressed as

$$A = \begin{bmatrix} A_t & 0 & 0 \\ 0 & A_t & 0 \\ 0 & 0 & A_| \end{bmatrix} \tag{3.10}$$

Thus the effective value of the property A along the Z-axis will given by

$$A_z = A_1 \cos^2 v + A_\tau \sin^2 v \tag{3.11}$$

$$= A_1 \cos^2 v + A_t \sin^2 u \tag{3.12}$$

$$= (A_| - A_t) \cos^2 v + A_\tau \tag{3.13}$$

$$= \frac{(A_| + 2A_t)}{3} + \frac{2}{3}(A_| - A_t)\frac{(3\cos^2 v - 1)}{2} \tag{3.14}$$

$$= A_{\text{iso}} + \frac{2}{3}(A_| - A_t)\left(\frac{3\cos^2 v - 1}{2}\right) \tag{3.15}$$

where A_{iso} is the isotropic or average value of the property A. Summing over all molecules:

$$A_\| = A_{iso} + \frac{2}{3}(A_| - A_t)\left\langle \frac{3\cos^2 v - 1}{2} \right\rangle \tag{3.16}$$

However

$$S_2 = \frac{1}{2}\langle 3\cos^2 v - 1 \rangle \tag{3.17}$$

so that

$$S_2 = \frac{3}{2}\frac{A_\| - A_{iso}}{(A_| - A_t)} = \frac{A_\| - A_\perp}{A_\| - A_t} \tag{3.18}$$

It is therefore possible if the single crystal property is known to determine the order parameter of the liquid crystalline phase.

3.7 Elastic Behaviour of Nematic Liquid Crystals

The *nematic* phase has properties that resemble those of a solid yet also those of a fluid. A very successful approach to modelling the behaviour of liquid crystals is based on fluid dynamics. These theories are based on the concept of the Leslie–Erickson theory of hydrodynamic flow.[50,51] The theory considers the use of a series of coefficients, the rotations between the coefficients describing the liquid. The state of a fluid at point r where $r = (x, y, z)$ at a given time is defined by the fluid velocity $v(r, t)$ and any two thermodynamic variables, such as pressure $P(r, t)$ or density $\rho(r, t)$. The fluid obeys the conservation laws:

- *Conservation of mass.* Consider a surface of unit area, through which a fluid flows. In a unit time, the mass of fluid that flows through that area is the component of ρv normal to that area. The variation of ρv is the rate of change of the density per unit time so that: $(\partial \rho/\partial t) + \Delta(\rho \mathbf{v}) = 0$.
- *Conservation of linear momentum.* The rate of change of momentum of a volume element in the fluid is equivalent to the total force $\mathbf{F}^T$ acting on the element $\mathbf{F}^T = \rho(\partial v/\partial t)$ and the acceleration is given by $\partial v_i/\partial t = (\partial v_i/\partial t) + v_j(\partial v_i/\partial x_j)$. The possible contributions to $\mathbf{F}^T$ may come from pressure P, viscous drag $\mathbf{F}^v$ or an external field $\mathbf{F}^f$ which could be gravitational, electrical or magnetic: $\rho(\partial \mathbf{v}/\partial t) = \mathbf{F}^v + \mathbf{F}^f - \Delta P$. The force per unit area t acting on a normal to a surface v is given by $t_i = t_{ij}v_j$. The components of t_{ij} represent the ith component of force on a surface and x_j equals a constant. Where $i = j$, this is equivalent to a compression and where $i \neq j$, these are shear stresses. Consequently t_{ij} can be split into parts for an incompressible fluid: $t_{ij} = -P\delta_{ij} + \hat{t}_{ij}$, where $\hat{t}_{ij}$ is the extra stress tensor and δ_{ij} is the Kronecker delta. The conservation of linear momentum now takes the form $\rho(\partial v_i/\partial t) = F_i - P_i + \hat{t}_{ij,j}$. In the case of a transverse velocity gradient $\partial u/\partial x$, the induced shear stress in flow becomes $\hat{t}_{21} = \eta(\partial u/\partial x)$, where η is the viscosity coefficient. In the general

form $\hat{t}_{ij} = \eta[(\partial v_i/\partial x_j) + (\partial v_j/\partial x_i)]$, where $\partial u_i/\partial x_j$ is the transpose of $\partial u_j/\partial x_i$. It can be shown that every tensor can be split into a symmetric and an asymmetric part. Carrying out this operation on the viscosity gradient tensor, the symmetric part is

$$A_{ij} = \frac{1}{2}\left(\frac{\partial v_i}{\partial x_j} + \frac{\partial v_j}{\partial x_i}\right) = \frac{1}{2}(v_{i,j} + v_{j,i})$$

and the anti-symmetric part is

$$\omega_{ij} = \frac{1}{2}\left(\frac{\partial v_i}{\partial x_j} - \frac{\partial v_j}{\partial x_i}\right) = \frac{1}{2}(v_{i,j} - v_{j,i})$$

Thus A_{ij} is the stretching tensor or rate of strain and ω_{ij} is the rotation or vorticity tensor and are illustrated below.
- *Conservation of angular momentum.* The net torque about a given point which acts on the particle is a volume V, relating to the body forces acting throughout the volume and the stress acting on the surface. Consequently at least force couples may also act on the material and are characterized by the vector K_i, force per unit mass. Although force couples are generally not considered in hydrodynamics, they are important in this context as they allow the introduction of external forces: magnetic and electric fields, *etc.* By equating the net torque about a point with the rate of change of angular momentum and assuming that there are no stress couples, it can be shown that $\varepsilon_{ijk}t_{kj}+K_i=0$. This illustrates that when there is no force couple acting throughout the fluid ($K_i=0$) the stress tensor is symmetric.
- *Conservation of energy.* The conservation of energy is the equivalence of the material derivative of kinetic and internal energy with the rate at which mechanical work is done on the fluid. Thus work is concerned with the effects of external forces producing kinetic energy and viscous dissipation. If the rate of viscous dissipation is $\rho\dot{\Phi}$ then $\rho\dot{\Phi} = \hat{t}_{ij}v_{ij}$, and since the stress is symmetrical the above expression becomes $\rho\dot{\Phi} = \hat{t}_{ij}A_{ij} = 2\eta A_{ij}A_{ji}$ and the rate of increase of entropy $\dot{S}$ is equivalent to the heat supply divided by the temperature: $\dot{S} = (1/T)2\eta A_{ij}A_{ji}$. As $\dot{S}$ is positive, so the coefficient of the viscosity must be positive.

The above is a quick summary of the relevant hydrodynamic equations for a simple fluid. The behaviour of a *nematic* liquid crystal is more complex in that the stress tensor is now non-symmetric. Another variable that is introduced is the director, **n**, defined by a unit vector, where $\mathbf{n} \sim -\mathbf{n}$. Allied to these constraints, one defines the rate of rotation of the director with respect to the background fluid by

$$N_i = \dot{n}_i - \omega_{ij}n_j \qquad (3.19)$$

Applying the conservation laws summarized above gives

$$\text{Conservation of mass}: \quad v_{i,i} = 0 \qquad (3.20)$$

Conservation of linear momentum : $\rho \dot{v}_i = \rho F_i + t_{ij,i}$ (3.21)

Conservation of angular momentum : $o\varepsilon_{ijk} n_j \ddot{n}_k = K_i + \varepsilon_{ijk} t_{kj} + l_{ij,j}$ (3.22)

Conservation of energy : $\dot{E} = t_{ij} v_{i,j} + s_{ij} (\dot{n}_i)_j - g_i n_i$ (3.23)

where o is an internal constant associated with the director. This term is taken as being equal to a^2, where a is a molecular dimension. Since this is a constant it is often ignored. F is the external body force, K is the external body couple, l is the stress couple tensor, E is the internal energy per unit volume, s is the director stress tensor and g is the intrinsic *director* body force. The constitutive equations are[50,51]

$$t_{ij} = -P\delta_{ij} - \frac{\partial W}{\partial n_{k,j}} n_{k,i} + \tilde{t}_{ij}$$ (3.24)

$$s_{ij} = n_i \beta_j + \frac{\partial W}{\partial n_{i,j}} + \tilde{s}_{ij}$$ (3.25)

$$g_i = \lambda n_i - (n_i \beta_j)_j - \frac{\partial W}{\partial n_i} + \tilde{g}_i$$ (3.26)

where W is the free energy density relating to the elastic energy deformation, $\tilde{t}_{ij}$ is the viscous stress tensor, β is an arbitrary vector arising from the constrained director, ($n_i.n_i = 1$), $\tilde{g}_i$ is the intrinsic director body extra force and $\tilde{s}_{ij}$ is the director extra stress tensor, which equals zero. These extra terms relate to the dynamic behaviour of the system. The above equations lead to

$$\tilde{t}_{ij} = \alpha_1 n_k n_p A_{kp} n_i n_j + \alpha_2 n_i n_j + \alpha_3 n_j n_i + \alpha_4 A_{ij} + \alpha_5 A_{ik} n_k n_j + \alpha_6 A_{jk} n_k n_i$$ (3.27)

where $\gamma_1 = \alpha_3 - \alpha_2$ and $\gamma_2 = \alpha_6 - \alpha_5$ and α_i are the Leslie coefficients of the viscosity. Leslie has subsequently shown that these coefficients are related to the various components of the viscosity. The coefficient α_4 is the isotropic viscosity coefficient and does not involve the director. The coefficients α_1, α_4, α_5 and α_6 are related to the deformation tensor, whereas α_2 and α_3 involve only the vorticity and are related to the coupling between the director orientation and the flow.

The above equations can be related to measurable viscosity coefficients. The coefficients allow one to develop the concepts of stability of the flow in the applied field and have allowed examination of the dynamics of switching of the liquid crystal systems. The above theory illustrates that although the liquid crystals have many facets of liquids they exhibit coupled behaviour that is characteristic of solids.

3.8 Computer Simulations

Paralleling the developments of liquids, considerable interest has been shown in the ability to develop computer-based theoretical models that can describe the onset of the *mesophase*. As might be expected, much of the modelling is based on a Monte Carlo or a molecular dynamics approach. Zannoni[52] has reviewed some of the earlier simulations. The Monte Carlo approach involves the orientation of the molecules being determined by an averaged potential function and using the Metropolis technique,[53] which involves introducing a stochastic process in that asymptotically each configuration recurs with a frequency proportional to the Boltzmann factor for that state. In order to achieve calculations that describe the whole system it is usual not to impose boundary conditions and use the technique of periodic boundaries to simulate bulk behaviour. In the usual application, it is known that surface interactions are critically important and hence the simulation of these effects is particularly taxing in terms of computer time. To assist with the scaling of the calculations it is also appropriate to compute the orientational order parameter to assess the degree and type of alignment being created. Lattice models based on the density correlations have also been widely reported and demonstrate good qualitative correlation with experiment. The molecular dynamics simulations involve setting up and solving numerically the equations of motion for a system of N molecules contained in a box of given volume. The calculations involve the separation of rotational and translational motion and examination of the contribution each makes to the total energy.

Recent simulations work[54] has demonstrated the Gay–Berne (GB) potential,[55] developed more than twenty years ago as a model for describing the interactions between two elongated rigid molecules, provides a very useful approach to liquid crystal modelling. The GB potential owes much to the pioneering work of Corner[56] who noted that the Lennard-Jones 12–6 potential provided a good description of the potential between a pair of atoms and so could also be used to describe the interactions between larger molecules if their deviations from spherical symmetry were coded into the potential. Although his ideas worked reasonably well for small molecules such as the nitrogen dimer, the potentials were not suitable for more elongated molecules, especially those of length-to-breadth ratios characteristic of liquid crystals.

Gay and Berne[55] showed that the interaction potential between a pair of rigid, elongated molecules could be reasonably well represented in a Lennard-Jones 12–6 form by

$$U(\text{GB})(\hat{u}_i, \hat{u}_j, r) = 4\varepsilon(\hat{u}_i, \hat{u}_j, r) \left[\frac{\sigma_s}{r - \sigma(\hat{u}_i, \hat{u}_j, \hat{r}) + \sigma_s} \right]^{12} - \left[\frac{\sigma_s}{r - \sigma(\hat{u}_i, \hat{u}_j, \hat{r}) + \sigma_s} \right]^{6} \tag{3.28}$$

where $\hat{u}_i$ and $\hat{u}_j$ are unit vectors describing the orientations of the two molecules and $\hat{r}$ is a unit vector along the intermolecular vector r, with $r = |r|$. Unlike the scalar s in the well-known Lennard-Jones potential for atoms, the distance parameter $\sigma(\hat{u}_i, \hat{u}_j, \hat{r})$ for elongated molecules should be dependent on the orientations of the two molecules and the intermolecular vector.[57] This is written as

$$\sigma(\hat{u}_i, \hat{u}_j, \hat{r}) = \sigma_s \left[1 - \frac{\chi}{2} \left(\frac{(\hat{u}_i\hat{r} + \hat{u}_j\hat{r})^2}{1 + \chi(\hat{u}_i\hat{u}_j)} \right) + \frac{(\hat{u}_i\hat{r} - \hat{u}_j\hat{r})^2}{1 + \chi(\hat{u}_i\hat{u}_j)} \right]^{-1/2} \quad (3.29)$$

Although this looks complicated, it is just a way of coding the shape of the molecule into the orientation-dependent contact separation. A key component of the potential is the anisotropy in the shape of the molecule, χ, which is dependent on the length-to-breath ratio, κ ($= \sigma_e/\sigma_s$, where σ_e and σ_s are, respectively, the length and breadth of the molecule), and defined by $\chi = (\kappa^2 - 1)/(\kappa^2 + 1)$. Similarly, the energy parameter $\varepsilon(\hat{u}_i, \hat{u}_j\hat{r})$ should also be orientation dependent.[56] An important parameter entering this term is the ratio of the potential energy for a pair of molecules in the side-by-side arrangement to that in the end-to-end arrangement: $\kappa' = \varepsilon_s/\varepsilon_e$. Two further parameters are included, m and n, allowing some extra freedom for the potential to be fitted to a particular model. Using this approach, fairly accurate estimates of the temperature of the transitions have been obtained; however, some of the more detailed predictions tend to be less accurate.

3.9 Defects, Dislocations and Disclinations[20,58]

Whereas defects in molecular crystals are associated with dislocations, in liquid crystalline materials disclinations are discontinuities in orientation in the bulk material, *i.e.* within the *director* field. Disclinations are line singularities perpendicular to the layer. They appear as dark brushes when thin films (10 µm) are viewed under cross polarizers. On rotating the polarizers, the positions of the points remain unchanged but the brushes themselves rotate continuously. The sense of the rotation may either be the same sense as that of the polarizers (positive disclinations) or opposite (negative disclinations). The rate of rotation is about equal to that of the polarizers when the disclination has four brushes and is twice as fast when it has only two. The strength of the disclination is defined as $S =$ (number of brushes)/4. A number of these structures have been observed, $S = +1/2$, $-1/2$, $+1$ and -1. Neighbouring disclinations connected by brushes are of opposite signs and the sum of the strengths of all disclinations in the sample tends to zero. Normally, disclinations remain static in the field of view but as the temperature approaches the nematic–isotropic transition they tend to become mobile and disclinations of opposite sign are seen to attract one another and coalesce.

To understand the relationship to the director it is helpful to consider a planar sample in which the director orientation is parallel to the glass surfaces and not a function of the thickness. This assumption is of course not valid close to the singularities. We will firstly consider *nematic* liquid crystals with a curved structure, splay, twist and bend, as indicated below:

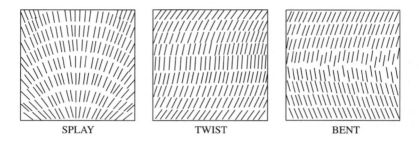

SPLAY   TWIST   BENT

The elastic free energy per unit volume is $\frac{1}{2}k_{11}(\partial\phi/\partial x)^2$, where $\phi = n_x$ is the tilt of the *director* and k_{11} the splay constant. The free energy density may be written for splay as $F = \frac{1}{2}k_{11}(\partial n_x/\partial x)^2$, for twist as $F = \frac{1}{2}k_{22}(\partial n_x/\partial y)^2$ and for bend as $F = \frac{1}{2}k_{33}(\partial n_x/\partial z)^2$. In the more general form it may be written as

$$F = \frac{1}{2}k_{11}(\nabla\mathbf{n})^2 + \frac{1}{2}k_{22}(\mathbf{n}\nabla x\mathbf{n})^2 + \frac{1}{2}k_{33}(\mathbf{n}x\nabla x\mathbf{n})^2 \qquad (3.30)$$

The elastic constants k_{11}, k_{22} and k_{33} are usually of the order of 10^{-6} to 10^7 dyn.

In the case of the disclinations, only two elastic constants are involved, k_{11} and k_{22}, that to a first approximation are taken as being equal to k. If we consider cylindrical coordinates (r, α) and now seek solutions for eqn (3.30) in which the director orientation ψ is independent of r, then the elastic free energy is

$$F = \frac{\frac{1}{2}k}{r^2}\left(\frac{\partial\psi}{\partial\alpha}\right)^2 \qquad (3.31)$$

Minimization of eqn (3.31) gives ψ a constant which describes the uniformly orientated nematic sample, or $\psi = s\alpha + c$, where c is a constant. In the nematic case, the orientational order is taken to be polar, and hence $S = \pm 1/2, \pm 1, \pm 3/2, \ldots$, with $0 < C < \pi$. The angle between two successive dark brushes is therefore $\Delta\alpha = \Delta\psi/S = \pi/2S$, and thus the number of dark brushes per singularity is $2\pi/\Delta\alpha = 4|S|$. A few examples of the types of molecular orientation in the neighbourhood of a disclination are shown below:

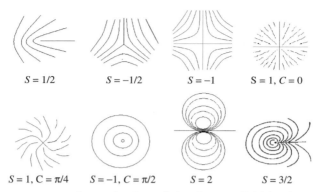

$S = 1/2$　　$S = -1/2$　　$S = -1$　　$S = 1, C = 0$

$S = 1, C = \pi/4$　　$S = -1, C = \pi/2$　　$S = 2$　　$S = 3/2$

The curves represent the projection of the *director* field in the xy plane. For $S \neq 1$, a change in C merely causes a rotation of the figure by $C/(1 - S)$, while for $S = 1$ the pattern itself is changed. The disclinations are characterized by their 'strengths', which are defined as the number of multiples of 2π that the director rotates in a complete circuit around the disclination core. A value of $+1$ indicates that the director is rotated through 2π. The micrograph of a liquid crystal material, shown below, illustrates the various disclinations described above.

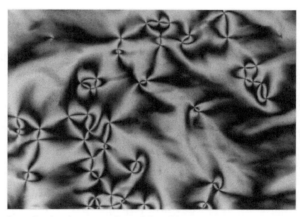

A variety of optical micrographs of liquid crystal systems are to be found at the following websites:

http://micro.magnet.fsu.edu/micro/gallery/liqcryst/liquidcrystal.html
www.cemes.fr/Oper_recherche/Polymer/LCP/Html_LCP/Liquid_Crystals_Physics.htm
www.warwick.ac.uk/fac/sci/Chemistry/jpr/lq/liqcry.html
www.lci.kent.edu/polmmicpic.hmtl

These collections of images illustrate the various types of liquid crystal that have been studied and the broad range of morphologies that can be observed.

3.10 Applications[59]

It is not appropriate to discuss the various ways in which liquid crystals are used, save to say that there are two principal effects that are employed. With chiral nematics the pitch of the rotation of the light will influence the colour that is observed. The interlayer spacing that is itself controlled by the temperature influences the pitch. As a consequence changing the temperature will lead to a change in the colour. This is used in liquid crystal thermometers. Displays, in contrast, use the fact that the alignment of the liquid crystal can be influenced by an external electric or magnetic field. In the case of a twisted nematic the application of an external electric field can change the alignment of the director as viewed through the plane of the liquid crystal film and this will hence change the effects it can have on polarized light. Most optical displays use the fact that liquid crystals can be aligned by interaction with a substrate. As a consequence it is relatively easy to achieve a twisted configuration. Application of an external field can be arranged to rotate the alignment of the liquid crystals so that they will now change their interaction with the polarized light. The results is that colour contrasts can be achieved between those areas where the electric field exists and those where it does not. The reader is refered to specialist texts for a detailed discussion of this topic.[60-64]

3.11 Polymeric Liquid Crystals

There has been considerable interest in recent years in the study of liquid crystalline order in polymeric materials. Following on from the use of small molecules in display applications, the possibility of creating polymers with similar characteristics became attractive. Onsager[65] and Flory[66] predicted that rigid rod-like macromolecules should form liquid crystalline phase. However, it was not until 1975[67] that the first observation of a thermotropic liquid crystalline polymer was reported. Several reviews have been published on polymeric liquid crystals.[68-79]

Before examining the properties of polymers which exhibit liquid crystalline characteristics it is appropriate to mention briefly that it is possible to disperse low molecular mass liquid crystals in a polymer matrix and hence achieve an immobilization of the liquid in a solid, yet still achieve many of the desirable optical characteristics of the pure media. These dispersed phases are called *polymer dispersed liquid crystals* and will not be discussed here. These materials, however, are being shown to have many applications and clearly may be technologically important in the future.[80-86]

3.12 Polymeric Liquid Crystalline Materials

In general, polymeric liquid crystal materials are classified according to the dominant features of their backbone structure into backbone, side chain or combined liquid crystalline materials:

| main chain | side chain | combined |

These characteristic structures can describe most situations found in polymer systems.

3.12.1 General Factors Influencing Polymeric Liquid Crystalline Materials

As would be expected, molecular mass is one of the important factors that will influence the formation of the liquid crystalline phase. Upon increasing the molar mass from a monomer to a polymer, the entropy of the liquid phase (S_i) decreases. The decrease of the entropies of mesomorphic and crystalline phases is less than that of the isotropic phase. The decrease in S_i and that of the liquid crystal phase (S_{lc}) tends asymptotically to zero with increasing molar mass. It follows that the free energies G_i and G_{lc} increase with increasing polymer molecular weight, again becoming asymptotic above a certain molecular weight, and we may consider both parameters as remaining constant. Consider a series of special cases.

Case 1. Both monomeric structural unit and polymer display an enantiotropic mesophase. Upon increasing the molar mass from dimer to trimer, *etc.*, S_i decreases and therefore G_i increases. Beyond a certain molar mass G_i remains for all practical considerations constant. However, the slope of the increase of T_{lc-i} is steeper than that of T_{c-lc}. The difference between these two slopes determines the relative thermodynamic stabilities of the mesomorphic versus the crystalline phases at different polymer molar masses. As the molar mass is increased, the liquid crystalline regime is increased for main chain[87-89] and side chain[90-100] liquid crystalline polymers.

Case 2. The structural unit displays a virtual or a monotropic mesophase; the polymer displays a monotropic or an enantiotropic mesophase. The slope of the T_{lc-c} versus molar mass (M) is steeper than that for T_{c-I}, and as a consequence there will exist a critical value of M below which a mesophase is *not* observed. This effect has been observed in main chain[101-103] and side chain[104] liquid crystalline polymers.

Case 3. The structural unit displays a virtual mesophase; the polymer displays a virtual mesophase. This corresponds to the situation where the slope of T_{lc-c} versus M is greater than that of T_m versus M. The latter lies above the former throughout, and hence the two curves do not cross. Therefore the resulting polymer displays a virtual mesophase. This thermodynamic situation has been observed in polyethers containing both flexible mesogens and flexible spacers.[105]

Case 4. Rigid rod-like polymers. The discussion from cases 1 to 3 refers to semi-flexible or semi-rigid and flexible polymer systems that exhibit first-order transition temperatures (*i.e.* melting, isotropic liquid phase) that are molecular mass dependent only up to a certain degree of polymerization. In these systems the melting temperature is determined by the length of the chain fold. Rigid rod-like polymers such as poly(*p*-phenylene)s, poly(*p*-phenylenebenzobisthiazole) (**PBT**) and poly(*p*-phenylbenzobisoxazole) (**PBO**) exhibit a linear dependence of their melting transition over the entire range of molar mass.[106–108] Theory predicts that the axial ratio ($x = L/d$) of rod-like molecules reaches a value of about 6.2, and the compound should exhibit a mesophase. This is the case for poly(*p*-phenylene)s where nematic mesophase behaviour is observed for polymers with a degree of polymerization *n* between 6 and 7. However, for oligomers with *n* larger than 7 the nematic phase cannot be observed since the decomposition temperature overlaps the melting temperature and then becomes lower than the melting temperature. Other rigid rod-like polymers such as **PBT** and **PBO** and even semi-rigid systems like fully aromatic and unsaturated polyamides and polyesters decompose before melting.

3.12.2 Main Chain Crystalline Polymers

Certain phenyl-linked polymers (**32**) that exhibit mesophase behaviour will have elements of the structures shown below:[109–112]

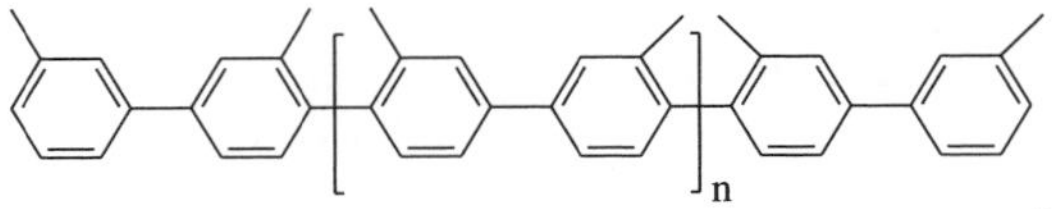

(32)

n	Melting temperature (°C)	Isotropic–mesophase temperature (°C)
1	142	–
2	273	311 (nematic)
3	242	260 (smectic)
4	298	–
5	171	–
6	219	–

These oligomeric methyl-substituted *p*-phenylenes have an axial ratio (x) equal to 0.8 times the number of benzene rings. According to theory predictions hexamethylhexphenyl ($n = 1$) which has $x = 4.8$ and lower oligomers should not exhibit a mesophase, whereas octamethyloctphenyl ($n = 2$) has $x = 6.4$, *i.e.* larger than the critical value of 6.2, and exhibits a nematic phase. A variety of other systems with phenylene structures have been explored; whilst they have rigid rod-like character, they do not meet the criteria for the breadth to length to make them liquid crystalline.

Substituted polyesters[113] **(33)** as a consequence of the conjugation through the ester function can exhibit the required characteristics of liquid crystalline behaviour.

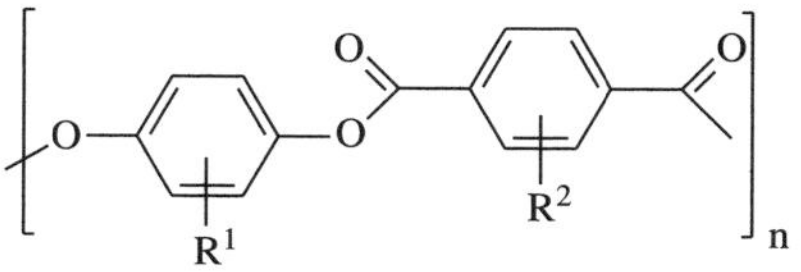

(33)

R^1	R^2	T_g (°C)	T_m (°C)	Mesophase
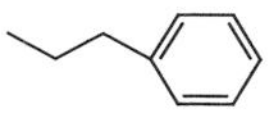 –CH₂–C₆H₅	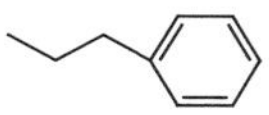 –CH₂–C₆H₅	60	195	Nematic
Bu^t	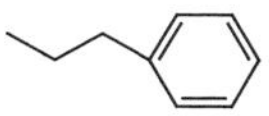 –CH₂–C₆H₅	100	225	Nematic
Bu^t	Br	131	–	–
Bu^t	CF_3	141	228	Nematic
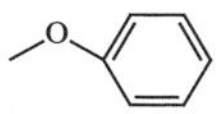 –CH₂–O–C₆H₅	Cl	85	241	Nematic
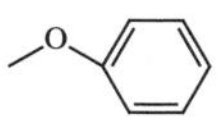 –CH₂–O–C₆H₅	Br	90	220	Nematic
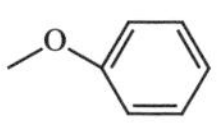 –CH₂–O–C₆H₅	CF_3	88	–	Isotropic
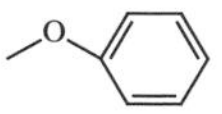 –CH₂–O–C₆H₅	 –C₆H₅	112	–	Isotropic

This series of polymers illustrates the breadth to length criteria and demonstrates that only those systems that form a crystalline phase exhibit mesophase behaviour.

Rod-like soluble polyimides **(34)** have been shown to exhibit liquid crystalline characteristics.[114–116] All polymers containing alkyl side chains of identical length are isomorphic within their crystalline phase and form solid solutions regardless of the nature of their polymer backbone. For instance (35) is isomorphic with the polyester series (36).[116] The rigid rod-like polyimides form layered mesophases.

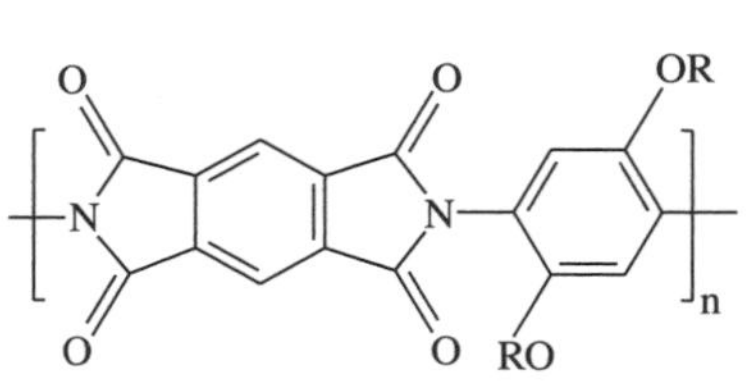

Most of the rigid rod systems form a liquid crystalline phase when cast from a suitable solvent and hence exhibit lyotropic properties.

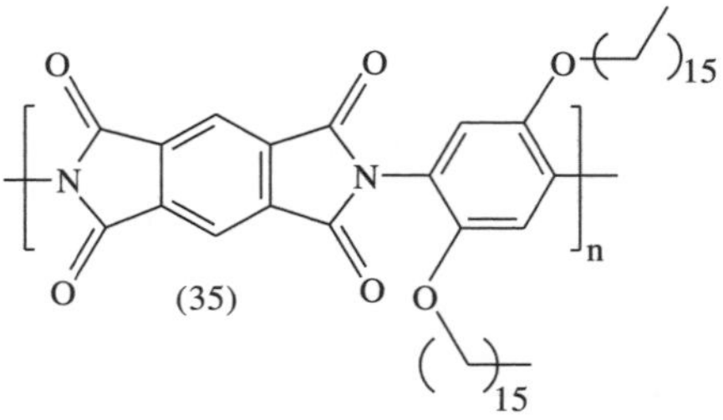

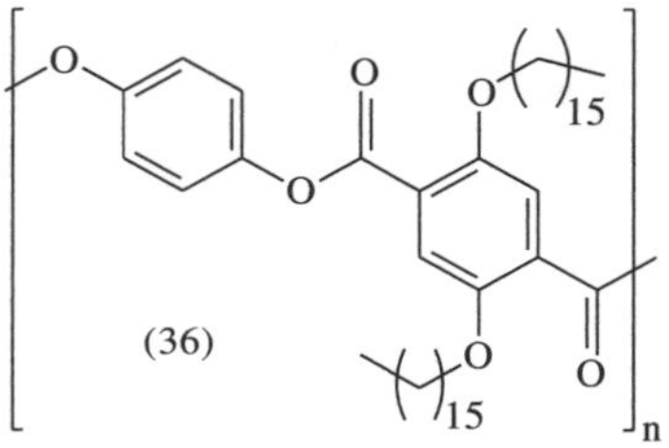

Polyurethanes with liquid crystalline behaviour have been produced from bischloroformates of 4,4'-alkylenedioxydiphenols by polycondensation with piperazine (37), trans-2,5-dimethylpiperazine (38) and 4,4'-bipiperidine (39). These molecules exhibit smectic properties.[117–119] In general the hydrogen bonding between the neighbouring polymer chains is sufficiently strong to suppress the formation of mesophases; however, some other systems have been reported.[120]

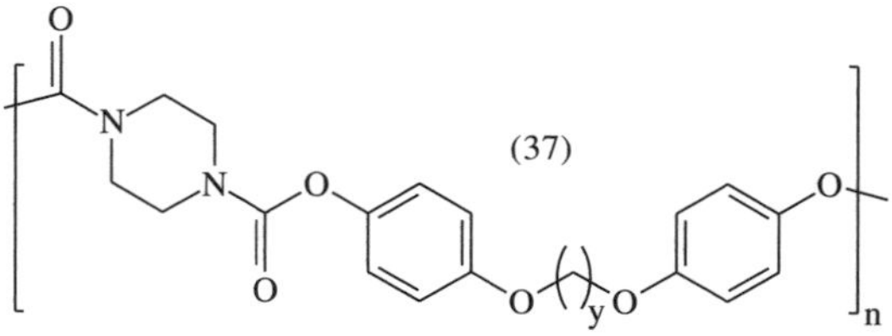

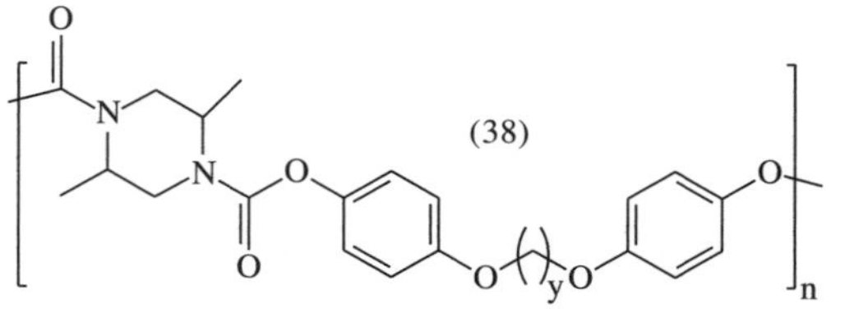

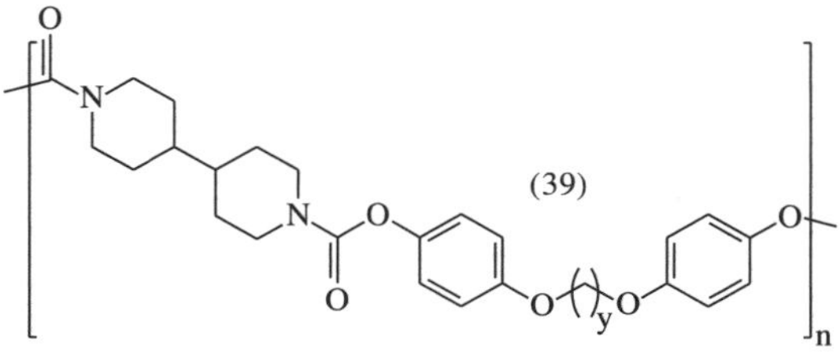

[Structure (39)]

Poly(ester anhydrides) of the type **(40)** exhibit nematic phase behaviour over a broad range of temperatures.[121]

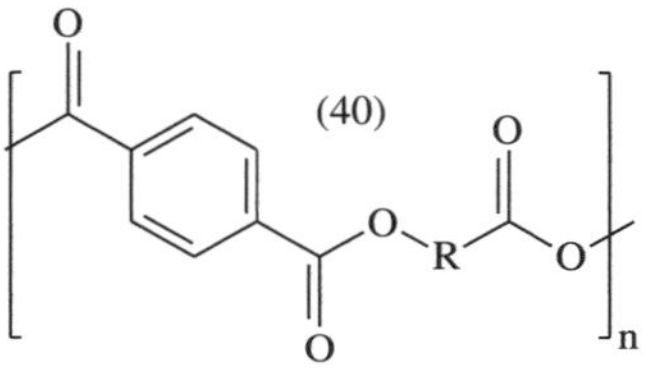

[Structure (40)]

R	T_g (°C)	T_i (°C)
 [1,4-phenylene]	136	360
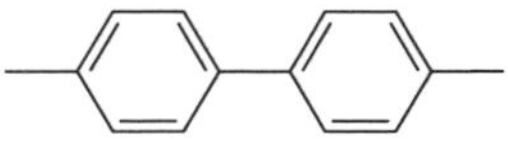 [4,4'-biphenylylene]	160	450
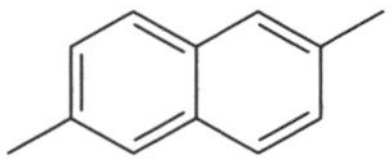 [2,6-naphthylene]	162	440

A number of poly(esterimide)s **(41)** have been shown to exhibit mesophase behaviour. The products of the reaction between trimellitic anhydride and α,ω-diaminoalkanes containing from 4 to 12 methylene units and containing various 4,4' aromatic entities crystallize in a layered morphology but show certain smectic properties.[122,123] The polymer with Ar = 4,4'-biphenylyl melts into a smectic mesophase. The melting point varies from 297 °C for the polymer with $n = 12$ to 393 °C for the polymer with $n = 4$. The liquid crystalline phase becomes an isotropic liquid at 386 °C for the polymer with $n = 12$ and at 467 °C for the polymer with $n = 4$. The rate of crystallization of all these polymers is usually fast with the exception of the polymers based on Ar = 1,4-phenylene, 1,4-tetrachlorophenylene and 2,6-naphthalene that exhibit smectic mesophases; the other polymers melt directly into an isotropic phase.[124]

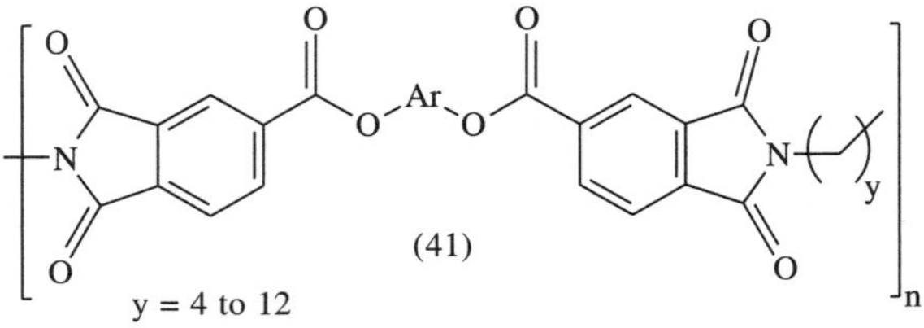

(41)
y = 4 to 12

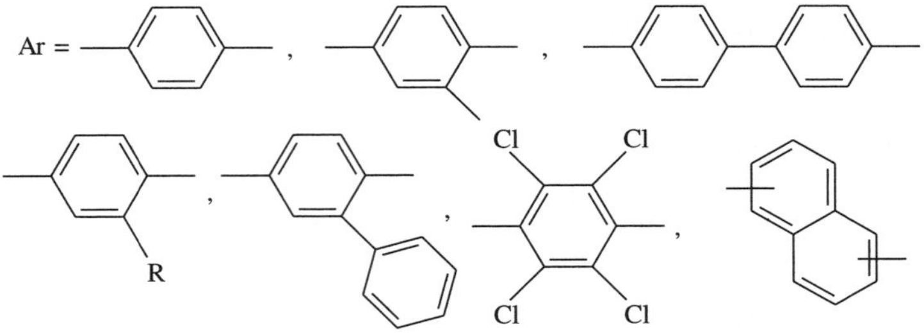

Ar = [1,4-phenylene], [2-substituted phenylene], [4,4'-biphenylyl], [substituted phenylene with R], [biphenyl], [tetrachlorophenylene with Cl at 4 positions], [naphthalene (1,4 : 1,5 : 2,6 : 2,7)]

The reaction of pyromellitic dianhydride or benzophenone-3,3',4,4'-tetracarboxylic anhydride with amino acids or lactams followed by condensation of the resulting diacids with diacetates of hydoquinone, 2,6-dihydroxynaphthalene or 4,4'-dihydroxybiphenyl produced layer polymers **(42)**. With the exception of the polymer with $y = 11$, all the polymers based on Ar = 4,4'-biphenylyl exhibit a smectic mesophase. The polymer with $y = 10$ and Ar = 1,4-phenylene exhibits a smectic mesophase. Polymers based on Ar = 1,4-phenylene or 2,6-naphthalene melt directly into an isotropic phase.[124]

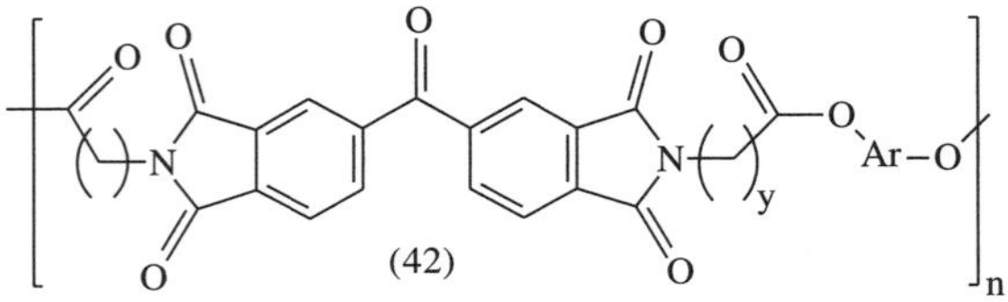

(42)

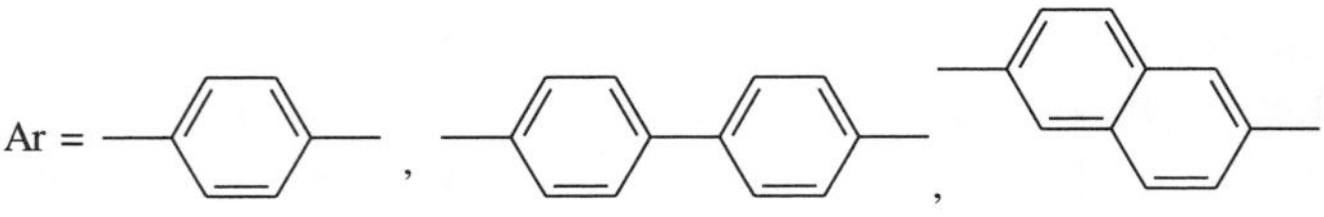

Ar = [1,4-phenylene], [4,4'-biphenylyl], [2,6-naphthalene]

A number of other polymer systems have been shown to exhibit liquid crystal characteristics. In all cases the critical length rule is a very useful guide as to whether such behaviour will be observed. Clearly in these materials, the flexible alkyl element allows the disorder to be created that is the essential feature of liquid crystallinity.

3.12.3 Side Chain Liquid Crystalline Polymers

An extensive study of possible side chain liquid crystalline materials has been undertaken.[80,125] The majority of materials that have been studied have structures in which a mesogenic low molar mass entity is flexibly attached to a polymer backbone. The usual backbone is a flexible polymer and interestingly small-angle neutron scattering (SANS) experiments[126–128] and X-ray scattering experiments[129,130] have shown that the statistical random coil conformation of the polymer backbone is slightly distorted in the nematic phase and highly distorted in the smectic phase.

3.12.4 Nature of Flexible Spacer and Its Length

According to the previous discussion it is found that increasing the degree of polymerization decreases the entropy of the system and if the monomeric structural unit exhibits a virtual or monotropic or mesophase, the resulting polymer should most probably exhibit a monotropic or mesophase. Alternatively, if the monomeric structural unit displays a mesophase, the polymer should display a mesophase that is broader. It is also possible that the structural unit of the polymer exhibits more than one virtual mesophase and therefore at high molecular weights the polymer will increase the number of its mesophases. All these effects have been observed in various systems.[131]

The length of the flexible spacer determines the nature of the mesophase. Long spacers favour smectic phases while short spacers favour nematic phases. This effect is similar to that observed in low molar mass liquid crystals.

3.12.5 Nature of the Backbone

At constant molecular weight the rigidity of the polymer backbone determines the thermodynamic stability of the mesophase. Accordingly a polymer with a more rigid backbone should have a more stable mesophase. However, the experimental situation is reversed and is explained by assuming that a more flexible backbone uses less energy to get distorted and therefore generates a more decoupled polymer system. In fact, the more flexible backbones do not only generate higher isotropic temperatures but also greater ability towards crystallization. However, contrary to all expectations the entropy change of the isotropic transition is higher for these polymers that are based on more rigid backbones and therefore they exhibit lower isotropic temperatures. This contradiction between the values of the entropy change and the isotropic temperatures can be accounted for by a different mechanism of distortion of the polymer backbone. In the case of a flexible chain the backbone can become sandwiched between the smectic layers, whereas in a less flexible system the chains have to weave in and out of the liquid crystalline layers. Hence the difference observed reflects the ability of the backbone and side chains to accommodate each other in the compacted matrix. There will always be a driving force for the elements to phase separate and microphase separation has been reported in a number of systems. The molar mass at which the isotropic temperature becomes independent of molar mass depends on the flexibility of the polymer backbone; the temperature for the polysiloxanes[132,133] is higher than that for poly(methyl acrylates).[133] As pointed out recently,[134] for monomers, based on the ability to form mesophases, it is possible to predict whether they will form mesophase polymer systems. The vast majority of liquid crystal dimers studied consist of two identical rod-like mesogenic units linked in terminal positions via a flexible alkyl chain normally containing between 3 and 12 methylene units.[135–138] Three possible structures are conventionally found in the dimers: a liner molecule, an H-shaped dimer in which the mesogenic units are laterally attached and a T-shaped dimer containing a terminal and a laterally linked mesogenic unit. Examples are shown below:

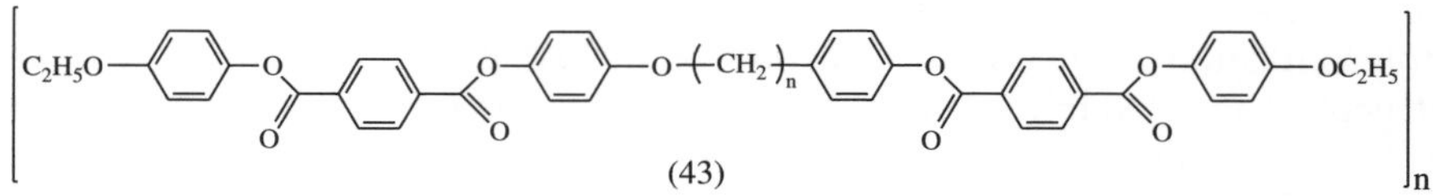

(43)

Linear dimer

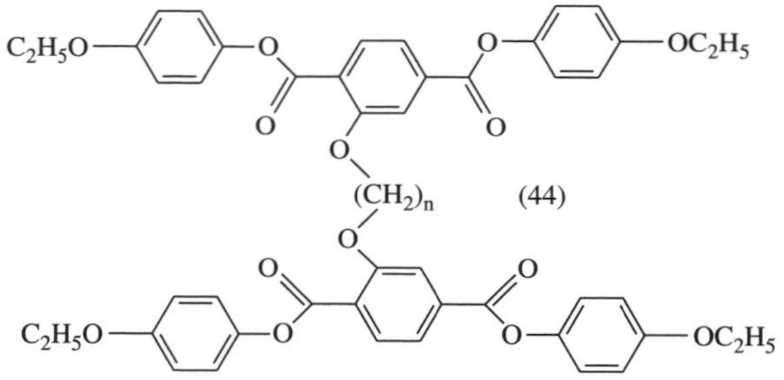

(44)

H-shaped dimer in which the mesogenic units are laterally attached

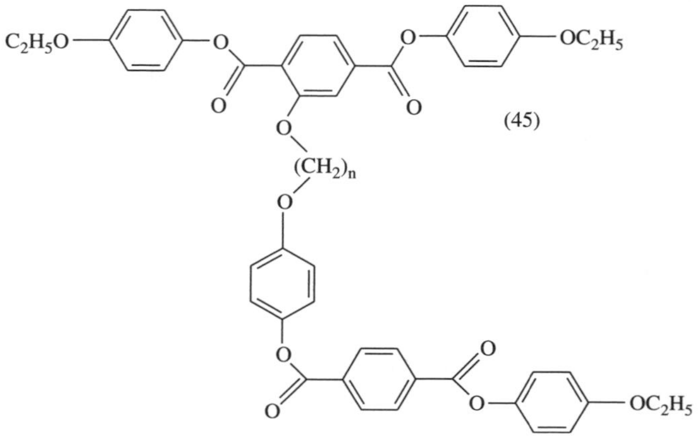

(45)

T-shaped dimer containing a terminal and a laterally linked mesogenic unit.

The linear dimer exhibits the highest clearing temperature and the lateral dimer the lowest. Smectic behaviour is observed only for the lateral dimers while the linear and T-shaped dimers exhibit solely nematic behaviour. This study considered only even spacers ($n = 4$, 8 and 12) and for all three series the clearing temperature decreases with increasing spacer length. It has been suggested that the strong nematic tendencies of the T-shaped dimers imply that the spacer adopts conformations for which the two mesogenic units are held more or less co-parallel. A wide variety of different systems have been studied and these support the general contention that whilst the rules for simple liquid crystals apply it is important to understand in detail the subtle influences which ultimately control the conformation of the chain and hence the ability to pack into either a crystalline or a mildly disordered form which is consistent with liquid crystalline behaviour.

3.12.6 Polymer Network Stabilized Liquid Crystal Phase

Polymer network stabilized ferroelectric liquid crystals with homogeneous alignment have been produced in cells without a surface alignment layer.[139–141] Normally an alignment layer is required to achieve the desired orientation of the molecules for them to exhibit ferroelectric properties. In this approach, a crosslinkable monomer is mixed with the liquid crystal and polymerized in a magnetic field to form a polymer network that will stabilize the alignment of the ferroelectric liquid crystal.[142] A concentration of the monomer of less than 10 wt% is required to avoid disturbance of the morphology of the low molecular weight material. The type of monomers which have been studied are shown below:

(46)

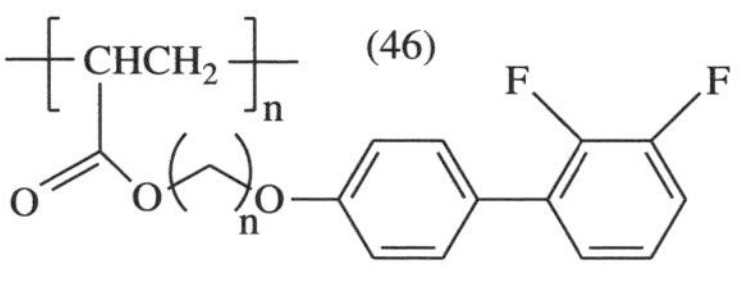

n = 0, 6-16

(47)

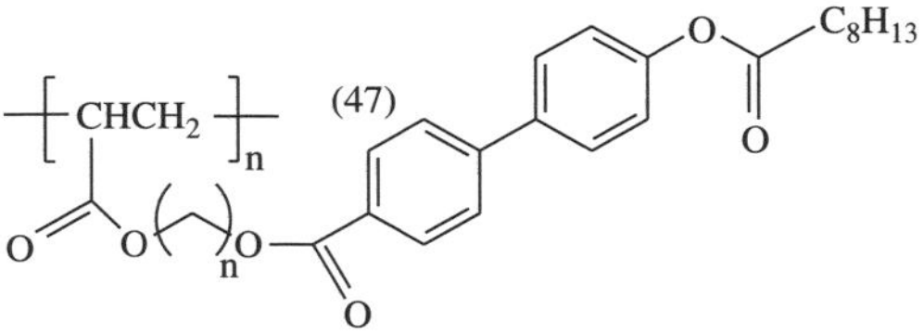

(48)

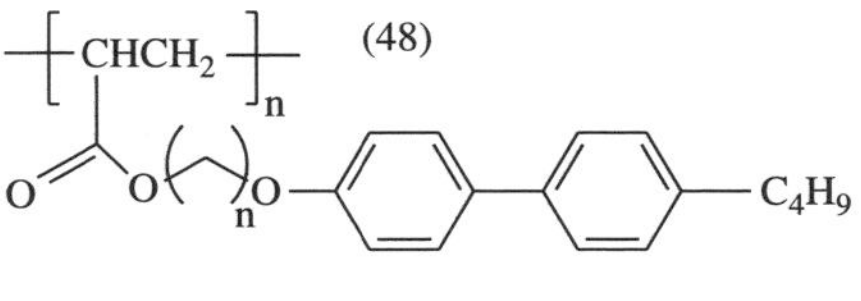

3.13 Structure Visualization

It will be appreciated that visualization of the structure of liquid crystalline materials is a particularly difficult task as the phase being studied has liquid-like character.[143] The technique of freeze fracture transmission electron microscopy allows examination of most systems; however, lyotropic materials which contain greater than 85% water still prove to be difficult. The technique involves the fast freezing of the material and then examination of the fracture surface. Despite the obvious attraction of this method it appears to be still in its infancy. Studies of cholesteric,[144] smectic[145] phases have been reported and show that it is possible to identify stacks of well-ordered materials which are often banana-shaped but do conform to the concepts that have been developed above.

3.14 Conclusions

Liquid crystals both as small molecule and polymeric materials will remain as fascinating materials and their useful physical properties will be used in many applications. We have not in this chapter considered two very important classes of liquid crystalline materials: lyotropic and naturally occurring materials. Lyotropic liquid crystals are formed when the concentration of the material is raised from dilute solution. They will often retain solvent as part of the phase structure and a wide range of phases can be observed depending on the concentration. They are discussed more fully in Chapter 11. Many celluloses, peptides, polysaccharides, *etc.*, have rigid backbone structures and hence can and do exhibit liquid crystalline behaviour. An example of one such system is hydroxypropylcellulose (HPC) that is a modified cellulose structure.[146] It is a semi-flexible polymer that can easily be aligned and exhibits the banded structure that is characteristic of the liquid crystalline state. Morphological features in naturally occurring systems are considered further in Chapter 11.

Recommended Reading

W.H. De Jeu, *Physical Properties of Liquid Crystalline Materials*, Gordon and Breach, London, 1980.
P.G. DeGennes, *The Physics of Liquid Crystals*, Oxford Science Publications, 1974.
G.R. Luckhurst and G.W. Gray, *The Molecular Physics of Liquid Crystals*, Academic Press, London, 1979.
I.W. Stewart, *The Static and Dynamic Continuum Theory of Liquid Crystals*, Taylor and Francis, London, 2004.

References

1. F. Reinitzer, *Monatsch Chem.*, 1888, **9**, 421–441 (English translation *Liq. Cryst.*, 1989, **5**, 7–18).

2. G. Friedel, *Ann. Phys. (Paris)*, 1922, **18**, 273–474.
3. H. Sackmann, *Liq. Cryst.*, 1989, **5**, 43–55.
4. P.G. de Gennes and J. Prost, *The Physics of Liquid Crystals*, Clarendon Press, Oxford, 1993.
5. D. Demus, G. Kunicke, J. Neelson and H. Sackmann, *Z. Naturforsch.*, 1968, **23a**, 84; S. Diele, P. Brand and H. Sackmann, *Mol. Cryst. Liq. Cryst.*, 1972, **17**, 1963.
6. A. De Vries and D.L. Fishel, *Mol. Cryst. Liq. Cryst.*, 1972, **16**, 311.
7. J. Doucet, A.M. Levelet and M. Lambert, *Phys. Rev. Lett.*, 1974, **32**, 301.
8. G.W. Gray and P.A. Windsor, in *Liquid Crystals and Plastic Crystals*, ed. G.W. Gray and P.A. Windsor, Ellis Horwood, Chichester, UK, 1974, vol. 1.
9. K. Markau and W. Maier, *Chem. Ber.*, 1962, **95**, 889.
10. W.H. de Jeu, *J Phys. (Paris)*, 1977, **38**, 1265.
11. G.W. Gray. *J. Phys. (Paris)*, 1975, **36**, 337; G.W. Gray, K.J. Harrison and J.A. Nash, *Electron Lett.*, 1973, **9**, 130.
12. J.W. Goodby and G.W. Gray, *J Phys. (Paris)*, 1976, **37**, 18.
13. G.W. Gray and D.G. McDonnell, *Mol. Cryst. Liq. Cryst.*, 1976, **37**, 189.
14. G.W. Gray, *Molecular Structure and the Properties of Liquid Crystals*, Academic Press, New York, 1962, p. 131.
15. D. Coates and G.W. Gray, *J. Chem. Soc., Chem. Commun.* 1975, 514.
16. D. Coates and G.W. Gray, *Mol. Cryst. Liq. Cryst.*, 1976, **37**, 249.
17. D. Coates and G.W. Gray, J. Chem. Soc., *Perkin Trans. 2*, 1976, **7**, 863.
18. G.W. Gray, K.J. Harrison and J.A. Nash, *J. Chem. Soc., Chem. Commun.*, 1974, 431.
19. H.B. Burgi and J.D. Dunitz, *Helv. Chim. Acta*, 1970, **53**, 1747; J. van der Veen and A.H. Grobben, *Mol. Cryst. Liq. Cryst.*, 1971, **15**, 239.
20. G.W. Gray, in *The Molecular Physics of Liquid Crystals*, ed. G.R. Luckhurst and G.W. Gray, Academic Press, 1979, ch. 1.
21. J.M. Robertson and I. Woodward, *Proc. R. Soc. London, Ser. A*, 1937, **162**, 568.
22. M.J.S. Dewar and R.S. Goldberg, *J. Org. Chem.*, 1970, **35**, 2711; *J. Am. Chem. Soc.*, 1970, **92**, 1582.
23. M.J.S. Dewar, A.C. Griffin and R.M. Riddle, in *Liquid Crystals and Ordered Fluids*, ed. J.F. Johnson and R.S. Porter, Plenum Press, New York, 1973, vol. 2, p. 733; M.J.S. Dewar and R.M. Riddle, *J. Am. Chem. Soc.*, 1975, **97**, 6658; M.J.S. Dewar and A.C. Griffin, *J. Am. Chem. Soc.*, 1975, **97**, 662.
24. H.J. Deutscher, F. Kuschel, H. Schubert and D. Demus, *Deutsche Demokratische Republik Pat.*, DOS 24 29 093, 1975.
25. R. Eidenschink, D. Erdmann, J. Krause and L. Pohl, *Angew. Chem.*, 1977, **89**, 103.
26. C. Weygand and R. Gabler, *Z. Phys. Chem.*, 1940, **B46**, 270.
27. H. Schubert, R. Dehne and V. Uhlig, *Z. Chem.*, 1972, **12**, 219.
28. W.R. Young, I. Haller and D.C. Green, *IBM Res. Rep.*, 1972, RC3827; *J. Org. Chem.*, 1972, **37**, 3707; W.R. Young and D.C. Green, *IBM Res. Rep.*, 1972, RC4121; *Mol. Cryst. Liq. Cryst.*, 1974, **26**, 7.

29. A.C. Griffin, D.L. Wertz and A.C. Griffin Jr, *Mol. Cryst. Liq. Cryst.*, 1978, **44**, 267.
30. W.H. de Jeu, *J. Phys. (Paris)*, 1977, **38**, 1265.
31. J. van der Veen, W.H. de Jeu, A.H. Grobben and J. Boven, *Mol. Cryst. Liq. Cryst.*, 1972, **17** 291; W.H. de Jeu and J. van der Veen, *Philips Res. Rep.*, 1972, **27**, 172; J. van der Veen, W.H. de Jeu, M.W.M. Wanninkhof and C.A.M. Tienhoven, *J. Phys. Chem.*, 1973, **77**, 2153; W.H. de Jeu and Th.W. Lathouwers, *Z. Naturforsch.*, 1974, **A29**, 905.
32. G.W. Gray and P.A. Winsor, in *Liquid Crystals and Plastic Crystals*, ed. G.W. Gray and P.A. Winsor, Ellis Horwood, Chichester, UK, 1974, vol. 1.
33. G.W. Gray, in *Advances in Liquid Crystals*, ed. G.H. Brown, Academic Press, New York, 1976, vol. 2, p. 1.
34. G.W. Gray, *Mol. Cryst. Liq. Cryst.*, 1969, **7**, 127.
35. M. Leclerq, J. Billard and J. Jacques, *Mol. Cryst. Liq. Cryst.*, 1969, **8**, 367.
36. J.W. Goodby, G.W. Gray and D.G. McDonnell, *Mol. Cryst. Liq. Cryst.*, 1976, **34**, 183.
37. G.W. Gray, in *The Molecular Physics of Liquid Crystals*, ed. G.R. Luckhurst and G.W. Gray, Academic Press, 1979, ch. 12.
38. J.E. Lydon and C.J. Coakley, *J. Phys. (Paris)*, 1975, **36**, 45.
39. G.W. Gray and J.E. Lydon, *Nature (London)*, 1974, **252**, 221.
40. A.J. Leadbetter, J.L.A. Durrant and M. Rugman, *Mol. Cryst. Liq. Cryst. Lett.*, 1977, **34**, 231.
41. G.W. Gray and K.J. Harrison, *Symp. Faraday Soc.*, 1971, **5**, 54.
42. D.J. Byron, G.W. Gray, B.M. Worrall. *J. Chem. Soc.*, 1965, 3706.
43. M.A. Cotter, *Phys. Rev. A*, 1974, **10**, 625.
44. M.A. Cotter, *J. Chem. Phys.*, 1977, **66**, 1098.
45. M.A. Cotter, in *The Molecular Physics of Liquid Crystals*, ed. G.R. Luckhurst and G.W. Gray, Academic Press, 1979, ch. 12.
46. C. Zannoni, in *The Molecular Physics of Liquid Crystals*, ed. G.R. Luckhurst and G.W. Gray, Academic Press, 1979, ch. 3.
47. G.R. Luckhurst, in *The Molecular Physics of Liquid Crystals*, ed. G.R. Luckhurst and G.W. Gray, Academic Press, 1979, ch. 4.
48. W. Maier and A. Saupe, *Z. Naturforsch. A*, 1958, **13**, 564.
49. W. Maier and A. Saupe, *Z. Naturforsch. A*, 1959, **14**, 882.
50. F.M. Leslie, *Quart. J. Mech. Appl. Math.*, 1966, **19**, 357.
51. F.M. Leslie, *Arch. Rat. Mech. Anal.*, 1986, **28**, 265.
52. C. Zannoni, in *The Molecular Physics of Liquid Crystals*, ed. G.R. Luckhurst and G.W. Gray, Academic Press, 1979, ch. 9.
53. N. Metropolis, A.W. Rosenbluth, M.N. Rosenbluth, A.H. Teller and E. Teller, *J. Chem. Phys.*, 1953, **21**, 1087.
54. M.A. Bates, *Liq. Cryst.*, 2005, **32**(11), 1365–1377.
55. J.G. Gay and B.J. Berne, *J. Chem. Phys.*, 1981, **74**, 3316.
56. J. Corner, *Proc.R. Soc. London, Ser. A*, 1948, **192**, 275.
57. M.A. Bates and G.R. Luckhurst, *Struct. Bond.*, 1999, **94**, 65.
58. S. Chandrasekar, in *Polymers Liquid Crystals and Low Dimensional Solids*, ed. N. March and M. Tosi, Plenum Press, New York, 1984, ch. 7.

59. G. Durand, in *Polymers, Liquid Crystals and Low Dimensional Solids*, ed. N. March and M. Tosi, Plenum Press, New York, 1984, ch. 11.
60. I.W. Stewart, *The Static and Dynamic Continuum Theory of Liquid Crystals*, Taylor & Francis, London, 2004.
61. B.S. Scheuble, *Kontakte (Darmstadt)*, 1989, **1**, 34–48.
62. F. Schneider and H. Kneppe, in *Handbook of Liquid Crystals*, ed. D. Demus, J. Gooby, G.W. Gray, H.W. Spiess and V. Vill, Wiley VCH, Weinheim, 1998, vol. 1, pp. 454–476.
63. V. Sergan and G. Durand, *Liq. Cryst.*, 1995, **18**, 171–174.
64. M.J. Stephen and J.P. Straley, *Rev. Mod. Phys.*, 1974, **46**, 617–704.
65. L. Onsager, *Ann. N.Y. Acad. Sci.*, 1949, **51**, 627.
66. P.J. Flory, *Proc. R. Soc. London, Ser. A*, 1956, **234**, 73.
67. A. Roviello and A. Sirigu, *J. Polym. Sci., Polym. Lett. Ed.*, 1975, **13**, 455.
68. M. Gordon and N.A. Plate (eds), *Adv. Polym. Sci.*, 1984, 60/61.
69. A. Blumstein, *Mesomorphic Order in Polymers and Polymerization in Liquid Crystalline Media*, ACS Symposium Series 74, American Chemical Society, Washington, DC, 1978.
70. A. Blumstein, *Liquid Crystalline Order in Polymers*, Academic Press, New York, 1978.
71. A. Blumstein, *Polymeric Liquid Crystals*, Plenum Press, New York, 1985.
72. L.L. Chapoy, *Recent Advances in Liquid Crystalline Polymers*, Elsevier, London, 1985.
73. A. Ciferri, W.R. Krigbaum and R.B. Meyers, *Polymer Liquid Crystals*, Academic Press, London, 1982.
74. C.B. McArdle, *Side Chain Liquid Crystal Polymers*, Blackie, Glasgow, 1989.
75. A. Ciferri (ed.), *Liquid Crystallinity in Polymers. Principles and Fundamental Properties*, VCH, New York, 1991.
76. R.A. Weiss and C.K. Ober, *Liquid Crystalline Polymers*, ACS Symposium Series 435, American Chemical Society, Washington, DC, 1990.
77. E. Chiellini and R.W. Lenz, in *Comprehensive Polymer Science*, ed. G. Allen and J.C. Bevington, Pergamon Press, Oxford, 1989, vol. 5, p. 701.
78. R. Zentel, in *Comprehensive Polymer Science*, ed. G. Allen and J.C. Bevington, Pergamon Press, Oxford, 1989, vol. 5, p. 723.
79. A.M. Donald and A.H. Windle, *Liquid Crystalline Polymers*, Cambridge University Press, Cambridge, 1991; C. Noel and P. Navard, *Prog. Polym. Sci.*, 1991, **16**, 55.
80. V. Percec and D. Tomazos, in *Comprehensive Polymer Science*, ed. S. Aggawal, S. Russo and G. Allan, Pergamon Press, 1992, 1st supplement, ch. 14, p300.
81. J. Moon, J.H. Ford and S. Yang, *Polym. Adv. Technol.*, 2006, **17**(2), 83–93.
82. J. Moon, J.H. Ford and S.J. Yang, *Macromol. Sci. Polym. Rev.*, 2005, **C45**(4), 351–373.
83. J. Qi and G.P. Crawford, *Displays*, 2004, **25**(5), 177–186.
84. J. Gu, Y. Xu, Y.S. Liu, J.J. Pan, F.Q. Zhou and H. He, *J. Opt. A: Pure Appl. Opt.*, 2003, **5**(6), S420–S427.

85. J.M. Mucha, *Prog. Polym. Sci.*, 2003, **28**(5), 837–873.
86. T.J. Bunning, L.V. Natarajan, V.P. Tondiglia and R.L. Sutherland, *Annu. Rev. Mater. Sci.*, 2000, **30**, 83–115.
87. L. Bouteiller and P. Lebarny, *Liq. Cryst.*, 1996, **21**(2), 157–174.
88. A. Blumstein, S. Vilasagar, S. Ponrathnam, S.B. Clough, R.B. Blumstein and G. Maret, *J. Polym. Sci., Polym. Phys., Ed.*, 1982, **20**, 877.
89. V. Percec, H. Nava and H. Jonsson, *J. Polym. Sci., Polym. Chem. Ed.*, 1987, **25**, 1943.
90. J.L. Feijoo, G. Ungar, A.J. Owen, A. Keller and V. Percec, *Mol. Cryst. Liq. Cryst.*, 1988, **155**, 487.
91. V. Percec and D. Tomazos, *Polymer*, 1990, **31**, 1658; V. Percec, B. Hahn, M. Ebert and J.H. Wendorff, *Macromolecules*, 1990, **23**, 2092.
92. S.G. Kostromin, R.V. Talrose, V.P. Shibaev and N.A. Plate, *Makromol. Chem. Rapid Commun.*, 1982, **3**, 803.
93. Y.K. Godovsky, I.I. Mamaeva, N.N. Makarova, V.P. Papkov and N.N. Kuzmin, *Makromol. Chem. Rapid Commun.*, 1985, **6**, 797.
94. V. Perec and B. Hahn, *Macromolecules*, 1989, **22**, 1588.
95. V. Perec, M. Lee and H. Johnsson, *J. Polym. Sci., Polym. Chem. Ed.*, 1991, **29**, 327.
96. V. Perec and M. Lee, *J. Macromol. Sci. Chem.*, 1991, **A28**, 651.
97. V. Perec, M. Lee and C. Ackerman, *Polymer*, 1992, **33**, 703.
98. V. Perec and M. Lee, *Macromolecules*, 1991, **24**, 2780.
99. V. Perec, A.D.S. Gomes and M. Lee, *J. Polym. Sci., Polym. Chem. Ed.*, 1991, **29**, 1615.
100. V. Percec, C.S. Wang and M. Lee, *Polym. Bull.*, 1991, **26**, 15.
101. V. Percec, Q. Zheng and M. Lee, *J. Mater. Chem.*, 1991, **1**, 611.
102. J. Majnusz, J.M. Catala and R.W. Lenz, *Eur. Polym. J.*, 1983, **19**, 1043.
103. Q.F. Zhou, X.Q. Duan and Y.L. Liu, *Macromolecules*, 1986, **19**, 247.
104. V. Percec and H. Nava, *J. Polym. Sci., Polym. Chem. Ed.*, 1987, **25**, 405.
105. H. Stevens, G. Rehage and H. Finkelmann, *Macromolecules*, 1984, **17**, 851.
106. P.J. Flory and G. Ronca, *Mol. Cryst. Liq. Cryst.*, 1979, **54**, 311.
107. P.A. Irvine, W. Dacheng and P.J. Flory, *J. Chem. Soc., Faraday Trans. 1*, 1984, **80**, 1795.
108. V. Percec, D. Tomazos and C. Pugh, *Macromolecules*, 1989, **22**, 3259.
109. W. Kern, W. Gruber and H.O. Wirth, *Makromol. Chem.*, 1960, **37**, 198.
110. J.K. Stille, F.W. Harris, R.O. Rakutis and H. Mukamal, *J. Polym. Sci., Polym. Lett. Ed.*, 1966, **4**, 791.
111. H. Mukamal, F.W. Harris and J.K. Stille, *J. Polym. Sci., Part A-1*, 1967, **5**, 2721.
112. J.K. Stille, R.O. Rakutis, H. Mukamal and F.W. Harris, *Macromolecules*, 1968, **1**, 431.
113. (*a*) W. Heitz and H.W. Schmidt, Makromol. Chem., *Macromol. Symp.*, 1990, **38**, 149; (*b*) H. Kromer, R. Kuhn, H. Pielartzik, W. Siebke, V. Eckhardt and M. Schmidt, *Macromolecules*, 1991, **24**, 1950; (*c*) B.S. Hsiao, R.S. Stein, N. Weeks and R. Gaudiana, *Macromolecules*, 1991, **24**, 1299.

114. M. Wenzel, M. Ballauff and G. Wegner, *Makromol. Chem.*, 1987, **188**, 2865.
115. F.H. Metzmann, M. Ballauff, R.C. Schulz and G. Wegner, *Makromol. Chem.*, 1989, **190**, 985.
116. R. Duran, M. Ballauff, M. Wenzel and G. Wegner, *Macromolecules*, 1988, **21**, 2897.
117. F.W. Harris and S.L.C. Hsu, *High Perform. Polym.*, 1989, **1**, 3.
118. H.R. Kricheldorf and J. Awe, *Makromol. Chem., Rapid Commun.*, 1988, **9**, 681.
119. H.R. Kricheldorf and J. Jenssen, *Eur. Polym. J.*, 1989, **25**, 1973.
120. H.R. Kricheldorf and J. Awe, *Makromol. Chem.*, 1989, **190**, 2579.
121. P.J. Stenhouse, E.M. Valles, S.W. Kantor and W.J. MacKnight, *Macromolecules*, 1989, **22**, 1467.
122. H.R. Kricheldorf and D. Lubbers, *Makromol. Chem., Rapid Commun.*, 1990, **11**, 303.
123. H.R. Kricheldorf and R. Pakull, *Macromolecules*, 1988, **21**, 551.
124. H.R. Kricheldorf and R. Pakull, *Polymer*, 1987, **28**, 1772.
125. H.R. Kricheldorf, R. Pakull and S. Buchner, *Macromolecules*, 1988, **21**, 1929.
126. L. Noirez, J.P. Cotton, F. Hardouin, P. Keller, F. Moussa, G. Pepy and C. Strazielle, *Macromolecules*, 1988, **21**, 2889.
127. P. Davidson, L. Noirez, J.P. Cotton and P. Keller, *Liq. Cryst.*, 1991, **10**, 111.
128. F. Hardouin, S. Mery, M.F. Achard, L. Noirez and P. Keller, *J. Phys.*, 1991, **1**, 511.
129. H. Mattoussi, R. Ober, M. Veyssie and H. Finkelmann, *Europhys. Lett.*, 1986, **2**, 233.
130. F. Kuschel, A. Madicke, S. Diele, H. Utschik, B. Hisgen and H. Ringsdorf, *Polym. Bull.*, 1990, **23**, 373.
131. P.W. Morgan, *Macromolecules*, 1977, **10**, 1381; S.L. Kwolek, P.W. Morgan, J.R. Shaefgen and L.W. Gulrich, *Macromolecules*, 1977, **10**, 1390; T.I. Bair, P.W. Morgan and F.L. Killian, *Macromolecules*, 1977, **10**, 1396; M. Panar and L. Beste, *Macromolecules*, 1977, **10**, 1401.
132. R.D. Richards, W.D. Hawthorne, J.S. Hill, M.S. White, D. Lacey, J.A. Semiyen, G.W. Gray and T.C. Kendrick, *J. Chem. Soc., Chem. Commun.*, 1990, **95**.
133. V. Perec and C. Pugh, in *Side Chain Liquid Crystal Polymers*, ed. C.B. McArdle, Chapman and Hall, New York, 1989, p. 30.
134. G.W. Gray, in *Side Chain Liquid Crystal Polymers*, ed. C.B. McArdle, Blackie, Glasgow, 1989, p. 106.
135. C.B. McArdle (ed.), *Side Chain Liquid Crystalline Polymers*, Blackie, Glasgow, 1989.
136. C.T. Imrie and P.A. Henderson, *Curr. Opin. Colloid Interf. Sci.*, 2002, **7**, 298–311.
137. C.T. Imrie and G.R. Luckhurst, in *Handbook of Liquid Crystals*, ed. D. Demus, J.W. Goodby, G.W. Gray and H.W. Spiess, Wiley VCH, Weinheim, 1998, vol. 2B, pp. 801–833.

138. C.T. Imrie, *Struct. Bond.*, 1999, **95**, 149–192.
139. W.-S. Bae, J.-W. Lee and J.-I. Jin, *Liq. Cryst.*, 2001, **28**, 59–67.
140. W. Zheng and G.H. Milburn, *Liq. Cryst.*, 2000, **27**, 1423–1430.
141. R.A.M. Hikmet, H.M.J. Boots and M. Michielsen, *Liq. Cryst.*, 1995, **19**, 65–76.
142. A.C. Guymon, E.N. Hoggan, D.M. Walba, C.N. Clark and C.N. Bownam, *Liq. Cryst.*, 1995, **19**, 719–727.
143. O. Mondain-Monval, *Curr. Opin. Colloid Interf. Sci.*, 2005, **10**, 250–255.
144. D.W. Berrman, S. Meiboom, J.A. Zasadzinski and M.J. Sammon, *Phys. Rev. Lett.*, 1986, **57**, 1737–1740.
145. J. Fernsler, L. Hough, R.F. Shao, J.E. Maclennan, L. Navailles, O. Madhusudham, O. Mondain-Monval, C. Boyer, J. Zasadzinski, J. Rego, D.M. Walba and N.A. Clark, *Proc. Natl Acad. Sci. USA*, 2005, **4**, 14191–14196.
146. C. Viney and W.S. Putnam, *Polymer*, 1995, **36**, 1731.

CHAPTER 4
Plastic Crystals

4.1 Introduction

In the previous chapters the properties of ordered and disordered low molecular weight organic materials have been considered, so as to help to develop our understanding of the factors which will influence the organization in high molar mass materials. Smectic liquid crystalline materials exhibit a higher degree of order, yet still retain at least one degree of freedom. Plastic and incommensurate phase materials similarly exhibit a high degree of order, yet, over a limited temperature range, retain a degree of freedom. As we shall see when we consider polymers, the ability to exhibit some level of motional freedom, yet still have solid characteristics is a very important and useful characteristic of polymers.

4.2 Plastic Crystalline Materials

In Chapter 2, the structural characteristics of a range of organic crystals were considered. Using appropriate conditions it is possible to grow large single crystals of materials such as carbon tetrabromide,[1] adamantine,[2] bicyclooctane,[2] camphene,[2] norbornylene,[2] succinonitrile,[2] pivalic acid,[2] hexamethylethane,[2] *etc*. An example of such a crystal is shown in Figure 4.1.

Superficially these materials show sharp melting transitions that would indicate that they are behaving as simple crystals. However, if one studies carefully a number of their physical properties it becomes evident that there is a transition that allows a degree of molecular motion to be achieved in these systems below the melting temperature. Examination of the molecular structure of theses molecules indicates that they all have a common feature, a spherical shape (Figure 4.2). The lattice structure of carbon tetrabromide is a face-centred cubic material. Examination of the melting process using differential scanning calorimetry (DSC) indicates that instead of the single melt peak expected, there is a second lower temperature peak occurring at 320.9 °C, which is significantly below the melting point at 366.4 °C (Figure 4.3). The DSC trace clearly shows the two shape peaks. The ultrasonic velocity decreases in an

Figure 4.1 Single crystal of carbon tetrabromide grown from ethanol solution (scale bar = 1 cm).[1]

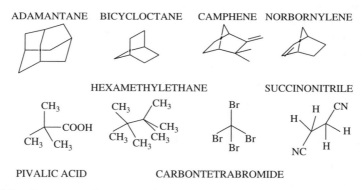

Figure 4.2 Structures of some plastic crystalline materials.

approximately linear manner as the temperature is increased, consistent with the expected decrease in the elastic modulus. At 320.9 °C, the velocity drops to a value below 1100 m s^{-1} and it becomes difficult to measure the value because the sound waves have become highly attenuated. This behaviour is unexpected as the material is still a solid.

This sudden drop in the velocity is associated with the carbon tetrabromide molecule suddenly gaining the freedom to rotate on its axis without the ability to undergo translation. The onset of translation is marked by the melting transition at 366.4 °C. The tetrahedral molecule below 320.9 °C is locked into the crystal structure and the C–Br bonds point in well-defined directions. At 320.9 °C the lattice expansion allows the molecule to rotate on its lattice point and being a pseudo-spherical molecule, the average force field has an

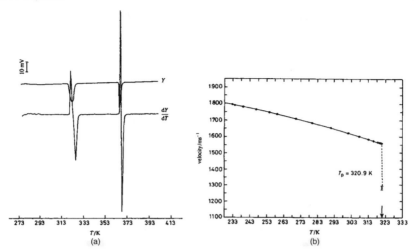

Figure 4.3 Differential scanning calorimetry trace (a) and ultrasonic velocity scan (b) as a function of temperature for carbon tetrabromide.

approximately spherical form. Studies of so-called 'globular molecules'[3] indicated that for molecules with the formula (CXYZW), where the groups involved are CH_3, NO_2, and Cl for X, Y and Z and W may be CN or Br, then rotor–plastic phase behaviour is observed. Another sequence identified was based on the simple polar derivatives of camphene and cyclohexane derivatives. Measurements of the dielectric permittivity as a function of temperature exhibited unusual behaviour[4] (Figure 4.4).

It would be expected that when a polar liquid is frozen, the dipoles would become immobilized and the permittivity would decrease. In the case of 2-chloro-2-nitropropane the reverse appears to occur with the magnitude of the permittivity increasing on freezing. Similar behaviour has been observed for all the rotator phase materials, the magnitude of the effect varying from system to system.[4–9] Normally the amplitude of the dipole relaxation will be governed by the distribution of states that it can occupy. In the liquid state the dipole is restricted by the shape of the potential formed by the neighbouring molecules (Figure 4.5). In the rotator phase, the potential has now become symmetric and more effective rotation of the dipole is possible.

In these simple molecules, the dielectric relaxation curves conform to the simple Debye form and hence it is relatively straightforward to determine activation energies from the variation of the relaxation frequency with temperature (Table 4.1). A surprising feature of the rotator phase is that in certain cases the activation for dipolar relaxation is smaller in the solid than it is in the liquid phase, when the solid conforms to a face-centred cubic lattice. When the solid melts, the local force field loses its symmetrical form and the result is that the activation energy for free rotation is observed to increase slightly (Figure 4.5). For most polar liquids, the activation energy for viscosity flow and for dipole

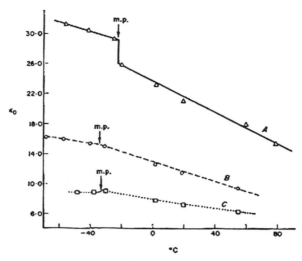

Figure 4.4 Permittivity values (ε_0) for 'spherical' methane derivatives in the liquid and solid phases. A: 2-chloro-2-nitropropane, $CH_3CClNO_2CH_3$; B: 2,2-dichloropropane, $CH_3CCl_2CH_3$; C: 1,1,1-trichloroethane, CCl_3CH_3.[4]

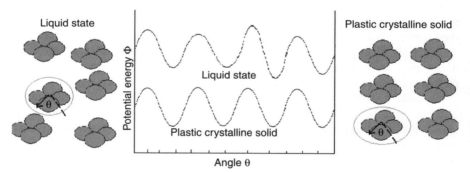

Figure 4.5 Variation of the potential energy with rotation angle: (a) the liquid state; (b) the plastic crystalline state.

relaxation are very similar, but in the case of the pseudo-spherical substituted methanes the activation energy of the rotational process is approximately half that for viscous flow. The reason for this difference is that translational motion of the neighbouring molecules in the fluid is coupled and higher energy is observed for motion. In the solid phase, the translational motion ceases and the activation energy is that for the dipole rotation in a pseudo-spherical force field.

Another group of dielectrically interesting solids are the polysubstituted aromatics having at least four halogen, nitro or methyl group substituents. These molecules have the symmetry of circular discs and exhibit ready rotation about one axis.[10] Two examples of such systems are presented in Table 4.1.

Table 4.1 Activation enthalpies for molecular rotation in liquids and solids.

Molecule	ΔH^*_E (kJ mol^{-1}), solid	ΔH^*_E (kJ mol^{-1}), liquid	ΔH^*_E (kJ mol^{-1}), viscosity	Ref
$(CH_3)_2CCl_2$	5.8	5.2	10.5	4
CCl_3CH_3	4.6	4.7	10.7	4
$(CH_3)CNO_2$	2.1	3.4	14.0	4
$(CH_3)CClNO_2$	5.5	6.1	13.6	4
Succcinonitrile	8.8	13.0	–	8
Camphene	9.2	–	–	4
Camphor	7.5	–	–	4
Bornyl chloride	10.4	–	–	4
Isoborneol	23.0	–	–	4
Pentachlorotoluene	49.7	–	–	9
1,2,4-Trimethyl-3,5,6-trichlorobenzene	44.3	–	–	9

A variety of other systems have been studied and include organic ions such as the butylammonium salts,[11,12] substituted cyclohexanes,[13] cyanoadamantane,[14] chloroadamantane,[14] tetrachloro-m-xylene,[15] pentachloronitrobenzene,[16] caffeine[17] and various zinc chloride salts ((n-$C_5H_{11}NH_3)_2ZnCl_4$ and (n-$C_{12}H_{25}NH_3)_2ZnC_{l4}$).[18] All these systems support the concept that in the rotator–plastic phase motion is easier than in the liquid and that it only ceases when the strength of the intermolecular interactions is sufficiently strong to impose an asymmetric profile on the pseudo-spherical cavity.

4.3 Alkanes and Related Systems

Although the initial studies of the rotator phase were concerned with molecules which have a pseudo-spherical shape, it was found that other molecular systems were able to exhibit rotational motion in the solid state. Liquid crystals possess mesophase characteristics because of the rotational freedom about one of their axes. The alkanes have received considerable attention over the years and are precursors for the study of polymers. Methane and propane are pseudo-spherical and will in principle have rotator phases but these occur at very low temperature and are difficult to access. In Chapter 1, the possibility of *gauche–trans* isomerism was discussed in the context of n-alkane chains. Analysis of the dynamic distribution of molecular shapes at the melting point temperature has been carried out in order to estimate the length of the alkane chain exhibiting the first premelting transition.[19,20] This study indicates that nonane ($n=9$) is the first linear hydrocarbon to exhibit a premelt transition. However, the majority of studies have been carried out with longer chain lengths.

X-Ray scattering studies of octadecane, $CH_3(CH_2)_{16}CH_3$, and other even chain length alkanes in their triclinic phase formed by the supercooling of melt, have shown that a transient metastable rotator phase exists. Such metastable and transient phases are likely to have analogues in polymeric systems.[21–23]

Molecular dynamics simulations have been used to provide a better understanding of the processes occurring on cooling n-nonadecane.[24] A molecular dynamics simulation of the free surface of the melt have been carried out using three layers of lamellas and studied over the a temperature range (385–410 K). In this range the middle layer prefers to be in the melt state with both surface layers remaining crystalline and reflects the experimental observation of surface freezing found in n-alkanes. It is found that the molecules in the surface monolayer align their axes nearly perpendicular to the surface and form well-defined hexagonal packing. It is also found that the molecules in the surface monolayer show large centre-of-mass fluctuations, translational and transverse, along the surface normal and parallel to the surface, respectively. The chains in the middle are therefore exhibiting characteristics of a rotator phase.

Investigations of mixtures of alkanes for $C_{19}H_{40}$–$C_{20}H_{42}$ indicate that the temperature range of the rotator phase widens due to chain mixing.[25] Eicosane has also been studied and X-ray scattering measurements indicate the existence of a rotator phase.[26] Studies of the ultrasonic attenuation and DSC measurements[27] (Figure 4.6) and Raman scattering[28] have indicated that between 33.3 °C and 36.6 °C there exists a rotator phase as evidenced by a marked drop in the velocity from the expected linear behaviour. The DSC traces show a clear shoulder prior to the main melting peak which is indicative of the existence of the rotator phase.

Raman scattering studies[28] indicated the concentration of *gauche–trans* defects increased continuously with increasing temperature and these defects can be associated with the ends of the chain.

The diffusion coefficients in a number of n-alkanes have been investigated,[29,30] which show the influence of the rotator phase assisting the motion of the polymer chains. Rotator phases have been reported in chains with as many as 36 carbon atoms.[31,32] Measurements of DCS and X rays scattering[33–36]

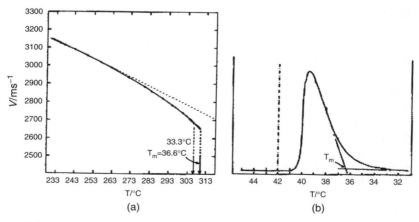

Figure 4.6 Plot of the acoustic velocity in the chain direction (a) and the DSC trace (b) for the melting of eicosane. The dotted line in (a) indicates the linear extrapolation of the velocity from low temperature.

of a number of *n*-alkane systems have confirmed the existence of the rotator phase.

4.4 Conclusions

Rotator phase materials are related to liquid and molecular crystals, discussed in Chapters 2 and 3. The rotator phase is characterized by the following:

- in low molecular mass materials the molecules exhibit almost free rotation yet are fixed on their lattice points, and
- in the short-chain alkanes the chains are free to rotate about their molecular axis and also exhibit enhanced diffusion coefficients due to their ability to move down the cylinders as a consequence of the free rotation of the chains.

This mobility close to the melt temperature is very important in understanding the behaviour of polymer materials in the crystalline state and is considered in subsequent chapters.

Recommended Reading

J.N. Sherwood (ed.), *Plastically Crystalline State*, Wiley, Chichester, UK, 1970.

References

1. C.S. Yoon, J.N. Sherwood and R.A. Pethrick, *J. Chem. Soc., Faraday Trans.*, 1989, **85**(9), 2867.
2. M. Eldrup, D. Lightbody and J.N. Sherwood, *Chemical Physics*, 1981, **63**(1-2), 51–58.
3. J. Timmermans, *Les Constantes Physiques des compos'es organiques crystallise's*, Masson, Paris, 1953.
4. C. Clemett and M. Davis, *Trans. Faraday Soc.*, 1962, **58**, 1705.
5. J.G. Powles, *J. Chem. Phys.*, 1952, **20**, 1048.
6. A.H. White and W.S. Bishop, *J. Am. Chem. Soc.*, 1940, **62**, 16.
7. C.P. Smyth, *Trans. Faraday Soc.*, 1946, **42A**, 175.
8. C. Clemett and M. Davis, *J. Chem. Phys.*, 1960, **29**, 1347.
9. A. Turney, *Proc IEE*, 1953, **100**, 46.
10. A. Gerschel, I. Darmon and C. Brot, *Molecular Physics*, 1972, **23**(2), 317.
11. M. Hattori, S.I. Fukada, D. Nakamura and R. Ikeda, J. Chem. Soc., *Faraday Trans.*, 1990, **86**, 3777–3783.
12. M. Hattori, Y. Onoda, T. Erata, M.E. Smith, M. Hattori, H. Ohki and R. Ikeda, *Z. Naturforsch. A*, 1994, **49**, 291–296.

13. T.M.R. Maria, F.S. Costa, M.L.P. Leitao and J.S. Redinha, *Thermochim. Acta*, 1995, **269**, 405–413.
14. J. F. Willart, M. Descamps and N. Benzakour, *J. Chem. Phys.*, 1996, **104**, 2508–2517.
15. J.J.M. Ramos, R.J.C. Sousa, N.T. Correia and M.S.C. Dionisio, *Ber. Bunsen-Ges. Phys. Chem.*, 1996, **100**(5), 571–577.
16. J.J. M. Ramos, N.T. Correia and M.J. Teixeira, *Mol. Cryst. Liq. Cryst. C*, 1996, **6**(3), 205–214.
17. M. Descamps, N.T. Correia, P. Derollez, F. Danede and F. Capet, *J. Phys. Chem. B*, 2005, **109**(33), 16092–16098.
18. K. Horiuchi, H. Takayama, S. Ishimaru and R. Ikeda, *Bull. Chem. Soc. Jpn.*, 2000, **73**(2), 307–314.
19. G.A. Arteca, *J. Phys. Chem. B*, 1997, **101**(20), 4097–4104.
20. B. Zgardzinska, J. Wawryszczuk and T. Goworek, *Chem. Phys.*, 2006, **320**(2–3), 207–213.
21. E.B. Sirota and A.B. Herhold, *ACS Symp. Ser.*, 2000, **739**, 232–241.
22. I.V. Filippova, E.N. Kotelnikova, S.Y. Chazhengina and S.K. Filatov, *J. Struct. Chem.*, 1998, **39**(3), 307–317.
23. I. Denicolo, J. Doucet and A.F. Craievich, *J. Chem. Phys.*, 1983, **78**(3), 1465–1469.
24. H.Z. Li and T. Yamamoto, *J. Chem. Phys.*, 2001, **114**(13), 5774–5780.
25. A.B. Herhold, H.E. King and E.B. Sirota, *J. Chem. Phys.*, 2002, **116**(20), 9036–9050.
26. B.M. Ocko, E.B. Sirota, M. Deutsch, E. DiMasi, C.S. Coburn, J. Strzalka, S.Y. Zheng, A. Tronin, T. Gog and C. Venkataraman, *Phys. Rev. E*, 2001, **63**, 33.
27. C.S. Yoon, J.N. Sherwood and R.A. Pethrick, *J. Chem. Soc., Faraday Trans.1*, 1989, **85**(10), 3221–3232.
28. M. Maroncelli, S.P. Qi, H.L. Strauss and R.G. Snyder, *J. Am. Chem. Soc.*, 1982, **104**, 6237.
29. T. Yamamoto, H. Aoki, S. Miyaji and K. Nozaki, *Polymer*, 1997, **38**(11), 2643–2647.
30. H. Yamakawa, S. Matsukawa, H. Kurosu, S. Kuroki and I. Ando, *J. Chem. Phys.*, 1999, **111**(15), 7110–7115.
31. H. Honda, S. Tasaki, A. Chiba and H. Ogura, *Phys. Rev. B*, 2002, **65**(10), 104112.
32. F. Mina, T. Asano, D. Mondieig, A. Wurflinger and C. Josefiak, *J. Phys. IV*, 2004, **113**, 35–38.
33. S.L. Wang, K. Tozaki, H. Hayashi, S. Hosaka and H. Inaba, *Thermochim. Acta*, 2003, **408**, 1–2.
34. K. Kato and T. Seto, *Jpn. J. Appl. Phys. Part 1*, 2002, **41**(4A), 2139–2145.
35. K. Nozaki, T. Yamamoto, T. Hara and M. Hikosaka, *Jpn. J. Appl. Phys. Part 2*, 1997, **36**(2A), L146–L149.
36. H. Honda, S. Tasaki, A. Chiba and H. Ogura, *Phys. Rev. B*, 2002, **65**(10), 104112.

CHAPTER 5
Morphology of Crystalline Polymers and Methods for Its Investigation

5.1 Introduction

Polymers when they solidify will form a crystalline, partially ordered or totally disordered structure, depending on the regularity of the polymer backbone and the strength of the polymer–polymer interactions. In the melt, the flexible chain will normally contain a number of higher energy *gauche* conformations and adopts a *random* coiled form consistent with its inherent entropic disorder. On cooling, the lower energy *trans* form becomes predominant and chains prefer a more extended structure. The all-*trans* sequences produce linear sections of chain that are able to interact with other chains and initiate crystal growth. The extent to which the molecule eliminates the higher energy conformations on cooling will dictate its ability to crystallize. If the interactions opposing adoption of the *trans* conformation are insufficiently strong, then the disorder is retained in the solid state and an *amorphous* material is formed. For a polymer to exhibit crystallinity, the backbone structure must have a regular structure, strong interchain interactions or alternatively a specific chain conformation. The presence of chain stereochemical defects, e.g. atactic sequences and/or chain branches at high concentrations, makes it impossible for the polymer chain to form a close packed structure and an *amorphous* structure is formed. Large side chains will inhibit packing and promote *amorphous* structure formation. Specific interactions, such as hydrogen bonds or strong dipole–dipole interactions, may be sufficiently strong to create order between neighbouring polymer chains and promote microcrystal formation.

5.2 Crystallography and Crystallization

To help us understand the factors that influence crystal formation in polymers, it is appropriate to consider particular polymer systems.[1]

(a) Polyethylene

In Chapter 3 the basics of crystallography were introduced. Although polymers have long chains it is still possible to identify a unit cell that often corresponds to the monomer or groups of monomers and which is repeated in space. In the case of polymers, it is conventional to select the chain axis as the c-axis, except in the case of the monoclinic cell when the chain axis is the unique axis (b-axis).[1] A central postulate of polymer crystallography is that the polymer chains are in their lowest energy state. The most extensively investigated polymer system is polyethylene. As discussed in Chapter 1, the chain can exist in two conformations: the *trans* state that is the lowest energy state and a doubly degenerate higher energy *gauche* state. Despite the single lowest energy *trans* conformational structure, polyethylene shows two different crystal forms at normal pressure, pointing to the possibility that the all-*trans* chains may pack differently. Packing of the extended polyethylene chains will create a 'sheet' of molecules with a one-dimensional repeating structure. The next stack coming alongside the first will 'see' small differences in the potential energy surface, leading to a slightly displaced form of the normal packing of the chains. Because this potential surface contains only small differences in energy between similar states, more than one form is possible. The existence of more than one crystal form for a specific compound is referred to as *polymorphism*. For polyethylene, the orthorhombic structure shown in Figure 5.1 is the most stable. The all-*trans* backbone does not present a circular cross-section and the preferred packing is the orthorhombic structure. The deviation from the ideal hexagonal packing is not large. A monoclinic structure has been observed when crystallization occurs in stressed samples. When considering the crystallization process we must remember two factors: the odd–even effects in terms of the enthalpy of fusion (Chapter 1) and also the mobility of the alkane molecules in the plastic crystals (Chapter 4).

All these factors will influence the initial nucleation process and hence the form of the crystal structure that is observed.

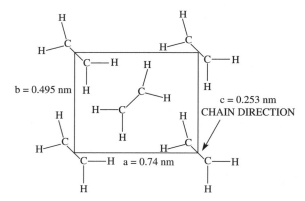

Figure 5.1 View along the c-axis, along the chain direction, for the orthorhombic polyethylene crystal.[1]

(b) Polypropylene

Whilst polyethylene with its simple regular structure is easy to visualize, forming an ordered crystalline structure, it is sometimes a little more difficult to appreciate the factors that influence crystallization in other polymer systems such as polypropylene. The methyl groups in the isotactic polymer would be expected to inhibit the packing of the chains; however, these groups impose a helical twist on the backbone and crystallization can occur (Figure 5.2).

X-Ray analysis indicates that fibres have a monoclinic unit cell structure containing 4 chains and 12 monomer units and cell parameters $a = 6.65$ Å, $b = 20.96$ Å, $c = 6.50$ Å, $\beta = 99.4°$. This structure is consistent with the chains having a three-fold helical conformation: 3_1 helix. The helical structures can register and pack to form a regular crystal structure. In contrast, the atactic material does not form a helical structure and crystallization is inhibited. As a consequence, atactic polypropylene is a soft flexible solid and is used as an additive to lubrication oils, whereas the isotactic polymer is rigid and used for fabrication of hot water pipes used in plumbing applications. Isotactic polypropylene has two helical forms, both having the same energy. The torsion angle about the $[-CH_2-CH[CH_3]-]_n$ bond is denoted ϕ_1 and the torsion angle associated with the $[-CH[CH_3]-CH_2-]$ bond is denoted ϕ_2. The two low-energy conformational states are repeats of $\phi_1 = 120°$, $\phi_2 = 0°$ (right-handed helical structure) and $\phi_1 = 0°$, $\phi_2 = 240°$ (left-handed helical structure). The isotactic polymer is a chiral molecule and hence can be either right or left handed. In the isotactic crystal of polypropylene there are four different possible helices: a right-handed helix pointing upwards and another pointing downwards and a complementary pair of left-handed helices. Although at first sight the upward and downwards helices would appear the same, a closer inspection indicates that they are different. It is therefore not surprising that isotactic polypropylene exhibits a number of polymorphs, referred to as α, β and γ forms. Packing adjacent chains with opposite senses of the helices creates the α form of isotactic polypropylene. In fact, it is generally found that the packing helices of an opposite form are better than if the same forms are used. The γ structure was first discovered in high-pressure crystallized isotactic polypropylene and X-ray

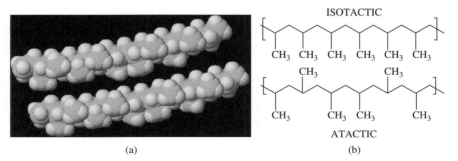

Figure 5.2 A space filling model of the isotactic structure of polypropylene (a) and symbolic structures for the polymers (b).

analysis suggests a triclinic structure with $a = 0.854$ nm, $b = 0.993$ nm and $c = 4.241$ nm.[2] If we compare these dimensions to those of the individual chains it becomes clear that the best fit of the structure is achieved with a structure in which alternative layers are each two chains thick, and with each of the bilayers at an angle of 80° to the adjacent bilayers.

The β form is a hexagonal form and produces spherulitic structure and is usually induced using nucleating agents. In addition to these three crystalline forms, isotactic polypropylene can exhibit a so-called smectic or mesomorphic phase on rapid cooling. An X-ray scattering pattern with two broad peaks centred at 14.8° and 21.2° (2θ, Cu K_α) showed that the density of this phase was 880 kg m^{-3}, compared with the density of 850 kg m^{-3} for the fully amorphous polypropylene[3] and it has been suggested that this corresponds to a *smectic* phase built up from bundles of straight chains of left- and right-handed helices with less than perfect order in the direction perpendicular to the chain helix axis.

(c) Polyoxymethylene

Replacement of one of the methylene groups of polyethylene by an oxygen creates polyoxymethylene. In this molecule, the lone pairs of electrons on the oxygen will interact with the hydrogen atoms on the adjacent carbon atoms and the preferred conformation is the nearly all-*gauche* structure which produces a stable trigonal form (I) and a less stable orthorhombic form (II). The unit cell contains chains with the same handedness, left and right-handed molecules appearing in different crystal lamellae.

(d) Poly(ethylene oxide)

In the crystalline state, the poly(ethylene oxide) chains form a structure which has seven monomer elements; $-(CH_2CH_2O)_n-$ and two helical turns per unit cell. The crystallographic unit cell contains four molecular chains and is monoclinic with $a = 7.96$ Å, $b = 13.11$ Å, $c = 1939$ Å and $\beta = 124.8°$.[4-6] The chains have dihedral symmetry, two fold axes, one passing through the oxygen atoms and the other bisecting the carbon–carbon bond. The chain conformation assigned to internal rotation about the $-O-CH_2-$, $-CH_2-CH_2-$ and $-CH_2-O-$ bonds is *trans*, *gauche*, *trans* respectively. Although poly(ethylene oxide) is essentially very similar to polyethylene, it is therefore quite surprising the complexity of the chain conformation found in the solid and reflects the dominance of the local interactions in determining the unit cell structure.

Many isotactic forms of polymers or simple linear chains, such as polytetrafluoroethylene and poly(vinylidene chloride), have helical conformations. The pitch of the helix is determined by the influence of the nonbonding short-range interactions between adjacent atoms on the polymer backbone. Helical structures are frequently observed and reflect the subtle effects of these interactions.[7] The controlling factor is the enthalpy of the melt process and Table 5.1 summarizes values for some common polymer systems.

These polymers illustrate the complexity of predicting the crystal structure of these polymers. Although polystyrene, with its bulky phenyl side group, is normally considered to be an amorphous polymer, the isotactic form of the polymer forms a helix and a crystalline structure. Usually the density of the

Table 5.1 Heats of fusion, melting points and densities for some common polymers.

Monomer unit	Enthalpy of fusion, ΔH $(J\ g^{-1})$	Melt temperature (°C)	Density (g cm^{-3})	
			Amorphous	Crystalline
Ethylene (linear)	293	141	0.853	1.004
Propylene (isotactic)	79	187	0.853	0.946
Butene (isotactic)	163	140	0.859	0.951
4-Methylpentenene-1 (isotactic)	117	166	0.838	0.822
Styrene (isotactic)	96	240	1.054	1.126
Butadiene (1,4 polymer) (cis)	171	12	0.902	1.012
Butadiene (1,4 polymer) (trans)	67	142	0.891	1.036
Isoprene (1,4 polymer) (cis)	63	39	0.909	1.028
Isoprene (1,4 polymer) (trans)	63	80	0.906	1.051
Vinyl chloride (syndiotactic)	180	273	1.412	1.477
Vinyl alcohol	163	2165	1.291	1.350
Ethylene terephthalate	138	280	1.336	1.514
ε-Caprolactone	142	64	1.095	1.184
Ethylene oxide	197	69	1.127	1.239
Formaldehyde (oxymethylene)	326	184	1.335	1.505
Nylon 6	230	270	1.090	1.190
Nylon 6,6	301	280	1.091	1.241

crystalline form is higher than that of the amorphous solid; however, in the case of 4-methylpentenene-1 the reverse is the case. The crystal of 4-methylpentenene-1 has a 7_2 helix and this occupies a large volume leaving a hollow cylinder down the centre of the coil. Although nylon 6 and nylon 6,6 have similar melting temperatures they have significantly different enthalpies that reflects the differences in the hydrogen bonding in the two polymers and explains the significant difference in their densities and susceptibilities to moisture uptake.

In general, however, identification of the crystal cell is only part of the problem of characterizing the structure of crystalline polymers. Crystals are never perfect and the units cells do not infinitely duplicate through space even when they are grown very carefully from dilute solution using low molecular mass materials. As with the organic crystals considered in Chapter 3, a variety of defects can be observed and are associated with chain ends, kinks in the chain and jogs (defects where the chains do not lie exactly parallel). The presence of molecular (point) defects in polymer crystals is indicated by an expansion of the unit cell as has been shown by comparison of branched and linear chain polyethylene. The c parameter remains constant, but the a and b directions are expanded for the branched polymer crystals. Both methyl and ethyl branches induce expansion, whereas larger pendant groups, propylene or longer homologues, are largely excluded from the crystals.

5.3 Single Crystal Growth

Our knowledge of the structure of crystalline polymers relies on a very large volume of research carried out on the crystal growth of polymers from dilute solution.[8–12] The majority of the early research focused on polyethylene and involved crystallization from 0.01% solutions in xylene. This method of crystal growth produces single crystals that are 10 nm thick platelets. In order to explain the shapes of these single crystals it was necessary to propose that the long polymer chains folded back and forward rather than being infinitely extended in space. Storks[13] proposed folding of polymer chains in *trans*-polyisoprene. An example of a transmission electron micrograph[12] showing that the crystals of polyethylene have a regular shape which are pyramidal in form is shown in Figure 5.3. Considering the polyethylene chains to be predominantly in an extended *trans* form, then the thickness of the platelets, ~10 nm, can only be explained by assuming that in some way the chains of the polymer fold back on themselves.

Careful examination of single crystals grown from solution at 70 °C in xylene indicated that they are lozenge shaped and have surfaces that are {110} planes.

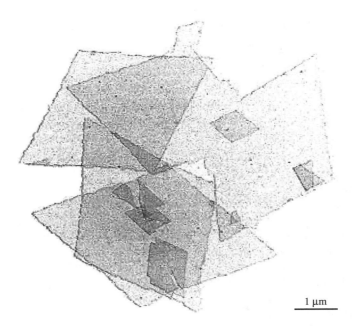

Figure 5.3 Transmission electron micrograph of a solution-grown single crystal of linear polyethylene showing lozenge-shaped crystals.[12]

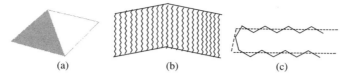

Figure 5.4 Chain packing to form pyramidal-shaped single crystal of polyethylene: (a) pyramidal shape; (b) chain alignment within pyramid; (c) close-up of the (110) chain fold with chain axes parallel but slightly displaced and the resulting inclined edge.

Crystals grown at 80 °C are truncated with two new sectors with {100} facets. The existence of crystallographic facets indicates that the growth process is controlled and that the chains are deposited in a regular manner. It must be recognized that the higher the temperature used for the growth, the greater will be the frequency of the higher energy *gauche* sequence in the chain. It is possible to rationalize the observed effects in terms of the conformation of the polymer chain in solution. In the {110} sectors, the folds of the chain lie along the [110] direction, whereas for {100} sectors, the chain folding is expected to occur along [010]. The difference in fold types affects the thermal stability of the sectors and it is noted that {100} sectors possess a lower melting point than the {110} sectors.

The single crystal images are obtained by lifting the crystal from the solution using a supporting metal grid. Corrugations of the crystal are sometimes observed and are a consequence of collapse of the pyramidal single crystals. The hollow pyramid shape arises from the chain axis not being parallel with the normal to the lamella direction. The chain axis is often found to be at an angle of $\sim 30°$ with respect to the normal of the lamella and the tilt is a consequence of a small vertical displacement when the chains fold (Figure 5.4).

Similar displacements to those shown in Figure 5.4 are found in many types of crystal and indicate that the (110) fold is consistent with the chains being placed on a diamond lattice structure. The pyramidal crystal arises from the regular chain folding and constraints of the (110) fold. Further studies[13] have shown that the fold surfaces in the {110} sectors were parallel to the {312} planes and that the fold surfaces in the {100} sectors were parallel to {201} planes. A match between {312} and 201} is obtained only for certain fixed ratios of {100} and {100} growth.[14] A preference for such a growth ratio and deviations from this ideal shape may occur and are associated with distortions of the lattice. Crystals grown from the melt exhibit similar structure with a rooftop shape and has implications for the alignment of crystalline polymers when the melt is drawn below the melt temperature.[15]

5.3.1 Habit of Single Crystals

As with simple organic crystals, the temperature at which crystallization occurs will influence the habit of the crystal formed. Whereas in simple organic crystals the growth can be envisaged as occurring from the primary unit cell, in the case of a polymer system the nucleus is formed by the alignment of sections of all-*trans* or similar chain structures. The conformational entropy of the polymer molecule in the melt will influence the length of these low-energy sections. The higher the temperature the more frequent the occurrence of the *gauche* sequences and the larger the B/A ratio with the consequence that a rounding of the crystal {100} surface occurs[16] (Figure 5.5). At the melt temperature, 130 °C, only a very small part of the periphery of the crystal lamellae is {110} facets, the remaining parts being rounded {100} surfaces. The variation in the shape, or *habit*, of the crystals is a direct reflection of the influence of chain folding on the growth mechanism. The changes in shape follow a smooth variation and the ratio B/A follows a continuous function with temperature. Crystals with well-defined {100} and {100} facts are typical of the crystallization of polyethylene at relatively low temperatures, 70 and 90 °C, from *p*-xylene.

Despite the apparent complexity of the growth behaviour of polyethylene, a number of characteristic shapes have been identified for particular polymer systems (Table 5.2).

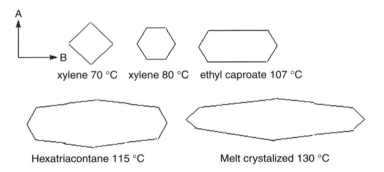

Figure 5.5 Schematic of the change of crystal shape with growth temperature for polyethylene obtained from different solvent systems.[15,16]

Table 5.2 Characteristic shapes for some polymer crystals grown from dilute solution.

Polymer	Characteristic shape of crystal
Polyoxymethylene	Hexagonal hollow pyramid
Poly(4-methyl-1-pentane) (isotactic)	Square-based pyramid
Polytetrafluoroethylene	Irregular hexagonal platelets
Poly(but-1-ene)	Square or hexagonal platelets
Polystyrene (iostactic)	Hexagonal platelets
Poly(ethylene oxide)	Square or hexagonal platelets
Poly(ethylene terephthalate)	Flat ribbons of ~30 nm width
Polyamide 6	Lozenge-shaped lamellae
Polyamide 6,6	Irregular hexagonal platelets tending to flat ribbons
Polypropylene (isotatic)	Lath shaped platelets

Most polymer crystals will exhibit facets and some like polyoxymethylene form hollow pyramids. The occurrence of the hollow pyramid is for similar reasons to that proposed for polyethylene and is a direct consequence of the constraints on the chain folding. The smooth surfaces observed for many crystal systems are evidence of regular chain folding, but are not proof that this occurs.

As in the case of simple crystals discussed in Chapter 2, spiral growth can be observed stemming from dislocations. The same factors appear to operate, except that now the step size is controlled by the lamellae thickness rather than the single atom attachment.

5.4 Crystal Lamellae and Other Morphological Features

5.4.1 Solution-Grown Crystals

If crystallization is carried out slowly in dilute solution then during lamellae growth, additional layers may be formed through the action of screw dislocations. When growth is slow the number of such defects is small and well-ordered multilayered crystals are formed. If, however, the growth rate is fast then complex lamellae aggregates that are either sheaf-like or spherical in outline are formed. The nature of these aggregates is very important as they can have profound effects on the mechanical properties of the resultant bulk solid. During the melt crystallization process, the shorter chains are rejected and become concentrated between the spherulites.[17] As such, interspherulitic regions constitute sites of potential weakness and are the location for crack propagation and dielectric breakdown. For this reason the spherulitic size and form are important in determining bulk polymer properties.

Whilst much of the initial research has focused on the structures generated from crystals grown from solutions of 0.01–0.001 wt%, it is the solid-phase morphology that is the more challenging to understand. A quantitative discussion of crystallization from dilute solution is attempted in Chapter 6; as an introduction, the factors that influence crystallization will be considered.

5.4.2 Chain Folding

A common feature of most polymer crystals is chain folding to form the lamellae and is observed in both solution-grown crystals and those obtained from the melt. One of the questions which arise when one considers chain folding is whether the folds are tight, occurring within four or so main-chain atoms, or loose and extending over many more. Although there may be complete folding in solution-grown single crystals, in the bulk a proportion of the molecules in adjacent lamellae must pass from one lamella to the next, *tie molecules*, and/or fold back into the lamellae from which they originate. Many mechanical properties are determined by the connectivity between the lamella structures and rely on the tie molecules to transfer the stress.

5.4.3 Crystal Habit

As indicated above, a wide variety of *habits* are found in solution-grown polymer crystals. Although it is possible to identify the 'ideal' crystal habit, it is also often found that in addition to the general rounding of the crystal faces, twinning of crystals can occur with the result that objects with six arms are common for {110} because (110) and (1$\bar{1}$0) faces are inclined at 67.5°, close to 60°. Moreover, lath-type structures are prone to develop with twin boundaries along their centre lines because of the accelerated growth provided by the notch at the tip, which aids molecular attachment.[18,19] This is a similar effect to that observed in small-molecule crystals where defects form the sites for rapid growth.

5.4.4 Sectorization

Polymers are long chains and so far in the discussion there has not been obvious evidence of the effects of chain length on the crystal structure, chain folding, removing the obvious effects of the chain length. However, sectorization is the first effect that can be related to the length of the polymer chains. It is observed as a surface texture in which the lamellae are divided into discrete regions bounded by a growth face. This phenomenon is a direct consequence of chain folding along the growth face and would not occur if the chain formed pre-folded blocks before attachment to the crystal. Sectorization arises because folding of the chains transforms a long molecule into a pleated sheet which can extend across several successive lattice planes but essentially lies along the relevant growth surface denoted the *fold* plane.[20] The so aligned chain breaks the symmetry and slightly distorts the repetitive unit of chain packing within the lamellae. The fold plane in a given sector differs from nominally equivalent ones along which there is no folding and so transforming single lamellae into a multiple twin. A four-sided polyethylene lozenge (Figure 5.5) bounded by four {110} growth surfaces is, accordingly, a four-fold twin with four equivalent sectors in each of which the two nominally equivalent {110} surfaces will be parallel to the fold plane, the other is not. At higher growth temperatures lozenges become six sided with truncating {100} surfaces, and then there are six corresponding sectors. The two new ones denoted {100} sectors are twinned with respect to each other but are different in character from the four twinned {110} sectors and have a lower melting point. The size of the sector is governed by that of its growth face and can be very small, e.g. dendritic growth, when individual facets, each of which has its own micro-sector, may be as little as 20 nm wide.

5.4.5 Non-planar Geometries

One consequence of the distortions which are associated with the sectorization is to make the lamellae non-planar. Folding along the growth faces makes nominally equivalent planes in the subcell become non-equivalent adopting slightly different spacings. For the three systems in which measurements have been reported, polyoxymethylene, isotactic poly(4-methylpent-1-ene)[21] and low molar mass polyethylene[22] which have, respectively, hexagonal, tetragonal and orthorhombic subcells, the distortion reported is of the order of ~ 0.001 nm. As a consequence, the transverse axes of the subcell no longer meet at precisely 60° or 90° and distortion of the crystal results. Such small effects are probably universal in sectored polymer lamellae. The pyramidal structures identified in Figure 5.4 can invert and appear as chair crystals or form derivative structures such as half hollow pyramidal/half ridged lamellae with a central hollow pyramidal portion.

5.5 Melt Grown Crystals

5.5.1 Melt-Crystallized Lamellae

Development of various specimen preparation methods[20] has allowed some important features for melt crystallization to be identified. It is found that the bulk material is full of lamellae that often have different profiles.[23] In a typical crystalline polymer, such as polyethylene, spherulitic growth is responsible for the spatial variation in physical properties. The structures that are observed are the results of the growth of dominant lamellae that branch and diverge. The space created between the lamellae is, except in the case of very low molecular weight materials, filled by *subsidiary* or *infilling* lamellae. This type of growth will produce lamellae with different characteristics for two reasons. Firstly, fractional crystallization in which the chains with different molar mass (chain length) segregate from one another allows the shorter chains to form lamellae which have a lower modulus and melting point. Secondly, the different orientations of the lamellae help to develop isotropic properties, a proportion of the lamellae being in the direction of the applied force and a proportion being perpendicular with a statistical distribution being in all other directions. If the material is cold drawn, then the applied stresses will develop physical characteristics in the material that are the result of alignment of the dominant lamellae and the development of an enhanced modulus in the draw direction. The natural draw ratio is influenced by the 20° alignment of the subsidiary lamellae to the dominant structures.

Changes in growth temperature can lead to the profile of the lamellae varying from planar to S or C shaped. The change in shape to an S profile {201} fold structure occurs at a critical growth rate and decreases for longer and/or more branched molecules.[24,25] The time taken for a single 5 nm thick molecule layer to be added to the crystal is $\sim 10^{-1}$ s for linear polyethylene of 100 000 molecular mass. For slower growth, {201} surfaces occur and nucleate as such, in contrast to faster growth, when lamellae nucleate with {001} surfaces and subsequently transform behind the growth front, adopting S profiles and thickening isothermally. For slower growth a molecule is able to achieve inclined fold packing as it adds to the crystal, but not for faster growth when

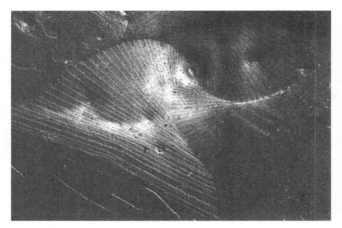

Figure 5.6 Electron micrograph of a multilayered crystal of polyethylene displaying regular rotation of successive growth terraces.[26]

improvements in fold packing must be accomplished within the constraints of surrounding lamellae. When a region of inclined packing is produced in the central part of the lamellae, *i.e.* the oldest portion of the lamellae, this leads automatically to an S profile in cross-section. Once more the habit can be significantly influenced by the occurrence and distribution of defects in the lamellae surface.

Keller[26] has shown that a terrace-like structure is observed reflecting the stacking of the lamellae structure in a melt-grown crystal (Figure 5.6).

5.5.2 Polymer Spherulites

The supermolecular structure exhibited by many polymers has features that are in the range of 0.5 μm to several millimetres. Such features can be observed using polarized light optical microscopy. The common form of structure observed is called a spherulite and, as its name implies, it is a circular crystalline object. The first spherulites found were in vitreous igneous rocks and the name comes from the Greek word for a ball or globe. A spherulite is an 'object' with spherical optical symmetry. Two unique refractive indices may be determined, namely the tangential (n_t) and radial (n_r) refractive indices. As an example, consider crystalline polyethylene. The single crystal or any orientated structure of this polymer may be considered to be uniaxially birefringent, with the unique direction (largest refractive index) along the chain axis—down the *stem*. Negative spherulites with $n_t > n_r$ have a higher proportion of the chains in the circumferential planes than along the radius of the spherulites. The direction of growth of polyethylene spherulites is always close to [010], *i.e.* the radius of the spherulite is parallel to the crystallographic b-axis (Figure 5.7).

Other polymers have other growth directions, e.g. monoclinic isotatic polypropylene grows faster along the a-axis. Similar behaviour is observed with

Morphology of Crystalline Polymers and Methods for Its Investigation 119

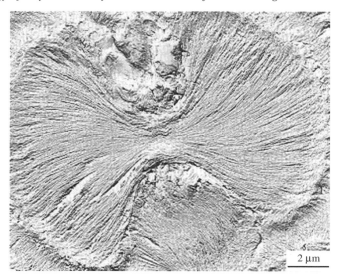

Figure 5.7 Sheaf-like lamellar aggregates crystallized from the melt in a blend of linear and low-density polyethylene at 125 °C.[27]

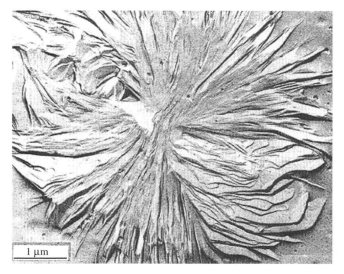

Figure 5.8 Micrograph of isotactic polystyrene crystallized at 200 °C showing sheaf-like spherulites.[27]

isotactic polystyrene crystallized at 200 °C that shows the characteristic sheaf-like spherulite structure (Figure 5.8). The size of the spherulites is controlled by the nucleation process which is almost invariably heterogeneous, *i.e.* growth starts from extraneous material such as dust or residual catalyst particles. Most of the information on bulk structure has been obtained either from fracture surfaces or strained microtomed thick sections of polymer. As a result, the structures that are observed are never as clear and definitive as those for solution-grown single crystals. It is generally agreed, however, that spherulitic structure is controlled by the dominant lamella growth that usually adopts a circular format. Starting from an individual lamella, the growth progression is to first create the dominant lamellae that then create a multilayered axialite, a parallel organized set of lamellae. Out of this axialite will emerge several fast growing lamellae that may splay and present a sheaf-like appearance down the principal axis of the splay (Figures 5.7 and 5.8). From these splays will emanate lamellae which will eventually be at right angles to the original axis as well as growth occurring parallel to the original structures. A schematic representation is shown in Figure 5.9. The axialite is a non-spherical and irregular superstructure. Axialites are primarily found in low molar mass polyethylene at essentially all crystallization temperatures and in intermediate molar mass polyethylene crystallized at higher temperatures at undercooling less than 17 K.

Very high molar mass samples, typically with molar mass of 10^6 g mol^{-1} or greater, form so-called random lamellar structures. The entanglement effect is so extensive that the crystalline mass becomes very low and regular lamellar stacking is absent in these samples.

The gross morphology is a consequence of the intersection of the growth fronts from the various nucleation centres and the boundaries are often hyperbolic reflecting the intersection of the spherical structures. It must always be remembered that these are three-dimensional structures and apparently distorted shapes are often structures viewed at a different angle. All the structures in Figure 5.9b are created by rotation or skewing of the original spherulite (Figure 5.9a). When viewed with polarized light, a Maltese cross structure is usually observed and spherulites appear to have different shades. This behaviour is exemplified in the case of the optical microscopic images for ethylene–propylene shown in Figure 5.10.

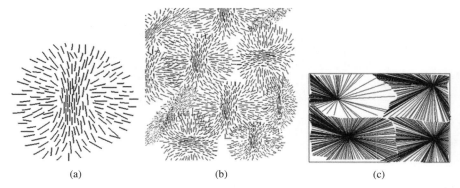

Figure 5.9 Schematic of lamellae growth (a) from an axialite to a spherulite and (b) packing into a solid. (c) Schematic of spherulites viewed with polarized light.

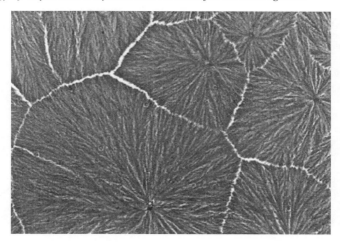

Figure 5.10 Optical micrograph of the spherulites produced by a propylene–ethylene copolymer.[28]

A scanning electron micrograph of a fracture surface of polypropylene shows the characteristic rings of an underlying spherulitic structure[29] (Figure 5.11). The spherulitic crystal structure can be subjected to low-temperature drawing. During this process the lamellae remain intact but are rotated into the line of the applied stress. As a consequence the lines of weakness that are the points at which the spherulites intersect open and form voids between the developing fibres. The spherulites elastically respond up to ~30% strain, followed by permanent inhomogeneous changes which involve combinations of slip, twinning and phase changes.

The morphologies discussed so far have all been formed from isotropic melts. In practice melts are often subjected to shear forces and this can induce alignment in the chains, which aids crystallization in particular directions. Orientation of the melt causes an increase in the free energy and this itself constitutes an important factor in practical processing. Shish-kebabs[30] are formed from solutions that are subjected to elongational flow, which induces orientation of the solute molecules (Figure 5.12).

A central core of orientated bundles of fibres is formed at first as a direct consequence of the orientation. The shish-kebab structure consists of a central group of highly orientated fibrils from which lamellar crystals can grow and form the kebab structures (Figure 5.12). The central fibrils are formed from high molar mass material. Similar structures have been obtained during extrusion/injection moulding of the melt when extreme conditions are used and the melt is subjected to high elongational flow in combination with a high pressure or a high cooling rate. The orientated melt solidifies in a great many fibrous crystals from which lamella overgrowth occurs. The radiating lamellae nucleating in adjacent fibrils are interlocked, a fact that is considered to be important for the superior stiffness properties of the melt-extruded fibres. The fibrillar structure is present in ultra-orientated samples. It consists of highly

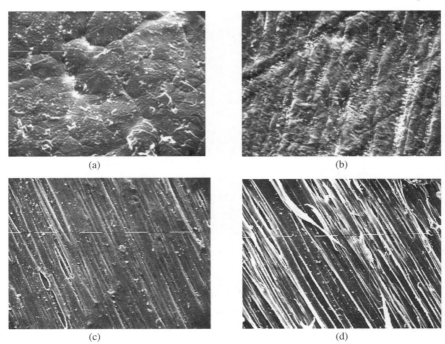

Figure 5.11 Scanning electron micrographs of the fracture surfaces of undrawn (a) and drawn (b), $\lambda = (l/l_0) = 2.6$; (c) $\lambda = 6$; (d) $\lambda = 14$ (where l is the drawn length and l_0 is the undrawn length of the sample).[29]

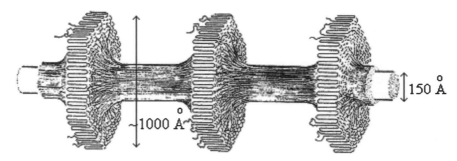

Figure 5.12 Schematic of a section through a shish-kebab structure showing extended core with radial lamellar growth.[30]

orientated microfibrils. These microfibrils are sandwiches of alternating sequences of amorphous and crystalline regions. A great many taut interlamellar tie chains are present and the resulting mechanical properties are excellent. The fibres are formed from stacks of the lamellae and have a very high strength consistent with the crystal structure of the polymer. At high draw ratios the fibres part and significant void structure can be observed between the fibres.

Figure 5.13 Thin film of polypropylene, part of spherulite crossed polars.[31]

Recent studies of highly linear metallocene-synthesized polyethylene of low molecular weight have shown the sheaf-like morphology found in solution is prevalent in the melt phase. The highly linear regular polymers are able very effectively to grow lamellae. The growth of typical lamellae is nicely illustrated in the optical microscopy of Saville[31] (Figure 5.13).

If growth occurs in contact with a nucleating surface then *trans*-crystalline structure is observed. The nucleating object will normally have a flat surface or a fibril. These surfaces become covered with a high density of nucleations resulting in a one-dimensional (columnar) growth in a direction parallel to the normal of the surface. The thickness of the *trans*-crystalline layers depends on the balance between surface nucleation and 'bulk' nucleation.

5.6 Annealing Phenomena

Annealing of monolayer solution-grown single crystals at high temperature leads to crystal thickening and formation of cylindrical holes in the crystals. This 'Swiss cheese' structure of annealed monolayer single crystals is thermodynamically more stable than the original 'continuous' single crystals. This is because the specific surface energy of the hole surface (~ 15 mJ m^{-2}) is much less then the specific surface energy of the fold surface (~ 90 mJ m^{-2}). Holes do not develop on annealing of mats of overlapping single crystals or melt-crystallized samples.

Polyethylene crystals are observed to thicken at temperatures greater than 110 °C, and if grown under isothermal conditions show this thickening gradually with time. This thickening process can have a profound effect on the data obtained from scanning differential calorimetry (DSC) measurements. On heating the sample an endotherm is observed associated with the melting process. Observation of the melting temperature as a function of the applied heating rate indicates that the melting point of the crystal is dependent on the heating rate used. Thin crystals have a tendency to thicken as the temperature is raised, prior to melting. The melting temperature decreases with increasing heating rate until finally a constant melting point is attained at the highest heating rate. In Chapter 4 the ability of eicosane chains to move led to the observation of a plastic phase just below the melt temperature. The thickening of the lamellar structure must involve a similar type of motion that allows movement of the folds and an increase in the length of the linear all-*trans* content of the crystal.

It appears that the thickening process occurs in abrupt and discontinuous steps,[32] crystals doubling or trebling their thickness in discrete steps. In melt-crystallized samples with a broad crystal thickness distribution this 'integer' crystal thickness change is not as clearly observed. Experiments on melt-crystallized samples commonly show a linear increase in the average thickness with the logarithm of annealing time. Polyethylene and a few other polymers form extraordinarily thick (micrometres) crystal lamellae after crystallization at elevated pressure, typically at 4–6 kbar. The polyethylene is transformed into a hexagonal phase with appreciable axial disorder at these elevated pressures consistent with the formation of a pseudo-plastic phase. The longitudinal mobility of the chains in such crystals is extremely high in the hexagonal phase as opposed to that of the conventional orthorhombic phase and linear elements of the chain in the lamellae grow past the length of the folded molecules.

Polyethylene has to a large extent dominated morphological studies in polymer systems; however, it is a good model for other systems. Optical microscopy of acetal shows the same characteristic patterns found in polyethylene when a spherulite is viewed under cross-polar imaging (Figure 5.14).

5.7 Experimental Techniques for the Study of Polymer Crystals

Whilst it is not appropriate to discuss the detail of the methods used it is appropriate to mention the principal methods that can be used. More detail of the methods can be found elsewhere.[33,34] The main techniques used are X-ray diffraction for determination of the lattice structure and scanning transmission electron microscopy together with optical microscopy for visualization of the structure. Figure 5.3 was obtained by placing a single crystal on a carbon-coated metal grid and then shadowed with a heavy metal such as gold and viewed using the bright field technique.

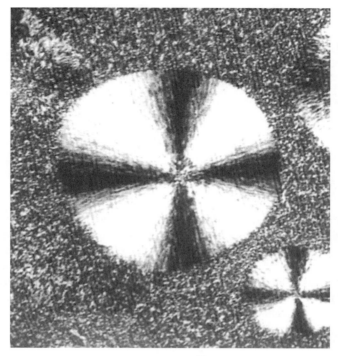

Figure 5.14 Section of acetal (crossed polars).[31]

5.7.1 Optical Microscopy

Polarized light microscopy is the most widely used method for characterizing the morphology of semi-crystalline polymers. Small variations in the local chain orientation produce variations of typically 0.01 or less in the birefringence. Using linear polarized incoming light and a crossed polarizer positioned after the objective lens allows the small differences in birefringence to be used to reveal the morphology. Crystals with their main axis at 45° to the polarizer/analyser pair will transmit the maximum of light; those with their axis parallel to the polarizer will transmit the minimum.

A spherullite is usually pictured as an array of lamellae radially disposed to one another and as a result is 'spherically' birefringent. A spherullite has two unique refractive indices: the radial (n_r) and the tangential (n_t). The refractive index ellipse can represent the variation in refractive index in the plane, where the length of the major axis of the ellipse is proportional to the maximum refractive index in the plane and the length of the minor axis is proportional to the minimum refractive index. If the larger refractive index is in the tangential direction, i.e. $n_r < n_t$, the spherulite is termed negative. Spherulites show a characteristic Maltese cross pattern with a maximum in the intensity in the direction at 45° to the polarizer/analyser pair.

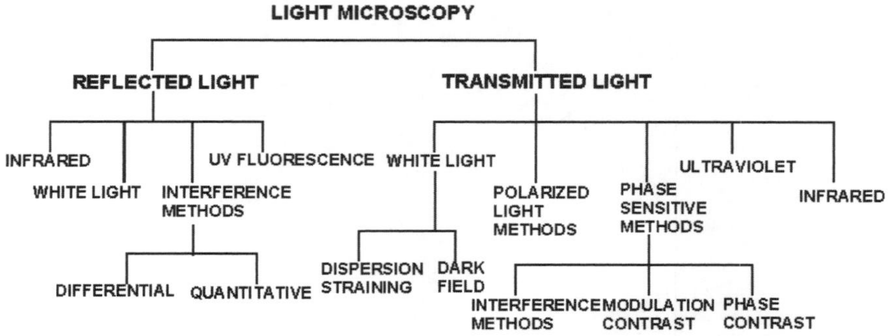

Figure 5.15 Summary of the various types of optical microscopy.

With dimensions on the micrometre scale it is possible for these features to scatter light. For this reason many crystalline polymers are opaque to the human eye. Small-angle light scattering (SALS) can be used to examine pseudo-crystalline material.[33] Using the birefringent nature of the material, studies are used with either H_v (crossed polarizers) or V_v (parallel polarizers). The scattering pattern is recorded, and from the scattering angle (θ_{max}) the average dimensions of the spherulite (R) are determined:

$$R = \frac{4.1\lambda_0}{4\pi n}\left[\sin\left(\frac{\theta_{max}}{2}\right)\right]^{-1} \quad (5.1)$$

where λ_0 is the wavelength of light *in vacuo* and n is the average refractive index of the polymer sample.

This scale of the dimension can be accessed by optical microscopy (Figure 5.15).

The structures themselves are often made up of orientated and organized substructures, as discussed above. The organization within the substructures can be used to aid their visualization. In addition to using direct observation of the specimen, it is possible to obtain images using reflected light. As indicated in Figure 5.15, the techniques that are available can be divided according to whether they are viewed using transmitted or reflected light. The transmitted light techniques require that the samples be obtained in the form of a thin section and require the use of microtome techniques to create these sections.

5.7.2 Microtomes[35]

For polymer work the best and most reliable microtome, in terms of consistently good results, is the base sledge type. Sections will be cut as thin as 0.5 µm, provided the samples are sufficiently rigid. The runners or glides on which the vice, which forms the moving element of the microtome, must also be rigid. The knife blade itself requires a precise edge and may be formed from a diamond edge. For soft samples, it may be necessary to cool the sample and cold working microtomes are available which can operate at liquid nitrogen temperatures. Alternatively, soft samples can be placed in a rigid resin and then microtomed at room temperature. It is important when carrying out any examination of a polymeric material to observe certain basic rules:

- Handle the sample as little as necessary and use forceps where possible.
- Define the problem and identify which part of the sample requires examination.
- Identify the problem associated with obtaining the specimen in the correct form. Clearly obtaining a thin section can introduce damage that could be misinterpreted as morphology unless it is correctly recognized.
- Assess the size of possible structural features to be examined and areas over which this size might reasonably be expected to vary. It may be necessary to cut 'sighting' section as an aid to making this assessment.

In order to increase the contrast in light microscopy, staining with selective dye molecules that absorb light or have a different refractive index is often used. This process is not widely used in the synthetic polymer field, but there are three specific instances that are worth noting:

- Unsaturated rubbers in sections can be selectively stained black with osmium tetroxide.
- The addition of a fluorochrome to methyl methacrylate used for embedding purposes can be useful in demonstrating impregnation of the specimen.
- Voiding or porosity exposed in cut surfaces can be more easily seen and the extent of penetration established if the surface is impregnated with drawing ink. The technique is to deposit the ink on the surface, subject it to a vacuum cycling routine and then draw off the excess ink with a pipette or tissue. When the remaining ink is dry the surface is polished gently on dry tissue or cloth after which the ink-filled voids will be clearly visible.

Good specimen preparation is the essential precursor of good microscopy. Because of the variability of polymeric materials, not just between types and grades but also between batches, and also the vagaries of processing techniques, each sample will be different from the next. Consequently there are no 'standard routines' or 'magic buttons' that will ensure good results. The secret of success is knowledge of what structural feature requires definition, an appreciation of how the sample will react to different procedures, an awareness and skill to perform as many varied preparation techniques as possible and to modify these as necessary, and patience.

5.7.3 Basic Light Microscopy[36–40]

The simplest approach to light microscopy is the use of a simple configuration in which light is passed through the sample and observed by direct observation of the image. The microscope will usually have a condenser system to focus the

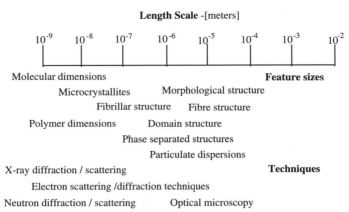

Figure 5.16 The length scale of the feature of interest dictates the appropriate technique for its study.

light into a narrow beam before it passes through the sample. The image is produced by an objective lens system which consists of a number of elements that act together to achieve the desired level of magnification. The observation of an object depends on detection of differences in refractive index between the object and its surroundings. If, as is often the case, the refractive index of the material is fairly uniform across the sample then the specimen may appear to be a uniform bright field. For a feature to be visible there has to be a significant difference in the refractive index of the object relative to the surrounding material.

The scale of the feature that is of interest dictates the preferred technique for its observation (Figure 5.16). Routine light microscopy of plastics and rubber seldom requires the microscope to be used at it limit of resolution.

Indeed, an especially low power ×1 objective lens may be more value in the objective set than a ×100 lens. If the use of higher magnification is necessary it is particularly important with polymeric specimens that the user commences the observation at low power and work up. The interrelationship between microstructural features in polymers can be complex and isolated observations at high magnification can be misleading.

5.7.4 Light versus Electron Microscopy

The microscopic examination of a specimen is usually embarked upon with a clear aim in mind. It is the problem that should dictate the choice of method and the examination will often justify the use of a range of microscopy techniques.

In very general terms, light microscopes are cheaper than their electron counterparts, less expensive to run and there are almost no problems of adverse interactions between the specimen and the radiation used. In the light microscope, the specimen is not exposed to vacuum and the specimen preparation procedures are fairly straightforward and flexible.

The great advantages offered by the electron microscope are substantially higher resolution in several modes of operation and, through accessories, elemental analysis of very small volumes. An increased depth of field is another advantage, at least in the case of conventional scanning microscopy. This allows easier image interpretation and a more precise understanding of the spatial interrelationship of features in the image. The disadvantages in relation to electron microscopy are electron beam damage and the use of high vacuum. A chapter later in this volume will consider the use of environmental scanning electron microscopy, when these problems are to some extent addressed.

5.7.5 Phase Contrast Microscopy

If a grating is inserted in the microscope it is possible to create diffraction effects in the back focal plane of the objective lens. The creation of a phase difference between the diffracted and undiffracted light can often increase image visibility. Specifically the aim is to increase the phase difference from around 90° to 180°. This can be accomplished by allowing the undiffracted light to pass through a thin plate (a 'phase' plate) of such a thickness that placing the plate in the back focal plane of the objective produces the necessary phase shift. At this point, the zero-order diffracted light can be located in a separate plane from the diffracted light, but to do this it is necessary to restrict the illumination cone from the condenser lens. The process of restriction of the cone of light by closing down the aperture iris gives a collimated beam. The zero-order diffraction maximum would then be clearly located at the centre of the back focal plane of the objective.

In practice, phase contrast systems may be 'positive' or 'negative'. For the former, higher refractive indices in the specimen will appear as being darker than the surrounding field; in negative phase contrast the situation is reversed. However, the contrast actually seen will also depend on the magnitude of the phase shift (or optical path differences) in the specimen. Examination of thin films of polymers may reveal knife marks since the knife creates regions of differing thickness and this will introduce phase shifts. In phase contrast, microscopy knife marks may be made substantially more visible than with straight transmission techniques.

5.7.6 Polarized Light Microscopy[40]

Additional information can be obtained on many polymeric materials by use of polarized rather ordinary white light. When a beam of unpolarized light is directed on a crystal of a material (such as calcite), it is found that there are two refracted beams instead of the usual one. This effect can lead to a duplication of the image, when examining an object through calcite crystal. These two beams can be shown to be polarized at right angles to one another by viewing the image through a sheet of polaroid. As the polaroid is rotated, first one image will be extinguished so that only one image remains, and on rotation through a further 90° the other image will disappear. The light can be described as being elliptically polarized, indicating that it consists of two components vibrating at right angles to one another. This concept and notation finds greatest use in situations where polarized light traverses more than one thin plate, in which case analysis by resolution of the beam into its individual components becomes quite complex.

5.7.7 Origins of Birefringence

Organized structures are often able to induce optical birefringence and when viewed using polarized light will have enhanced contrast. Imaging using crossed-polarizers or with the aid of birefringent fluids can assist with the visualization of the order within the structure. The substructure can usually be revealed by the selective use of solvent systems that will allow differential swelling of the sample, use of dyes which are selective for a particular type of chemistry or through the use of the structures themselves as revealed through the use of polarized light. In a polymer such as cellulose, it is possible to identify regions of high order within a fibre surrounded by areas of the reverse order or high disorder, and follow the changes that occur during the processing of the fibres.

5.7.8 Orientation Birefringence

In a system such as cellulose, addition of dibutylphthalate to the fibres will reveal the sub-micrometre order through changes in the birefringence of the fibres.[34] The magnitude of the birefringence can also provide information on the nature of the chain packing and the orientation of the molecules that produce the polarization changes. Orientation birefringence is produced by the alignment of molecules that are themselves optically anistropic, as is generally the case for polymer chains. Polymer chains can be considered to have a different refractive index parallel to the chain direction from that perpendicular to it. When the chains are randomly arranged as in the melt or in amorphous materials then the net refractive index is intermediate between the two separate values. When either crystallization or alignment due to drawing aligns the polymer chains then differences in the refractive index arise and birefringence can be detected. Many polymers are subjected to a drawing process during their fabrication, as in the manufacture of fibres and films, or as a by-product of the deformation that occurs in many processes such as compression or injection moulding. Measurement of the birefringence can in principle be used to indicate the extent to which orientation is being induced in that material.

5.7.9 Strain Birefringence[38]

Application of strain to a material or the presence of residual stress remaining from the production process can generate orientation that is detected as birefringence. The superimposed stress can alter the distance between atoms of the material, thus changing the polarizability of the bonds in the direction of the applied stress and hence creating a difference in refractive index in the material between the stress direction and directions perpendicular to it. This effect can occur equally well with molecules that are optically isotropic in the unstrained state and those that are naturally anistropic.

5.7.10 Form Birefringence[38]

Form birefringence is found in materials that have two or more phases with different refractive indices. A number of polymers are able to undergo phase separation and create on a nano-dimensional scale regions of homogeneity but distinctly different dielectric properties. If one phase is in the shape of rods or plates with smallest dimensions less than the wavelength of light, the refractive index of the whole material parallel to the rods is different from that perpendicular to the rods, even when both phases are themselves isotropic. Effects of this type are found in styrene–butadiene–styrene block copolymers.[34] It is possible in some cases to estimate the contribution of form effects to the total birefringence by selectively swelling one of the phases in various liquids of different refractive index. The form birefringence will in theory fall to zero when the refractive index difference between the phases is reduced to zero.

5.7.11 Polarization Colours[38]

When a wedge of material is viewed using monochromatic light a series of regularly spaced dark bands can sometimes be observed corresponding to points where the product of the thickness and the birefringence is a whole number of wavelengths. However, white light, which is more usually used to examine specimens, contains a continuous spread of wavelengths between about 400 nm and 750 nm, so destructive interference will occur at a different path difference for each part of the spectrum. Therefore, at no point can there be total darkness, for a specimen viewed in white light. For a specimen with a given (non-zero) optical path difference there will be subtraction from the incident white light of that wavelength for which the path difference is an integral multiple. Other wavelengths will also be reduced in intensity to a decreasing extent the further they are removed from this particular wavelength. With thicker or more birefringent specimens the path difference can be a multiple of more than one wavelength at the same time. The resulting complex subtraction of wavelengths from the incident white light produces a characteristic colour that can be related by use of a chart to the optical path difference producing it. Thus identification is only possible at the lower end of the path difference scale. The refractive index differences and birefringence can in principle be measured using references such as the quartz wedge or the Babinet compensator.

5.7.12 Modulation Contrast Techniques[39]

It is desirable to attempt to enhance the contrast of an object to be imaged. A modulation contrast system (MCS) and differential interference contrast (DIC) system can significantly help in this respect. The MCS and DIC systems detect the optical gradients and convert them to intensity variations. Since the eye can only detect colour and intensity variation detail in non-absorbing (phase) objects, a piece of transparent polymer is normally invisible. The optical gradient occurs at the curved or sloped portion of an object and is the rate of change in optical path difference between the object and its surroundings. A changing gradient can in effect act as a prism and would lead to refraction. In the very local region of the edge of the object there will be a refractive index gradient that in principle should allow visualization of the object. However, if the edge is not sharp then the gradient is small and visualization will be difficult. The factors that influence the ability to see the object are its size, edges and curvature of the object. The quality of the image will in turn depend on the wavelength of light, the diffraction process and the relative coherence of the light just prior to its interaction with the object. Since the mid wavelength in the visible spectrum is about 0.5 µm, physical features around this size create strong diffraction effects. The image of these features depends on the recombination (interference) of the diffracted waves at the image plane. The visibility of this interference process is dependent on the relative coherence of light at the moment of interaction with the object. The greater the coherence the greater the visibility of the interference process and thus the more visible are the images of the diffraction sites in the object. In a microscope the aperture of the condenser system controls the coherence of the light that reaches the object plane. As the condenser aperture is decreased it is usually observed that the visibility of the edges and particles of an object may be increased, although the resolution is reduced.

In the MCS, an additional slit is added with a polarizer as part of the condenser lens system. The modulation contrast microscope is best discussed in terms of the object being imaged as a small prism. Light passes through the object (prism) and then through the objective. The light coming from the object is then examined using a special filter. The filter has three elements representing different degrees of attenuation of the light from the object. Diffracted light passing through the unattenuated region of the filter is combined with light that is attenuated by about 15% of its original intensity. The intensity of the image is controlled by the transmission factor of that portion of the modulator thorough that the light passes. After passing through the modulator the light forms a bright image of the prism at the real image plane. The overall result is that combination of the shifted and non-shifted waves leads to enhanced contrast and greater visibility of the object. The enhanced clarity that is possible without loss of resolution has enabled fine features associated with local stress concentrations, *etc.*, to be imaged in otherwise transparent materials.

5.7.13 Interference Microscopy[40]

If two waves of light are out of phase then they are capable of producing constructive and destructive interference. Although two waves may have identical wavelengths and amplitudes, if one lags behind the other along the time axis then they will be 'out of phase' and a change in amplitude will result in the combined wave. One of the simplest methods to create an interference situation is to use flat cover slips that are coated with a thin semi-reflecting layer of metal such as gold or aluminium. The thickness of the coating is not particularly critical but should be thin enough to allow approximately 50% of the light to be transmitted. The phase delay is introduced between the reflected and transmitted waves and the cover slips form the basis of a conventional interferometer structure. Closing down the aperture iris increases the visibility of the interference fringes but can restrict the lateral resolution by increasing fringe contrast. The fringes observed can be interpreted as variations in the path length through the sample and associated with thickness variations in morphological features or, in the case of observation of a surface, variation in roughness.

5.8 Electron Microscopy

Perhaps the most common method for the study of morphology is electron beam microscopy. For very thin samples it is possible to use transmission methods and in principle the elastically scattered electron image can be used to determine the lattice dimensions. To enhance the contrast and aid the image definition the sample will typically be coated with a layer of gold or conducting carbon. This process minimizes the charge build-up on the surface and in the case of gold shadowing enhances the scattering and makes features more visible.[34,35] For both optical and electron beam microscopy etching of the sample can help reveal the detail of the molecular organization which exists.

5.8.1 Sample Preparation: Etching and Staining

For solution-grown very thin crystals direct examination is possible. However, for bulk materials either thin sections have to be created with the use of a low-temperature microtome or internal surfaces can be exposed through fracture of the material. In both cases the process of sample preparation can induce damage and great care has to be taken that the features identified are not artefacts of the sample preparation method. To enhance the contrast it is sometimes appropriate to etch the material.

The mildest form of treatment involves solvent etching, which relies on low melting point being more easily dissolved than the bulk of the material. In crystalline materials phase segregation can occur and hence this process can be very useful in outlining spherulite boundary layers and similar vulnerable areas.

The solvent can be used as part of a polishing process and the creation of smooth surfaces reveals more clearly the underlying morphological features.

There are essentially two main etching techniques: vapour etching for only a few seconds; and controlled isothermal treatment with a liquid solvent for a considerably longer time (several hours). A solvent etching temperature (T_d) is selected on the basis of the melting point of the segregated species (T_m):

$$T_d = T_m - \Delta T \tag{5.2}$$

where ΔT is dependent on the solvent power. For etching polyethylene with p-xylene, ΔT is 31 K.

For optical and electron microscopy etching and selective degradation of material can be helpful. Various approaches are used. Plasma etching will create oxygen-containing functions as will etching with permanganic acid. Other favoured etchant systems include etching with chlorosulfonic acid and staining with uranyl acetate. Sulfur, chlorine, oxygen and uranium add selectively to the amorphous component. Unsaturated polymers like polybutadiene or polyisoprene can be stained with OsO_4 that adds selectively to the double bonds located in the amorphous phase. Ruthenium tetraoxide (RuO_4) has proven useful for preparing samples with other types of double bond, e.g. polystyrene and polyamides.

In the case of electron beam microscopy optimum contrast is obtained when the fold surface is parallel to the electron beam. Tilted crystals appear less sharp and areas without lamellar contrast are found over large areas of the electron micrographs. This is one of the disadvantages of these methods. Only a small fraction of the crystals can be viewed at the same time. The use of a tilting stage permits the assessment of more lamellae, *i.e.* lamellae of other tilt angles are included in the analysis. Another problem is shrinkage of the whole section that may occur in the sample due to dissipation of the electron beam energy. Samples etched with chlorosulfonic acid for only a short period of time showed significant shrinkage, whereas samples treated with acid for a long period of time showed only negligible shrinkage.

An alternative method is to use mixtures of concentrated sulfuric acid and potassium permanganate, giving contrast between crystals and amorphous domain that turns out to be applicable to many different polymers, e.g. polyethylene, polypropylene, poly(but-1-ene), polystyrene and poly(aryl ether ketone)s. The strong etchant degrades the amorphous phase more quickly than the crystals and the resulting topography is revealed by heavy metal shadowing. Replicates are prepared which are examined in the electron microscope.

5.9 X-Ray Diffraction[34]

X-Ray diffraction is used to provide information on the lattice dimensions but can also provide so-called long period data. The long period is the sum of the amorphous and the crystalline lamellae dimensions and is studied using small-angle X-ray scattering (SAXS). The periodic variation in the electron density along a line perpendicular to the fold planes gives rise to constructive interference at a very small scattering angle. The average crystal thickness (L_c) is given by

$$L_c = dv_c \tag{5.3}$$

where v_c is the volume fraction of the crystalline component and d is the SAXS long period. Wide-angle X-ray diffraction is used to determine the crystal size through the use of the Scherrer equation:

$$\overline{D_{hkl}} = \frac{K\lambda}{\beta \cos \theta} \tag{5.4}$$

where K is the Scherrer shape factor, which adopts values close to unity, λ is the wavelength of the X-rays and β is the breadth in radians at half the peak height of the diffraction peak associated with the $\{hkl\}$ planes, where $\overline{D_{hkl}}$ is the crystal thickness. This expression assumes that the main cause of broadening is the distribution of the crystal orientations but thermal vibrations and paracrystalline distortions can also contribute. These effects are accommodated in the following equation:

$$\beta_s^2 = \frac{1}{D_{hkl}^2} + \frac{\pi^2 g^4}{d_0^2} n^4 \tag{5.5}$$

where β_s is the breadth of the diffraction peak in scattering units ($s = (2\sin \theta)/\lambda$) after subtracting the instrumental broadening contribution, d_0 is the interplanar distance for the first-order reflection, n is the order of reflection and g is the degree of statistical fluctuation of the paracrystalline distortions relative to the separation distance of the adjacent lattice cell. If β_s^2 is plotted as a function of n^4 the thickness $\overline{D_{hkl}}$ is obtained as the square root of the reciprocal of the intercept. More details can be found elsewhere.[34]

5.10 Raman Scattering and Phonon Spectra[35]

The crystal thickness in polyethylene has been determined using Raman spectroscopy by measurement of the frequency of the longitudinal acoustic mode (which is inversely proportional to the length of the all-*trans* stems in the crystals) and size exclusion chromatography of samples etched with HNO_3 or O_3. Both the acid and ozone degrade the amorphous parts and leave essentially oligomers with chain lengths equal to the all-*trans* length ($L_c/\cos \theta$, where θ is the tilt angle).

5.11 Degree of Crystallinity[34]

Because of the complex manner in which the lamellae grow and interact, the crystalline content of bulk samples can vary significantly. Determination of the degree of crystallinity is a very useful indicator of the morphology of the material. For simplicity it is assumed that the bulk material exists either as a crystalline or an amorphous phase and that any intensive property (ϕ) is an additive function with contributions from the two components present:

$$\phi = \phi_c w_c + \phi_a(1 - w_c) \qquad (5.6)$$

where ϕ_a and ϕ_c are, respectively, properties of the crystallinity of the amorphous and crystalline phases and w_c is the mass of the crystalline phase. This equation can be applied with intensive properties such as the enthalpy, specific heat or specific volume of the material.

5.11.1 Density and Calorimetric Methods

The most obvious difference between amorphous and crystalline regions is the density of chain packing. Accurate measurements of the density, with a precision of about 0.2 kg m^{-1}, are carried out in a density gradient column or alternatively by a floating method, and then very accurate measurement of the mixture density. The mass crystallinity can be obtained from the following equation:

$$\frac{1}{\rho} = \frac{w_c}{\rho_c} + \frac{(1 - w_c)}{\rho_a} \qquad (5.7)$$

where ρ, ρ_c and ρ_a are, respectively, the densities of the sample, crystalline component and amorphous component. The density of the crystalline component is usually determined from the X-ray unit cell data and the amorphous density is obtained by extrapolation of dilatometric data of molten polymer to lower temperatures. The principal problem with the density measurement is the possibility of the liquids used swelling the samples. It is usually possible by the correct selection of solvent system to minimize this problem.

The ready availability of differential scanning calorimetry (DSC) and in recent years moduluated DSC have significantly increased the use of the enthalpy of fusion of crystals as a method for crystallinity determination. An illustration of this method was shown for plastic crystals in Chapter 4. The enthalpy of the solid is a combination of components from the amorphous H_a and crystalline H_c regions. The enthalpy at any given temperature H is therefore

$$H = H_c w_c + H_a(1 - w_c) \qquad (5.8)$$

where w_c is the mass crystallinity. In practice, annealing processes will influence the distribution between crystalline and amorphous regions making determination of the value of w_c problematic. If, however, the internal stresses are negligible and it is assumed that the amorphous component is liquid-like, *i.e.* the amorphous phase enthalpy at $T < T_m$ can be obtained by extrapolation of data from temperatures greater than T_m, then enthalpy (H_1) at a temperature T_1 well below the melting temperature range is

$$H_1 = H_{c1} w_{c1} + H_{a1}(1 - w_{c1}) \qquad (5.9)$$

Morphology of Crystalline Polymers and Methods for Its Investigation 137

where w_{c1} is the crystallinity at T_1 and H_{c1} and H_{a1} are the crystalline and amorphous enthalpies, respectively, at T_1.

At T_2 ($\gg T_m$) the enthalpy is given by

$$H_2 = H_{a2} \tag{5.10}$$

The difference in enthalpy of the sample between temperatures T_1 and T_2 is given by

$$\Delta H_{21} = H_2 - H_1 = [H_{a2} - H_{a1}] + [H_{a1} - H_{c1}]w_{c1} = \Delta H_{a21} + \Delta H^0_{f_1} w_{c1} \tag{5.11}$$

That is,

$$w_{c1} = \frac{\Delta H_{21} - H_{a21}}{\Delta H^0_{f_1}} \tag{5.12}$$

where $\Delta H^0_{f_1} = H_{a1} - H_{c1}$ is the heat of fusion at T_1 and $\Delta H^0_{f_1}$ is temperature dependent:

$$\Delta H^0_{f_1} = \Delta H^0_{T^0_m} - \int_{T_1}^{T^0_m} [C_{pa} - C_{pc}] \, dt \tag{5.13}$$

where C_{pa} and C_{pc} are, respectively, the specific heats at constant pressure of the amorphous and crystalline components.

In order to determine $\Delta H_{21} - \Delta H_{a21}$, it is necessary to extrapolate the post-melting scanning baseline down to lower temperatures. Often the extrapolated baseline intersects the premelting scanning baseline and this helps selection of T_1. The area under the curve starts at T_1 and ends once melting has been completed. More details on the methods can be obtained elsewhere.[34]

5.11.2 X-Ray Scattering

The most direct method of determining the degree of crystallinity is X-ray scattering. The intensity of the structural lines (coherent scattering) in the total scattering spectrum arises from the unit cell and other ordered features. Thus the ratio of the coherent scattering to the total scattering is directly related to the crystalline fraction. The amorphous component will add a broad background to the scattering spectrum. Thus the total coherent scattering from N atoms is independent of the state of aggregation and the mass crystallinity can be defined as

$$w_c = \frac{\int_0^\infty s^2 I_c(s) \, ds}{\int_0^\infty s^2 I(s) \, ds} \tag{5.14}$$

where I is the total scattered intensity which contains amorphous and crystalline components and I_c is the scattered intensity associated with the crystalline regions and corresponds to the Bragg reflections. In eqn (5.14), s is the scattering parameter which is equal to $(2\sin \theta)/\lambda$, where λ is the wavelength.

The practical problems encountered in determining w_c arise from three principal sources. Often the coherent Bragg peaks are superimposed on a broad incoherent scattering peak and some form of subtraction has to be applied to separate the contributions. The denominator should contain the total scattering from all sources; however, it is sometimes difficult to have a sufficiently wide range of angles to be sure that all the contributions from incoherent scattering are included. As indicated above, the coherent scattering will also contain components from thermal vibrations and paracrystalline defects that are difficult to allow for in the estimation of the coherent scattering intensity. Various approaches have been developed but all have limitations.

5.11.3 General Observations

Usually X-ray and density data are in good agreement. The degree of crystallinity for most polymers will depend on the ability of the chains to pack regularly together. Disruption of the structure will introduce defects that will reduce the crystallinity. Ethyl branches on a linear polyethylene backbone can cause reduction in the degree of crystallinity to the extent of 20% per mol% of ethyl groups.[41] The ethyl groups are probably located in the fold and hence segregated where possible to the amorphous phase. Polymers with larger branches, propyl or longer homologues, are fully non-crystallizable unless the branch becomes of sufficient molar mass to look like a short polymer. The molar crystallinity depression of smaller groups, carbonyl and methyl groups, is less than that due to ethyl groups.

In general, the crystallinity increases with increasing molar mass. Very low molar mass polymers crystallize in extended chains or once- or twice-folded crystals and form almost perfect crystals. Increasing the molar mass of the polymer increases the possibility of chains leaving one lamella and entering another, with a consequent disruption of the perfect order. Chain entanglements play an important role controlling the extent to which various parts of the chain can be incorporated in particular lamellae. The effect of the entanglements can be to suppress the crystallinity so that for a linear polyethylene of $\bar{M}_w = 10^6 \text{ g mol}^{-1}$ the crystallinity is between 40 and 50%.

5.12 Conclusions

Although the structures that are observed with electron microscopy look complex, it is apparent that they are all related to the chain folded lamellae to a lesser or greater extent. Defects in the lamellar structure, as with simple organic crystals, will control the habit of the crystals form. The spherulitic structures observed are a consequence of the multiple growth of lamellar structures and their tendency to splay with growth. Naturally occurring polymer materials exhibit more complex morphology and this topic will be briefly considered later in this book. It is now appropriate to consider the factors that influence the growth of these lamellar structures.

Recommended Reading

D.C. Bassett, *Principles of Polymer Morphology*, Cambridge University Press, Cambridge, 1981.
D. Campbell, R.A. Pethrick and J.R. White, *Polymer Characterization*, Stanley Thornes, Cheltenham, UK, 2000.
U.F. Gedde, *Polymer Physics*, Chapman & Hall, London, 1995, ch. 7.
H. Hasegawa and T. Hashimoto, in *Comprehensive Polymer Science Second Supplement*, ed. G. Allan, Pergamon, Oxford, 1996, p. 497.
D.A. Hemsley (ed.), *Applied Polymer Light Microscopy*, Elsevier Applied Science, London, 1989.
M. Srinivasarao, in *Comprehensive Polymer Science Second Supplement*, ed. G. Allan, Pergamon, Oxford, 1996, p. 163.
S. Vaughan and D.C. Bassett, in *Comprehensive Polymer Science*, ed. G. Allan and J.C. Bevington, Pergamon, Oxford, 1989, vol. 2, p. 415.

References

1. U.F. Gedde, *Polymer Physics*, Chapman & Hall, London, 1995, ch. 7.
2. S. Brückner and S.V. Meille, *Nature*, 1989, **340**, 455.
3. G. Natta and P. Corradini, *Novovo Cimento Suppl.*, 1960, **15**, 40.
4. J.L. Koenig and A.C. Angood, *J. Polym. Sci., Part A2*, 1970, **8**, 1787.
5. T. Miyazawa, K. Fumushima and Y. Iideguchi, *J. Chem. Phys.*, 1962, **37**, 2764.
6. T. Miyazawa, *J. Chem. Phys.*, 1961, **35**, 693.
7. H. Tadokoro, in *Molecular Structure and Properties*, Physical Chemistry Series 1, ed. A. D. Buckingham and G. Allan, Butterworths, London, 1972, vol. 2, p. 45.
8. A. Keller, *Philos. Mag.*, 1957, **2**, 1171.
9. P.H. Till, *J. Polym. Sci.*, 1957, **24**, 30.
10. E.W. Fischer, *Z. Naturforsch.*, 1957, **12a**, 753.
11. A.S. Vaughan and D.C. Bassett, in *Comprehensive Polymer Science*, ed. G.Allan and J.C. Bevington, Pergamon, Oxford, 1989, vol. 2, p. 415.
12. L. Mandlekern, *Physical Properties of Polymers*, American Chemical Society, Washington, DC, 1984, ch. 4.
13. (*a*) D.C. Bassett, *Principles of Polymer Morphology*, Cambridge University Press, Cambridge, 1981; (*b*) K.H. Storks, *J. Am. Chem. Soc.*, 1939, **60**, 1753.

14. T. Kawai and A. Keller, *Philos. Mag.*, 1985, **11**, 1165.
15. A. Peterlin, in *Structure and Properties of Orientated Polymers*, ed. I. M. Ward, Wiley, Chichester, 1975, ch. 2, p. 36.
16. S.J. Organ and A. Keller, *J. Mater. Sci.*, 1985, **20**, 1571.
17. H.D. Keith and F.J. Padden, *J. Appl. Phys.*, 1964, **35**, 1270.
18. D.C. Bassett and A. Keller, *Philos. Mag.*, 1962, **7**, 1553.
19. F. Khoury and F.J. Padden, *J. Polym. Sci.*, 1960, **47**, 455.
20. D.C. Basset, *Morphology Encyclopaedia of Polymer Science and Technology*, John Wiley, New York, 2003.
21. D.C. Basset, *Philos. Mag.*, 1964, **10**, 595.
22. D.C. Basset, *Philos. Mag.*, 1965, **12**, 907.
23. A.M. Hodge and D.C. Bassett, *J. Mater. Sci.*, 1957, **12**, 2065.
24. M.I. Abo and D.C. Bassett, *J. Macromol. Sci. Phys.*, 2001, **B40**, 849.
25. M.I. Abo and D.C. Bassett, *Polymer*, 2001, **42**, 4957.
26. A. Keller, *Kolloid Z. Z. Polym.*, 1967, **219**, 118.
27. A.S. Vaughan and D.C. Bassett, in *Comprehensive Polymer Science*, ed. G. Allan, Pergamon, Oxford. 1989, vol. 2. p. 432.
28. D.A. Hemsley, in *Applied Polymer Light Microscopy*, ed. D.A. Hemsley, Elsevair Applied Science, London, 1989, p. 67.
29. P.K. Datta and R.A. Pethrick, *Polymer*, 1978, **19**, 145.
30. A.J. Pennings, *J. Polym. Sci. Polym. Symp.*, 1977, **59**, 55.
31. B.P. Saville, in *Applied Polymer Light Microscopy*, ed. D.A. Hemsley, Elsevier Applied Science, London, 1989, pp. 67–112.
32. G. Ungar, J. Stejny, A. Keller, I. Bidd and M.C. Whiting, *Science*, 1985, **229**, 386.
33. R.A. Pethrick and C. Viney, *Techniques for Polymer Organisation and Morphology Characterisation*, John Wiley, Chichester, UK, 2003.
34. D. Campbell, R.A. Pethrick and J.R. White, *Polymer Characterization*, Stanley Thornes, Cheltenham, UK, 2000.
35. A.D. Curson, in *Applied Polymer Light Microscopy*, ed. D.A. Hemsley, Elsevier Applied Science, London, 1989, p. 19.
36. B.P. Saville, in *Applied Polymer Light Microscopy*, ed. D.A. Hemsley, Elsevier Applied Science, London, 1989, p. 73.
37. M.J. Folkes and A. Keller, *Polymer*, 1971, **12**, 222.
38. R. Hoffman, in *Applied Polymer Light Microscopy*, ed. D.A. Hemsley, Elsevier Applied Science, London, 1989, p. 151.
39. D.A. Hemsley, in *Applied Polymer Light Microscopy*, ed. D.A. Hemsley, Elsevier Applied Science, London, 1989, p. 151.
40. P. Calvert and N.C. Billingham, in *Applied Polymer Light Microscopy*, ed. D.A. Hemsley, Elsevier Applied Science, London, 1989, p. 233.
41. T. Tranker, M. Hendenqvist and U.W. Gedde, *Polym. Eng. Sci.*, 1994, **34**, 1581.

CHAPTER 6
Polymer Crystal Growth

6.1 Introduction

The processes involved in the formation of a crystalline polymeric solid are complex and still open to some debate. Using the analogy of the short-chain hydrocarbon dodecane (Chapter 1), the crystallization process simplistically involves the creation of favourable interactions between neighbouring chains. However, such an approach does not recognize the unique features possessed by polymers: restricted chain mobility and the effects of entanglement and other long-range interactions.

6.1.1 Thermodynamics of Polymer Molecule in the Melt

The energy of a polymer molecule in the molten state will be determined by a combination of inter- and intramolecular forces. Within the polymer molecule, the atomic interactions that produce the barriers to internal rotation will also create the energy differences that will define the heat capacity. In Chapter 1, the energy profile for the rotation about a single bond was discussed and it was pointed out that molecules can occupy either the lower energy *trans* or the higher energy *cis* or *gauche* state. The success of the RISM theory[1] was in the prediction of the effects of temperature on the size and shape of the polymer molecule. The RISM predicts that as the temperature is raised the occupancy of the higher energy states is increased so the size of the polymer coil *shrinks*. In the melt state, the polymer molecule will attempt to adopt a minimum coil size and the collapse for higher molecular mass polymers is restricted by the effects of entanglement and so-called 'excluded volume' effects. The excluded volume effect describes the restrictions that have to be placed on the theoretical polymer chain to avoid it occupying the same point in space as another element of either the same chain or of that of another polymer.

Cooling a polymer chain will increase the proportion of the coil which exists in the extended all-*trans* form and this will facilitate the creation of the *stem* structures which are required for crystal growth. The polymer chains are not static entities and changes in the location of the *gauche* conformations can disrupt the formation of favourable interactions for crystal growth, suppress nucleation or alternatively assist the creation of *stems* of the correct length and aid nucleation. If two or more *gauche* sequences become located close together on the chain a *hair pin* or *folded* structure is generated which aids the growth process. The number of such folds will have an equilibrium value at a particular temperature. Although intuitively one might expect polyethylene chains to adopt an extended regular structure, both solution-grown and melt-grown crystals are chain *folded*. The attachment process may involve either the end of chains or a low-energy all-*trans* section interacting with a preformed surface. The ends of chains will have a higher energy than the central sections and hence ends tend to be located at surfaces. Clearly, the process of attachment of the polymer chain to the surface involves not just a change in enthalpy but also subtle changes in the total entropy of the system. The process of nucleation should ideally only involve interactions between polymer molecules, but in practice residual catalysts and other solids present in the melt help the process.

The concept of chain folding is core to understanding the nature of the crystalline polymeric state and this was first recognized by Keller.[2,3] It is easy to visualize that as the polymer chain is being laid down on the substrate, folding will occur and energetically the process would drive the folds to be located at a surface. As a consequence, the polymer tends to form stacks of extended polymer chains which we ascribe the term *lamellae* and these have a thickness which is indicative of the temperature at which they were grown. This observation is consistent with the polymer chain having a distribution of conformational states and entropy defined by that population. Trapped within the crystalline material will be disordered *amorphous* material that is in a non-equilibrium state.

6.1.2 Nucleation

The nucleation and growth processes, similar to the situation in low molar mass organic crystals, are dependent on the degree of supercooling of the melt or solution phase. The crystal thickness or alternatively the thickness of each new crystalline layer in a growing crystal is the one that grows fastest rather than the one that is at equilibrium. There is a wealth of information available on the crystallization of many polymers as well as several theories that aim to predict the crystallization rates, crystal shapes and lamellar thicknesses.

Once the chain has attached itself to the substrate it is possible that as neighbouring chains start to build subsequent layers of the crystal, the energy balance may change. As a consequence the possibility of the original structure changing to a new more stable structure exists. This leads to the phenomenon of *polymorphism* and is well known in both small organic crystals and in crystalline polymers.

In the melt phase, thermodynamics would indicate that chains with different chain lengths should have different energies. As a consequence there is the possibility that chains with a similar energy will tend to segregate themselves in space. *Phase separation* is well known in polymer blends (Chapter 8) and can be described by the process of spinodal decomposition. In the case of a broad molecular mass distribution polymer system, the individual chains will

Polymer Crystal Growth

fractionate according to their molar mass and the morphology of the crystals can reflect this effect. However, the viscosity of the melt during the crystallization process will usually stop an equilibrium state being achieved and hence phase diagrams often give little insight into the state of the material during the crystallization process.

Whilst it is relatively straightforward to describe the processes involved in polymer crystallization, it is far more difficult to articulate in terms of a model that allows the quantitative prediction of the crystallization process.[4]

In Chapter 5 it was identified that in crystalline polymeric material lamellae constitute a major fraction of the material. In Chapter 2 it was shown that crystal growth depends on the molecule becoming attached to the crystal surface at the lowest energy. In the case of polymer chains, the attachment is of a length, linear *stem*, in the case of polyethylene an all-*trans* structure being attached to a similarly organized set of molecules. As with organic crystals, the *habit* of the crystal that grows depends on the occurrence of defects and the relative growth rate of each of the faces of the lattice involved. The equilibrium shape of a crystal is therefore the result of molecules searching for the minimum energy. High-energy surfaces should be relatively small and the crystal dimensions perpendicular to those surfaces large because the *stems* will be much longer than they are wide with the result that the growth rate is much greater than attachment to the fold surface. It is not very surprising therefore to find that lamellar sheets dominate the discussion of crystal growth in polymer systems.

6.2 Minimum Energy Conditions and Simple Theory of Growth[4-6]

Simplistically, we can consider the relative energy of attachment to a cube with faces 1, 2, 3 (Figure 6.1).

The dimensions of the crystal will be determined by the relative energy of each surface and will be a minimum with respect to the melt (ΔG) at a given volume:

$$\Delta G = V\Delta G_m^0 + 2L_1 L_2 \sigma_3 + 2L_1 L_3 \sigma_2 + 2L_2 L_3 \sigma_1 \tag{6.1}$$

where ΔG_m^0 is the specific free energy of melting and σ_i are the specific surface free energies. This equation is true for any crystal growth process and is not

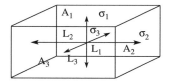

Figure 6.1 Equilibrium shape of a crystal with three different surfaces ($i = 1, 2, 3$) with different surface free energies σ_i.

specific for a polymer system. One of the L_i terms can be eliminated from the equation by considering that $V = L_1 L_2 L_3$ is constant, thus:

$$L_3 = V/(L_1 L_2) \tag{6.2}$$

Inserting eqn (6.2) into eqn (6.1) yields

$$\Delta G = V \Delta G_m^0 + \frac{2V}{L_3}\sigma_3 + 2L_1 L_2 \sigma_2 + \frac{2V}{L_1}\sigma_1 \tag{6.3}$$

Taking the derivatives of ΔG with respect to L_1 and L_3 and setting each component equal to zero, the following expressions are obtained:

$$\frac{\partial(\Delta G)}{\partial L_1} = L_3 - \frac{V}{L_1^2}\sigma_1 = 0 \Rightarrow \frac{L_1}{\sigma_1} = \frac{L_2}{\sigma_2} \tag{6.4}$$

and

$$\frac{\partial(\Delta G)}{\partial L_3} = L_1 \sigma_2 - \frac{V}{L_3^2}\sigma_3 = 0 \Rightarrow \frac{L_2}{\sigma_2} = \frac{L_3}{\sigma_3} \tag{6.5}$$

Combining eqn (6.4) and (6.5) gives

$$\frac{L_1}{\sigma_1} = \frac{L_2}{\sigma_2} = \frac{L_3}{\sigma_3} \tag{6.6}$$

Equation (6.6) indicates that the dimensions of the equilibrium crystal in different directions (i) are proportional to the surface free energies (σ_i) of the perpendicular surfaces. This result has already been shown to be appropriate for organic single crystals in Chapter 2. In the case of polymer crystals, two of the surfaces will contain chain folds having significantly different energies from the others; for polyethylene the specific surface energy (σ_e) of the folded surfaces is about 60–70 mJ m^{-2} which is about five times greater than the value for the surface containing the aligned chains (σ) which has a value of about 15 mJ m^{-2}. The ratio of the equilibrium thickness (along the chain axis direction) to the width perpendicular to the chain direction is consequently close to > 5. The predicted value of this ratio being > 5 is three to four orders of magnitude greater than that observed experimentally.[6] The difference between experiment and prediction could be a consequence of crystals not being in equilibrium with the melt and that crystals will rearrange when given enough 'thermal stimulation'. In Chapter 4 we noted that just below the melt temperature hydrocarbon chains have a relatively high degree of rotational mobility and this is associated with the plastic crystalline phase in these materials. It is therefore not unreasonable to expect that similar mobility would aid rearrangement of the crystal chains in the newly formed surface. In Chapter 5 it was noted that on thermal annealing, thickening of lamellae occurs. The thickness of solution-grown lamellae (L_c^*) depends on the degree of supercooling ($\Delta T = T_m^0 - T_c$, where T_m^0 is the equilibrium melting point and T_c is the crystallization temperature):[6]

$$L_c = \frac{C_1}{\Delta T} + \delta L \tag{6.7}$$

where C_1 and δL are constants for a particular polymer system. The effect of supercooling on the lamella thickness for the case of linear polyethylene[7–10] and similar regular polymers is shown in Figure 6.2.

The Thompson–Gibbs (TG) equation (4–6) allows the melting point to be related to the lamella thickness. The change in free energy on melting (ΔG_m) is given by

$$\Delta G_m = \Delta G^* + \sum_{i=1}^{n} A_i \sigma_i \tag{6.8}$$

where ΔG^* is the surface-independent change in free energy and σ_i is the specific surface energy of surface i with area A_i. At equilibrium:

$$\Delta G_m = 0 \Rightarrow \Delta G^* = \sum_{i=1}^{n} A_i \sigma_i \tag{6.9}$$

For a simple polymer crystal, the lamellae dominate the growth and the two surfaces that contain the folded chains dominate the total surface energy term. It can be assumed that there is a free energy term that is independent of this fold surface and represents the interaction between the all-*trans* linear chains, *stems*, and has the form

$$\Delta G^* = \Delta G_B^* A L_c \rho_c \tag{6.10}$$

where ρ_c is the density of the crystal plane. Both ΔH_B and ΔS_B can be regarded as temperature-independent bulk parameters, and the specific bulk free energy

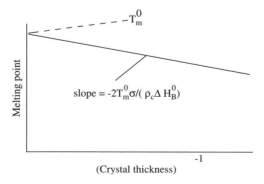

Figure 6.2 Schematic plot of the variation of the melting point with the reciprocal of the crystal thickness.

change (ΔG_B^*) is given by

$$\Delta G_B^* = \Delta H_B^0 - T_m \Delta S_B^0 = \Delta H_B \left(1 - \frac{T_m}{T_m^0}\right)$$
$$= \Delta H_B \left(\frac{T_m^0 - T_m}{T_m^0}\right) \qquad (6.11)$$

By inserting eqn (6.11) into eqn (6.10), the following equation is obtained:

$$\Delta G^* = \Delta H_B^0 (T_m^0 - T_m) \frac{AL_c \rho_c}{T_m^0} \qquad (6.12)$$

Inspection of the lamellae would indicate that the surfaces containing the chain *stems* are usually the short sides and have a small area compared with the sides that contain the folded chains. As indicated previously, it is the surfaces that contain the folded chains that make the major contribution to the surface free energy. It is therefore appropriate to assume that the total area of the four lateral surfaces is small compared to the area of the fold surfaces and the contribution to the total surface free energy of the crystal can be neglected:

$$\sum_{i=1}^n A_i \sigma_i \approx 2\sigma A \qquad (6.13)$$

Combination of eqn (6.9), (6.12) and (6.13) gives

$$\Delta H_B^0 (T_m^0 - T_m) \frac{AL_c \rho_c}{T_m^0} = 2\sigma A$$

$$T_m^0 - T_m = \frac{2\sigma A T_m^0}{A L_c \rho_c \Delta H_B^0} = \frac{2\sigma T_m^0}{L_c \rho_c \Delta H_B^0} \qquad (6.14)$$

which may be simplified to give the Thompson–Gibbs equation:

$$T_m = T_m^0 \left(1 - \frac{2\sigma}{L_c \rho_c \Delta H_B^0}\right) = \frac{-2T_m^0 \sigma}{\rho_c \Delta H_B^0} \qquad (6.15)$$

The Thompson–Gibbs equation predicts a linear relationship between the melting point and the reciprocal of the crystal thickness (Figure 6.2). In practice, it is difficult to obtain the melting point and crystal thickness data, as the experiments involved in determining the melting point allow crystal thickening to occur.

6.3 Nature of Chain Folding

If chain folding were regular and the chain *stems* were to lie adjacent to one another then the density would be close to that predicted from the crystal lattice structure. In Chapter 5, it was identified that small distortions of the lattice can be observed. Further, measurement of any crystalline polymer will indicate that the density is less that that from the predictions for the perfect crystalline material, indicating the presence of amorphous material. It is proposed that part of the amorphous content arises from disorder: irregular folding of the polymer chains (Figure 6.3).

The irregular folds in which the chains do not re-enter the lamellae at the next layer lead to the concept of the so-called switchboard model.[11] Early telephone switchboards were formed from a matrix of connections into which a connecting cable could be plugged. It was therefore possible to connect adjacent points—tight re-entry—or in a random fashion. Flory[11] argued that *random* re-entry was normal and regular, tight folding was rare. His argument was based on the melt behaviour where the polymer chains follow closely random coil statistics and the RISM model[1] has been every successful. It is only with the advent of small-angle neutron scattering (SANS) of blends of deuterated and protonated polymers, e.g. $(-CD_2-)_n$ and $(-CH_2-)_n$, that the validity of this model can be tested.[12]

For solution-grown single crystals of linear polyethylene,[13] the average radius of gyration ($<s>$) of the molecules is proportional to $M^{0.1}$. A much greater molar mass dependence of $<s>$ is observed for the polymer coil in solution, $<s> \propto M^{0.5}$, and indicates that the dimensions of the chains decrease markedly during crystallization. For the polymer chains to have a smaller value of $<s>$ they have to fold like an accordion and the *gauche* sequences have to be closely coupled in order to achieve close packing.

Flory based his ideas of re-entry on the tried and tested assumption that the coil follows random coil statistics in the melt. It is clear from the SANS data that the chains fold much more tightly than would be predicted by a random re-entry model.[14] Infrared spectroscopy showed that 75% of the folds

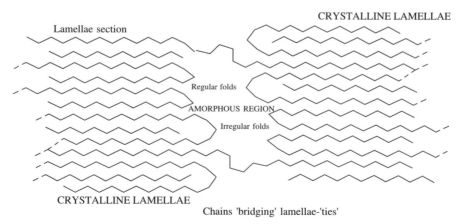

Figure 6.3 Schematic of a folded surface showing regular folds, irregular folds and bridging 'tie' molecules.

in solution-grown single crystals of polyethylene led to adjacent re-entry (tight folds) and that single molecules were diluted by 50% along the 110 fold plane. Both observations are consistent with the *super-fold* model where the chains behave more like the regular folds than the random re-entry model.

6.4 Crystals Grown from the Melt and Lamellae Stacks

The *super-folding* model indicates that in dilute solution the polymer molecules add to the growing face of a single crystal without significant competition from other polymer molecules. Crystal growth in concentrated solutions and the melt will be very different and chain entanglement will influence the growth process. When polymers are entangled, then the possibility arises of several chains adding to the surface at the same time and the process becomes more chaotic and the folding behaviour may tend more towards the switchboard model.[11] In general, low molecular weight polyethylene, because it is not entangled, can readily crystallize, exhibiting a well-defined lamellar morphology with a thickness-to-width ratio of 0.001–0.01. The lamellae often align parallel to one another forming regular stacks. This behaviour is particularly apparent with metallocene polyethylene, where there are very few branched chains to inhibit chain folding. The lamellae formed with low molecular weight materials can exhibit thickness-to-width ratios which are smaller than 0.001. Once entanglement can occur then growth of the large lamellae is inhibited and smaller less regular lamellae are usually observed. Branched and/or high molar mass polymers exhibit lamellae that tend not to be able to form regular stacks and exhibit much higher ratios. As discussed in Chapter 5, the lamella distortions can lead to characteristic shaped crystals. The regular tight-folded lamellae can form roof-shaped lamellae, pyramidal structures, whereas intermediate molar mass materials tend to form C-shaped lamellae and high molar mass polymers form S-shaped lamellae. Branching will introduce defects and, depending on its location and the length of the branches, different effects on the lamellae growth process will occur. The incorporation of the branches will impart curvature to the lamellar crystal sheets. Branched polyethylene shows mostly C- and S-shaped lamellae.

Polymers are always polydisperse with a distribution in molar mass and often contain chain branches, either introduced specifically during synthesis or as a consequence of synthetic defects, and both these effects will influence the observed morphology. As we shall see later, copolymers are a special case; however, the introduction of low levels of comonomers can lead to behaviour which is rather like that of random branched chains. Different molecular species crystallize in different stages indicating the thermodynamic control on the overall process, *i.e.* they are incorporated into the crystal structure at different temperatures and times. The intermediate and high molar mass component crystallizes early in the stacks of thick dominant crystals. Small pockets of rejected molten low molar mass material remain after crystallization of the dominant lamellae. The low molar mass species crystallize in separate crystal lamellae, favouring stacks of so-called subsidiary crystal lamellae. The process leading to separation of high and low molar mass material that accompanies crystallization is referred to as a molar mass segregation or fractionation. Some polymers show a preference for segregation of low molar mass species to the spherulitic boundaries.

For melt-grown polymers the amorphous density is 10–20% lower that the crystal density. Flory argued that all the chains entering the amorphous phase would take a random walk before re-entering the crystal lamellae. A significant fraction of the chains must, however, be fold directly back into the crystal in order to account for the observed low amorphous density.

An expression can be derived for the fraction of tight folds. Usually it is assumed that all of the chains can be divided into two groups: those that form tight folds and those that do not. The latter are assumed to exhibit random re-entry behaviour. In Figure 6.3 the tight regular folds have a density that is essentially that of the lamellae and the irregular folds are much bigger, entering further into the amorphous region and exhibiting random re-entry behaviour. If the length of chain between the lamellae is designated L_a, and there are n bonds in a typical Gaussian amorphous chain sequence, then

$$n = \frac{L_a^2}{Cl^2} \quad (6.16)$$

where C is a characteristic constant for a given polymer–temperature combination and l is the monomer bond length. Regular chain folding constitutes a fraction (f_{fold}) of the entries and these will *not* contribute to the fraction of the amorphous chains. The number of chain segments in an average amorphous entry is given by

$$n = \frac{L_a^2}{Cl^2}(1 - f_{\text{fold}}) \quad (6.17)$$

The number of chain segments (n^0) in a straight chain, *stem*, is given by

$$n^0 = \frac{L_a}{l} \quad (6.18)$$

The ratio of segments in the two phases will be in the ratio of the amorphous (ρ_a) to crystalline density (ρ_c):

$$\frac{\rho_a}{\rho_c} = \frac{n}{n^0} \quad (6.19)$$

The amorphous density is then obtained by a combination of eqn (6.17)–(6.19):

$$\rho_a = \rho_c \frac{L_a}{Cl}(1 - f_{\text{fold}}) \quad (6.20)$$

Careful X-ray diffraction studies have shown that the chains do not match exactly and as a consequence they are tilted by an angle θ.[15–17] To allow for this effect, eqn (6.20) can be modified to give

$$\rho_a = \rho_c \frac{L_a}{Cl}(1 - f_{\text{fold}}) \cos \theta \qquad (6.21)$$

For a typical linear polyethylene the ratio for the densities ρ_a/ρ_c is 0.85 and the amorphous distance is approximately 5 nm and l has a value of 0.127 nm. The value of C is 6.85 and $\theta = 30°$; using these data the value of f_{fold} is 0.83; *i.e.* 83% of all chain stems are expected to be tightly folded and 17% are expected to be statistically distributed chains in the amorphous layer. The amorphous phase can have a dominating influence on the physical properties of many semi-crystalline polymers, determining the ultimate strength, ductility and diffusion coefficients for small molecules. The strength and ductility of semi-crystalline polymers of high molar mass arise from the presence of interlamellar tie chains connecting adjacent crystal lamellae and are critical in determining the resistance to environmental stress cracking.

6.4.1 Location of Chain Ends

Any discussion of the lamellar structure raises the following question: where are the chain ends?

In Figure 6.3 some of the chain ends are shown to be close to the surface of the crystal. Keller and Preist have shown that about 90% of the chain ends are located in the amorphous phase, surface, of the lamellae. The concentration of interlamellar *tie* chains in a given sample depends on molar mass that in turn determines the spatial distribution of the chains and long period, *i.e.* the sum of crystal and amorphous layer thickness. If the polymer chains are shorter than the critical length for entanglement, then the chains can act independently and become separately incorporated in the lamellae. For low molar mass chains where their length is twice as long as the long period, the chains will fold into the same lamellae and there will be few bridging *tie* molecules and a brittle crystalline material will develop. Once more we must recall that the chains in the newly formed surface can probably undergo significant rearrangement and the lower energy state of a chain end in the surface can be rapidly achieved. For high molecular mass materials, an increasing proportion of the chains will leave one lamella and enter the adjoining lamella. The bridging section will often be entangled and the *tie* molecules will add to the strength of the material.

6.5 Crystallization Kinetics[6]

A core assumption when discussing the crystallization kinetics of polymers is that the theories must consider the effects of chain folding. In Chapter 2 the principles of crystallization were outlined and the importance of the differences

Polymer Crystal Growth 151

in the energy of the surface to which the molecules were being attached was demonstrated. The same basic principles apply to polymer crystals except that now the entity that is being attached is a polymer chain. In principle, Monte Carlo simulations can predict the growth process in a manner similar to that used for small molecules; however, some of the more empirical theories help the experimentalist gain an appreciation of the mechanism of the processes involved in practice.

As in the case of small molecules, the crystallization process involves nucleation and diffusion of the relevant entity to the surface site. The formation of the critical nucleus will be controlled by thermodynamics. The total change in free energy ΔG is the sum of contributions from the bulk and surface energies. Using the convention which we introduced in Figure 6.1, where σ_i is the specific surface energy of surface i and A_i is the area, then the free energy can be described by

$$\Delta G = \Delta G_B V_{\text{crystal}} + \sum_i A_i \sigma_i \qquad (6.22)$$

where ΔG_B is the change in the specific free energy in transformation of stems from the solid to the melt and V_{crystal} is the volume of the crystal. For simplicity it will be assumed that the crystal being considered has a spherical form. The free energy on crystallization (ΔG) is then given by

$$\Delta G = \frac{4\pi r^3}{3} \Delta G_B + 4\pi^2 \sigma \qquad (6.23)$$

where r is the radius of the spherical crystal and σ is the average specific free energy of the surface. The radius of the sphere (r^*) associated with the free energy barrier is obtained by setting the derivative of ΔG with respect to r equal to zero:

$$\frac{\partial \Delta G}{\partial r} = 4\pi^{*2} \Delta G_B + 8\pi r^* \sigma = 0 \qquad (6.24)$$

where

$$r^* = \frac{2\sigma}{\Delta G_B} \qquad (6.25)$$

The temperature dependence of this equation lies in ΔG_B:

$$\Delta G_B = \frac{\Delta H_B^0 \Delta T}{T_m^0} \qquad (6.26)$$

where ΔH_B^0 is the heat of fusion per unit volume, T_m^0 is the equilibrium melting point, $\Delta T = T_m^0 - T_c$ is the degree of *supercooling* and T_c is the crystallization temperature. Equation (6.26) is valid provided that ΔH_B^0 and the entropy of fusion, ΔS_B^0, are temperature independent, which is a good approximation over a limited temperature range near the equilibrium melting temperature.

Insertion of eqn (6.26) into eqn (6.25) yields

$$r^* = -\frac{2\sigma T_m^0}{\Delta H_B^0 \Delta T} \quad (6.27)$$

Since ΔH_B^0 is negative, the radius of the critical nucleus increases with decreasing degree of *supercooling*. Inserting eqn (6.27) into eqn (6.23) gives an expression for the free energy barrier (ΔG^*):

$$\Delta G^* = \frac{4\pi(-2\sigma T_m^0)^3}{3(\Delta H_B^0 \Delta T)^3} \frac{\Delta H_B^0 \Delta T}{T_m^0} + \frac{4\pi(-2\sigma T_m^0)^2 \sigma}{(\Delta H_B^0 \Delta T)^2} \quad (6.28)$$

which simplifies to

$$\Delta G^* = \frac{16\pi\sigma^3 (T_m^0)^2}{3(\Delta H_B^0)^2 \Delta T^2} \quad (6.29)$$

Equation (6.29) predicts that nucleation occurs more readily at lower crystallization temperatures because of the lower critical nucleus size and the lower free energy barrier associated with the crystallization process. As with the case of simple crystal growth, nucleation can take a number of forms:

- Primary nucleation that involves the formation of the first nuclei and involves six new surfaces being formed.
- Secondary nucleation will usually occur on a surface and typically involves four surfaces being formed.
- Tertiary nucleation involves typically two surfaces being formed and represents the stem attaching to an edge face.

The various forms of nucleation are schematically presented in Figure 6.4. By analogy with case of small crystals (Chapter 2), the free energy barrier is highest for primary nucleation and this seldom occurs in practice, heterogeneous nucleation being the normal mechanism.

It has been proposed that the following equation can be used to describe the temperature dependence of both diffusive transport and nucleation. The overall crystallization rate ($\dot{w}_c$) at a general temperature (T_c) is a combination of several factors. The crystallization process involves the diffusion (the first

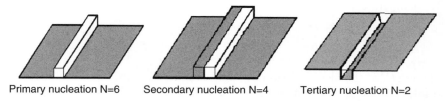

Primary nucleation N=6   Secondary nucleation N=4   Tertiary nucleation N=2

Figure 6.4 Schematic of the types of nucleation for stems adding to a crystal surface. The *stem* adding to the surface is shown in white except in the tertiary case where the addition is to the trough.

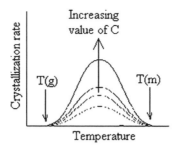

Figure 6.5 Variation of the crystallization rate with temperature. T_g and T_m are, respectively, the glass and melt temperatures of the material.

exponential term) of the stem to the surface and then its nucleation (the second exponential term) or attachment:

$$(\dot{w}_c) = C \exp\left(-\frac{U^*}{R(T_c - T_\infty)}\right) \exp\left(-\frac{K_g}{T_c(T_m^0 - T_c)}\right) \quad (6.30)$$

where C is a rate constant, U^* is an energy constant, R is the gas constant, T_∞ is the temperature at which all segmental mobility is frozen in and K_g is a kinetic constant for the secondary nucleation.

The pre-exponential factor C depends on the regularity and flexibility of the polymer segments and will be a large value for a flexible all-*trans* polymer, $C = 0$ for an atatic polymer and is low for an inflexible polymer such as isotactic polystyrene. The second term in eqn (6.30) describes the temperature dependence of the short-range motion of the stems to the surface and is expressed by the Williams–Landau–Ferry (WLF) equation.[17] The WLF equation describes the slowing down of the segmental motion of the chain backbone at the glass transition temperature and reflects the restriction on the volume available for motion to occur. At $T_c = T_\infty$ this term becomes zero. The third term describes the temperature dependence of the nucleation rate and has a zero value above the equilibrium melting temperature $T_c = T_m^0$. The form of the temperature dependence of the rate of crystallization is shown in Figure 6.5. It is a bell-shaped curve.

The form of the crystallization rate curve is common to many systems and similar expressions have been found for metals, inorganic compounds, sulfur, selenium, antimony, proteins and carbohydrates, graphite silicates and also polymers.

6.6 Equilibrium Melting Temperature

Central to most crystallization is the idea of an equilibrium melting temperature, T_m^0, above which crystallization does not occur. The rate of crystal growth is therefore related to the extent to which supercooling (ΔT) occurs and is defined by

$$\Delta T = T_m^0 - T_c \tag{6.31}$$

where T_c is the crystallization temperature. The equilibrium melting temperature for a polymer system normally refers to the growth of a theoretical crystal of infinite thickness in which the chains are fully extended. Homopolymers of intermediate or high molar mass can grow crystals of practically infinite thickness. The fully extended chain length of a polyethylene of $M = 100\,000$ g mol^{-1} is $100\,000/14 \times 0.127$ nm ~ 900 nm and is several orders of magnitude larger than the thickness of the lamellae.[6] The usual dimensions are ~ 10 nm, and for these dimensions to be achieved chain folding must occur. The melting point depression arising from the finite crystal thickness (900 nm) predicted by the Thompson–Gibbs equation (eqn (6.15)) is in this case negligible. However, low molar mass homopolymers can obviously grow crystals of only a limited thickness equal to their fully extended length (Chapter 2). The melting point of these materials does correspond to the predictions for the fully extended chain. Copolymers are more complicated, the change in the chemical sequence structure representing a disruptive element inhibiting chain packing. Depending on the nature of the comonomer, the 'foreign' moieties with a 'statistical' placement in the polymer chains may be fully non-crystallizable or they may be included to some extent in the crystals. As a consequence the equilibrium thickness will reflect the disruptive effects of these moieties. Very large lamellae are found in highly linear polymers such as certain metallocene-synthesized polyethylenes.

An example of the effect of different crystallization temperatures on the melt temperature for poly(trimethylene terephthalate) is shown in Figure 6.6. The value of T_m^0 is equal to 245.6 °C. The data follow a good linear variation of the melting temperature with the crystallization temperature.

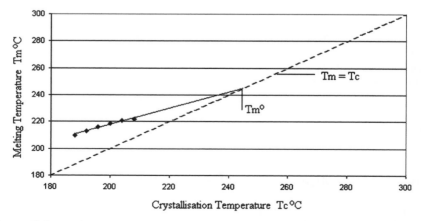

Figure 6.6 Variation of melting temperature with crystallization temperature for poly(trimethyl terephthalate) (unpublished data, A. Mckintosh and J. J. Liggat).

Polymer Crystal Growth

The equilibrium melting temperature can be determined in a number of different ways. The melting point (T_m) of samples with a well-defined crystal thickness (L_c) can be measured and the data extrapolated to $L_c^{-1} = 0$ using the Thompson–Gibbs equation:

$$T_m = T_m^0 \left[1 - \frac{2\sigma}{\Delta H_B^0 \rho_c L_c}\right] \qquad (6.32)$$

where ρ_c is the crystal density, ΔH_B^0 is the heat of fusion per unit mass and σ is the specific fold surface energy. Certain polymers, such as polyethylene, can be crystallized at elevated pressures to form extended chain crystals and these micrometre thick crystals have melting points that are close to the theoretical value of T_m^0.

An alternative approach is to derive the value of T_m^0 from a study of low molar mass analogues and then extrapolate to the infinitely thick crystal value. Using data for the enthalpy (ΔH_B) and entropy (ΔS_B) of fusion for oligomers, T_m^0 is obtained by extrapolation to infinite molar mass and for linear polyethylene has the form

$$T_m = 414.3 \left[\frac{x - 1.5}{x + 5.0}\right] \qquad (6.33)$$

where x is the degree of polymerization of the polymer. For high molecular mass materials this melting point has a limiting value of 414.3 K. Similar relationships have been observed for other polymer systems. Equilibrium melting points of a few selected polymers are presented in Table 6.1.[4-6]

It is clear from these examples that there is no correlation between the melting temperature and the enthalpy of fusion. The values of ΔH_B clearly do

Table 6.1 Equilibrium melting point and selected thermodynamic data for various polymers.[6,18]

Polymer	T_m^0 (K)	ΔS_B^0 (J K^{-1} mol^{-1})	ΔH_B^0 (kJ mol^{-1})
Polyethylene	414	9.6	4.01
Polytetrafluoroethylene	600	5.7	3.42
Isotactic polypropylene	463	7.5	2.31
Polyoxymethylene	457	10.7	4.98
Poly(ethylene oxide)	342	8.4	2.89
Nylon 6,6	553	10.2	4.85
Poly(1,4-*cis*-isoprene)	301	14.5	4.39
Poly(1,4-*trans*-isoprene)	347	36.6	12.70
Poly(1,4-*trans*-chloroprene)	353	23.7	8.36
Isotactic polystyrene	516	16.3	8.38
Poly(decamethylene adipate)	352.5	121.2	42.64
Poly(decamethylene sebacate)	353	142.1	50.16
Poly(tetramethylene terephthalate)	411	63.1	31.76
Cellulose tributyrate	480	33.8	12.54

not dictate the values of the melt temperature and it is rather the balance between the enthalpy and entropy that is the controlling factor.

Nylon 6,6 has a high enthalpy of fusion due to the strong hydrogen bonds between amide groups. In contrast, the higher melting point of polytetrafluoroethylene (PTFE) is due to its low entropy of fusion. At high temperature, PTFE crystals show considerable segmental mobility that leads to a relatively small increase in entropy on melting.

PTFE also exhibits a lattice change in the solid state just below its melting point, rather like the plastic phase transition discussed in Chapter 4. This lattice change is consistent with a high mobility in PFTE below its melting point. The high melting point of polyoxymethylene (POM) is due to the high enthalpy of fusion arising from intermolecular interactions involving the ether groups. The addition of the extra methylene group in poly(ethylene oxide) (PEO) leads to a reduction of the effects of the ether groups and a consequent reduction in the enthalpy of fusion. The introduction of a ring structure into a linear chain substantially increases the melting temperature relative to the aliphatic chain as would be expected from the decreased conformational entropy of the melt. Striking examples of this phenomenon are found in comparisons of the melting temperature of aliphatic and aromatic polyesters and polyamides. Cellulose derivatives usually have a highly extended structure with low entropy and consequently have high melting points. From the few examples presented it is apparent that the chain structure influences the melting temperature through its conformational properties.

6.7 General Avrami Equation

A generalized approach to the description of the crystallization process was proposed by Avrami.[19–21] Without prior knowledge of the molecular mechanism involved in the crystallization process, the Avrami equation gives a convenient means of empirically describing crystallization. The model assumes that crystallization starts randomly throughout the sample, which since nucleation is often heterogeneous, is a good approximation to reality. The theory attempts to accommodate the effects of the growth rate on the shapes of the crystals that are formed. It is assumed that the crystals that are seeded will grow smoothly in all three dimensions. All nuclei are formed and start to grow at time $t = 0$ and with a rate that is the same in all directions (spherical growth) and is equal to $\dot{\omega}$. The theory considers the growth front, $E(t)$, emanating from the central nucleus which will be described:

$$E(t) = \frac{4}{3}\pi(\dot{\omega}t)^3 q \qquad (6.34)$$

where q is the volume concentration of nuclei. Statistical analysis of the crystal growth problem can be considered in terms of an expanding circular wave front that mimics spherulite growth. The number of waves that pass a particular point at time t can be described in terms of a Poisson distribution that has the form

$$p(c) = \frac{\exp(-E)E^c}{c!} \tag{6.35}$$

where E is the average value of the number of waves passing a point P after some time t and the number of waves are constrained to be an exact number c. The probability that no fronts pass the point P is given by

$$p(0) = \exp(-E) \tag{6.36}$$

In the context of the crystallization process, $p(0)$ is equivalent to the volume fraction $(1 - v_c)$ of the polymer which is still in the molten state:

$$p(0) = 1 - v_c \tag{6.37}$$

where v_c is the volume fraction of crystalline material. Combination of eqn (6.34) with eqn (6.37) yields

$$1 - v_c = \exp\left(-\frac{4}{3}\pi\dot{\omega}^3 q t^3\right) \tag{6.38}$$

This equation indicates that in the case of spherical growth the growth rate depends on the cube of the time.

If now the growth rate is assumed to be constant and linear in space and time then the number of waves (dE) which pass the arbitrary point (P) for nuclei within the spherical shell confined between the radii r and $r + dr$ is given by

$$dE = 4\pi r^2 \left(t - \frac{r}{\dot{\omega}}\right) I^* dr \tag{6.39}$$

where I^* is the density, the number of nuclei per cubic metre per second. The total number of passing waves (E) is obtained by integration of dE between 0 and $\dot{\omega}t$:

$$E = \int_0^{\dot{\omega}t} 4\pi r^2 I^* \left(t - \frac{r}{\dot{\omega}}\right) dr = \frac{\pi I^* \dot{\omega}^3}{3} t^4 \tag{6.40}$$

which, after insertion into eqn (6.36) and (6.37) gives

$$1 - v_c = \exp\left(-\frac{\pi I^* \dot{\omega}^3}{3} t^4\right) \tag{6.41}$$

It should be noted that now the time is raised to the fourth power. In general it is found that crystallization based on different nucleation and growth mechanisms can be described by the same general formula, the general Avrami equation:

$$1 - v_c = \exp(-Kt^n) \tag{6.42}$$

where K and n are constants typical of the nucleation and growth mechanisms. The growth geometry describes the characteristics of the dominant process.

Equation (6.42) can be expanded according to $\exp(-Kt^n) \approx 1 - Kt^n + \ldots$ and for the early stages of crystallization where there is little restriction of crystallization due to impingement:

$$v_c = Kt^n \qquad (6.43)$$

The Avrami exponent (n) increases with increasing 'dimensionality' of the crystal growth (Table 6.2). Diffusion-controlled growth reduces the value of the exponent by a factor of 1/2 compared with the corresponding 'free' growth case. There are certain limitations and special considerations for polymers with regard to the Avrami analysis:

(i) The solidified polymer is always only semi-crystalline because, as discussed previously, the crystals are never 100% perfect or completely volume filling. This effect can be taken into account by a modification of eqn (6.43) to

$$1 - \frac{v_c}{v_{c\infty}} = \exp(-Kt^n) \qquad (6.44)$$

where $v_{c\infty}$ is the volume crystallinity finally reached.

(ii) The volume of the system studied changes during crystallization as a consequence of the difference in density between the melt and the solid:

$$1 - v_c = \exp\left\{-K\left[1 - v_c\left(\frac{\rho_c - \rho_l}{\rho_l}\right)\right]t^n\right\} \qquad (6.45)$$

where ρ_c is the density of the crystal phase and ρ_l is the density of the melt.

(iii) The nucleation is seldom either athermal or simple thermal. A mixture of the two is common.

(iv) Crystallization always follows two stages: (1) primary crystallization, characterized by radial growth of spherulites or axialites; and (2)

Table 6.2 Avrami exponent n for different nucleation and growth mechanisms.[4]

Growth geometry	Athermal	Thermal[a]	Thermal[b]
Linear growth	1	2	1
Two-dimensional, circular	2	3	2
Three-dimensional			
Spherical	3	4	5/2
Fibrillar	≤ 1	≤ 2	
Circular lamellar	≤ 2	≤ 3	
Solid sheaf	≥ 5	≥ 6	

[a] Free growth: $\dot{\omega} =$ constant.
[b] Diffusion-limited growth: $\dot{\omega} \propto 1/\sqrt{t}$.

secondary crystallization, *i.e.* the slow crystallization behind the crystal front caused by crystal thickening, and the formation of subsidiary crystal lamellae from secondary crystallization is associated with the fractionated low molar mass material and results in crystal imperfections.

The constants in the Avrami equation are obtained by taking the double logarithm of eqn (6.44):

$$\ln\left[-\ln\left(1 - \frac{v_c}{v_{c\infty}}\right)\right] = \ln K + n \ln t \qquad (6.46)$$

Differential scanning calorimetry (DSC) can be used to study the crystal growth kinetics. The nature of the experiment leads to data that are defined in terms of mass rather than volume. In order to convert the data the following expression is used to relate the mass crystallinity (w_c) to a volume fraction (v_c):

$$v_c = \frac{w_c/\rho_c}{\frac{w_c}{\rho_c} + \frac{(1-w_c)}{\rho_a}} = \frac{w_c}{w_c + \frac{\rho_c}{\rho_a}(1 - w_c)} \qquad (6.47)$$

where ρ_a is the amorphous density. Typically the theory is found to fit well the initial crystallization data but deviations are observed once the growth effects of neighbouring entities start to impinge.

6.8 Comparison of Experiment with Theory

Studies of low molar mass polyethylene, $\bar{M}_c \leq 10\,000$, show an exponent of 4 consistent with the theory for spherulitic growth.[22,23] The low molar mass samples display sheaf-like (axialitic) morphology and as expected a high value of the exponent is observed. Polymers with intermediate molar mass, $10\,000 < \bar{M}_w < 1\,200\,000$, have an Avrami exponent near 3 and for higher molar mass samples, $\bar{M}_w \geq 3\,000\,000$, the exponent is further reduced to a value of ~ 2. These observations are consistent with growth occurring predominately in one direction. The crystallization of high molar mass polymers is strongly influenced by chain entanglements and the slow and incomplete crystallization leads to small, uncorrelated crystals, *i.e.* to so-called random lamellar structures. A low value of n is expected for such a 'low-dimensional' platelet or fibrillar-like growth. It is common to observe that the Avrami exponent decreases with increasing molar mass reflecting the influence of the differences in morphology and crystal growth mechanisms on the observed behaviour.

6.9 Growth Theories

As in the case of small molecules, attempts have been made to develop theories that help to probe the mechanism of crystal growth. As with small molecules

(Chapter 2), theories can be divided into equilibrium and kinetic theories. Kinetic theories are principally of two types: enthalpic nucleation theories and entropic theories. Nucleation theories, e.g. the Lauritzen–Hoffman (LH) theory,[5,24,25] assume that the free energy barrier associated with nucleation has an energetic origin. The Sadler–Gilmer theory[12,26] regards the free energy barrier as predominantly entropic. Kinetic theories predict the temperature dependence of growth rate, initial crystal thickness (L_c^*) and other morphological parameters.

6.9.1 Lauritzen–Hoffman Theory[5,25]

The theory assumes that the spherulites or axialites grow radially at a rate which is a function of the degree of supercooling, $\Delta T = T_m^0 = T_c$, where T_m^0 is the equilibrium melting point and T_c is the crystallization temperature. The basic assumptions are similar to those of the Avrami theory: growth occurring at a linear rate. Secondary or tertiary nucleation is assumed and involves only relatively short-range diffusion of the crystallizing units (*stems*). The theory assumes that a secondary nucleus is firstly formed (Figure 6.7) and subsequent *stems* are attached to the surface at a rate ω_s. The thickness of the extended element of the polymer chain, *stem*, along the growth direction is b. As in the case of small molecules, the attachment may be initially to the crystal surface and it then diffuses to an edge, where stable growth will occur.

The rate at which the first *stem* is attached, A_0, will be different from the rest as indicated in Figure 6.5 and leads to a difference in the free energy change (Figure 6.8). The associated free energy change (a) is the sum of a contribution due to covering the crystal surface plus part of the free energy of crystallization. The first *stem* only effectively has interaction through one surface. The attachment of the second *stem* will involve the formation of the first fold and the interaction between the *stem* is between the crystal surface and the first *stem* which leads to a reduction in the free energy for formation. Each additional

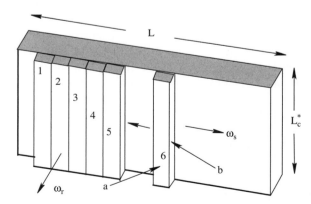

Figure 6.7 Growth processes on a crystal surface. Stem (6) is attached and will diffuse to the growing surface represented by stems (1)–(5). Stem (1) is assumed to be the first attached to the surface.

attachment will involve the creation of a further fold as indicate by the difference between (b) and (c).

The change in free energy involves the surface free energy ($2bL_c\sigma_L$) reduced by a fraction of the free energy of crystallization ($\psi 2bL_c\sigma_L$). The process is schematically described in Figure 6.8. The value of the free energy ΔG for the process is defined as being positive for $T_c < T_m^0$. The remainder of the free energy of crystallization ($(1 - \psi)2bL_c\sigma_L$) for the first *stem* is released on the other side of the maximum. The next process involves the formation of the first fold ($2ab\sigma$) that is accompanied by the release of a fraction of the free energy of crystallization ($\psi abL_c\Delta G$). On the rear side of the second maximum, the rest of the free energy of crystallization of the second *stem* is released. Later crystallizing *stems* exhibit the same energy barriers as the second stem (Figure 6.8).

The rate of deposition of the first stem is

$$A_0 = \beta \exp\left[-\frac{2bL_c\sigma_L - \psi abL_c\Delta G}{kT_c}\right] \tag{6.48}$$

where

$$\beta = \left(\frac{kT_c}{h}\right) J_1 \exp\left[-\frac{U^*}{R(T_c - T_\infty)}\right] \tag{6.49}$$

As indicated above, the term β allows for the diffusion of the *stem* to the site and is usually described by the WLF equation, h is Planck's constant, J_1 is a dimensionless scaling constant, U^* is a constant (dimensions J mol^{-1}) and T_∞ is the temperature at which diffusion is stopped. If the polymer is of sufficiently high molar mass for entanglement to take place, then instead of the WLF theory the reptation theory of de Gennes[27] should be used:

$$\beta = \left(\frac{kT_c}{h}\right)\left(\frac{1}{M_z}\right) \exp\left(-\frac{\Delta E_r}{RT_c}\right) \tag{6.50}$$

where ΔE_r is the activation energy for transport of a molecule undergoing reptation.

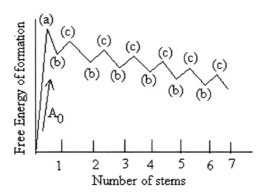

Figure 6.8 The free energy change associated with stem attachment.

The rate equations of the subsequent steps are

$$A = \beta \exp\left[-\frac{(2ab\sigma - \psi abL_c\Delta G)}{kT_c}\right] \tag{6.51}$$

$$B = \beta \exp\left[-\frac{[(1-\psi)abL_c\Delta G]}{kT_c}\right] \tag{6.52}$$

The ratios of A_0/B and A/B are independent of ψ. The absolute rate constants are dependent on ψ and the crystallization process involves subsequent addition of *stems* 1, 2, ..., n to the substrate surface. If it is assumed that a steady-state condition is developed then:

$$\begin{aligned}S &= N_0A_0 - N_1B = N_1A - N_2B = \dots \\ &= N_vA - N_{v+1}B\end{aligned} \tag{6.53}$$

Rearranging eqn (6.53) yields

$$N_{v+1} = \left(\frac{A}{B}\right)N_v - \frac{S}{B} \tag{6.54}$$

$$N_{v+1} = \left(\frac{A}{B}\right)^2 N_{v-1} - \frac{S}{B}\left(1 + \frac{A}{B}\right) \tag{6.55}$$

Further substitution of N_{v-1} finally leads to

$$S = N_0A_0\left(1 - \frac{B}{A}\right) \tag{6.56}$$

The equations describe the process of sequential addition of *stems* to the substrate. The initial *stems* are deposited and described by N_0A_0 and a balance will be achieved between the fraction of the *stems* that are permanently attached to the crystal and those that return to the melt by the ratio of the backward to the forward process, *i.e.* B/A.

By combining eqn (6.48)–(6.52) with eqn (6.56), the following expression is obtained:

$$S(L_c) = N_0\beta\left[\exp\left(-\frac{(2bl_c\sigma_L - \psi abL_c\Delta G)}{kT_c}\right)\right] \\ \times \left[1 - \exp\left(-\frac{(2ab\sigma - abL_c\Delta G)}{kT_c}\right)\right] \tag{6.57}$$

This equation relates the rate of crystallization to its dependence on crystal thickness (L_c) and temperature (T_c). The average crystal thickness L_c^* can be derived from the expression

$$L_c^* = \frac{\int\limits_{L_c=2\sigma/\Delta g}^{\infty} L_c S(L_c)\,dL_c}{\int\limits_{L_c=2\sigma/\Delta g} S(L_c)\,dL_c} \tag{6.58}$$

The lower limit of the above equation corresponds to the case of a crystal formed at the melting point:

$$L_{c,\min} = \frac{2\sigma}{\Delta G} = \frac{2\sigma T_m^0}{\Delta H^0 \rho_c (T_m^0 - T_c)} \tag{6.59}$$

and eqn (6.59) can be rearranged to give

$$T_c = T_m^0\left[1 - \frac{2\sigma}{\Delta H^0 \rho_c L_{c,\min}}\right] \tag{6.60}$$

It is interesting to note that eqn (6.60) has an identical form to that of the Thompson–Gibbs equation. Note that ΔH^0 is the mass-related heat of fusion (in kg^{-1}) and that $T_c = T_m$. Integration of eqn (6.58) gives the expression

$$L_c^* = \frac{2\sigma}{\Delta G} + \frac{kT}{2b\sigma_L}\frac{\left[2 + \left(1 - 2\psi a \frac{\Delta G}{2\sigma}\right)\right]}{\left(1 - \frac{a\Delta G \psi}{2\sigma_L}\right)\left(1 + \frac{a\Delta G(1-\psi)}{2\sigma_L}\right)} \tag{6.61}$$

For $\psi = 1$ this reduces to

$$L_c^* = \frac{2\sigma}{\Delta G} + \frac{kT}{2b\sigma_L}\left[\frac{\left(\frac{4\sigma_L}{a} - \Delta G\right)}{\frac{2\sigma_L}{a} - \Delta G}\right] = \frac{2\sigma}{\Delta G} + \delta L_c \tag{6.62}$$

The parameter δL_c becomes infinite when

$$\Delta G = \frac{2\sigma_L}{a} \tag{6.63}$$

The degree of super cooling (ΔT_s) corresponding to the singularity can be derived as follows:

$$\frac{\Delta H^0 \rho_c \Delta T_s}{T_m^0} = \frac{2\sigma_L}{a} \Rightarrow \Delta T_s = \frac{2\sigma_L T_m^0}{a\Delta H^0 \rho_c} \tag{6.64}$$

For linear polyethylene and putting $\psi = 1$ yields the value of ΔT as 55 K. The singularity (or δL; catastrophe as it is sometimes called) may be avoided by setting $\psi = 0$:

$$L_c^* = \frac{2\sigma}{\Delta G} + \frac{kT}{2b\sigma_L}\left[\frac{\left(\frac{4\sigma_L}{a} + \Delta G\right)}{\frac{2\sigma_L}{a} + \Delta G}\right] = \frac{2\sigma}{\Delta G} + \delta L_c \quad (6.65)$$

It must be remembered that the LH theory does not include any elements of crystal thickening and hence relaxation phenomena are not included. The LH theory introduces three possible growth regimes: I, II and III.

Regime I growth. It is assumed that in regime I (Figure 6.9), the linear growth rate is controlled by secondary nucleation and has a value ω_r.

The growth rate across the crystal surface (ω_s) is greater than the rate of formation of secondary nuclei (ω_i):

$$\omega_s \gg \omega_i \quad (6.66)$$

In this regime growth across the surface is assumed to occur as a monolayer until the whole of the substrate is covered. Monolayers are added to the substrate one by one and the linear growth rate (ω_r) is given by

$$\omega_r(\text{I}) = biL \quad (6.67)$$

where b is the monolayer thickness, i is the surface nucleation rate, the number of nuclei per length of substrate per second, and L is the substrate length ($= n_s.a$, where n_s is the number of *stems* adsorbed onto the substrate and a is the *stem* width). Since iL is the rate of the formation of secondary nuclei on the substrate and b is the monolayer thickness, then a linear growth rate must be thus a product of iL and b. The total rate of crystallization, *i.e.* the total flux (S) of polymer crystallizing is

$$S_T = -Na \quad (6.68)$$

where N is the total number of *stems*, surface nuclei per unit time:

$$S_T = \left(\frac{1}{L_u}\right) \int_{2\sigma/\Delta g}^{\infty} S(L_c)\,dL_c \quad (6.69)$$

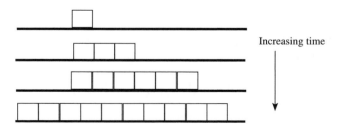

Figure 6.9 Regime I growth. The square boxes represent the cross-sections of the stems growing on the substrate.

Polymer Crystal Growth

where L_u is the length of the repeating unit. Combining eqn (6.68) and (6.69) gives

$$\omega_r(I) = bS_T \frac{n_s}{N} \quad (6.70)$$

Integration of eqn (6.69) gives

$$S_T = \left(\frac{N_0\beta}{L_u}\right) P \exp\left(\frac{2ab\sigma\psi}{kT_c}\right) \exp\left(-\frac{4b\sigma\sigma_L}{\Delta G k T_c}\right) \quad (6.71)$$

where

$$P = RT_c \left[\frac{1}{2b\sigma_L - \psi ab\Delta G} - \frac{1}{2b\sigma_L - (1-\psi)ab\Delta G}\right] \quad (6.72)$$

Using the WLF equation for short-range diffusion factor (β):

$$\beta = \left(\frac{kT_c}{h}\right) J_1 \exp\left(-\frac{U^*}{R(T_c - T_\infty)}\right) \quad (6.73)$$

Finally combination of eqn (6.70)–(6.73) yields

$$\omega_r(I) = \omega_{0r}(I) \exp\left(-\frac{U^*}{R(T_c - T_\infty)}\right) \exp\left(-\frac{4b\sigma\sigma_L}{\Delta G k T_c}\right) \quad (6.74)$$

where

$$\omega_{0r}(I) = b\left(\frac{kT_c}{h}\right) J_1 \exp\left(\frac{2ab\sigma\psi}{kT}\right) \quad (6.75)$$

Alternatively, using reptation to control the *stem* diffusion one obtains

$$\beta = \left(\frac{kT_c}{h}\right) \left(\frac{1}{M_z}\right) \exp\left(-\frac{\Delta E_r}{RT_c}\right) \quad (6.76)$$

where ΔE_r is the activation energy for reptation.

Regime II growth. As the temperature is lowered, multiple nucleation becomes possible and the condition for growth changes, *i.e.* $\omega_s < \omega_i$ (Figure 6.10), and the theory is modified accordingly to give the linear growth rate, $\omega_r(II)$:

$$\omega_r(II) = b\sqrt{i\omega_s} \quad (6.77)$$

where ω_s is the growth rate across the crystal surface $= a(A - B)$ given by

$$\omega_s = a\beta \left[\exp\left(-\frac{2ab\sigma}{kT_c} + \frac{\psi abL_c\Delta G}{kT_c}\right) - \exp\left(-\frac{(1-\psi)abL_c\Delta G}{kT_c}\right)\right]$$
$$\approx a\beta \exp\left(-\frac{2ab\sigma(1-\psi)}{kT_c}\right) \quad (6.78)$$

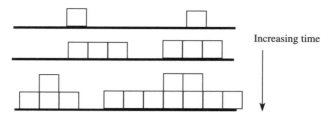

Figure 6.10 Regime II growth, showing that nucleation of a second layer occurs before the first is completed.

This expression is based on the concept that an isolated *stem* undergoes nucleation at time $t = 0$ and then grows equally in opposite directions. At a given time t, the length of the substrate covered by an additional monolayer is proportional to $2\omega_s \cdot t$. The rate at which new nuclei form on this surface is $2\omega_s t i$, where i is the nucleation rate. The number of nuclei formed during the incremental time dt is given by

$$N(t)dt = 2w_s t i dt \tag{6.79}$$

The total number of nuclei formed during a time period t is

$$\int_0^t N(t)\,dt \propto \int_0^t 2\omega_s i t\,dt = \omega_s i t^2 \tag{6.80}$$

The average time $<t>$ to form a new nucleus to grow on the surface is

$$\omega_s i \langle t \rangle^2 \propto 1 \Rightarrow \langle t \rangle \propto \frac{1}{\sqrt{i\omega_s}} \tag{6.81}$$

The rate at which new layers are formed is given by

$$\omega_r(\text{II}) = \frac{b}{\langle t \rangle} = b\sqrt{i\omega_s} \tag{6.82}$$

which is identical to eqn (6.77). Combining eqn (6.68), (6.71), (6.77) and (6.78) leads to an expression for the growth rate in regime II as:

$$\omega_r(\text{II}) = \omega_{0r}(\text{II}) \exp\left(-\frac{U^*}{R(T_c - T_\infty)}\right) \exp\left(-\frac{2b\sigma\sigma_L}{\Delta G k T_c}\right) \tag{6.83}$$

where

$$\omega_r 0(\text{II}) = b\left(\frac{kT_c}{h}\right) \exp\left(\frac{(2\psi - 1)ab\sigma}{kT_c}\right) \tag{6.84}$$

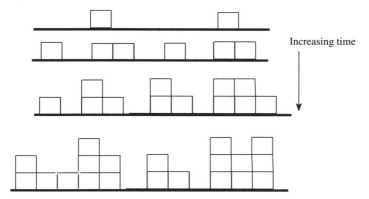

Figure 6.11 Regime III growth. The nucleation is faster than the growth across the surface and niches/grooves begin to be created.

Regime III growth. As the temperature is further lowered, prolific and multiple nucleation can occurs (Figure 6.11).

The niche separation becomes the same dimensions as the width of the *stem* and the growth rate, $\omega_r(\text{III})$, is given by

$$\omega_r(\text{III}) = biL \tag{6.85}$$

The growth rate in regime III may then be expressed by

$$\omega_r(\text{III}) = \omega_{r0}(\text{III}) \exp\left(-\frac{U^*}{R(T_c - T_\infty)}\right) \exp\left(-\frac{4b\sigma\sigma_L}{\Delta G k T_c}\right) \tag{6.86}$$

where $\omega_{r0}(\text{III}) = C(kT_c/h)$ and the last exponent, which is the nucleation factor, is the same as that for regime I. The LH theory implies that as the melt is cooled so there will be changes in the dominant growth mechanism.

The changes in the rate are a consequence of the change in balance between free energy of the melt and that of the solid and its influence on the nucleation processes. The LH theory has been used extensively to analyse polymer crystal growth data and is able qualitatively to describe processes that occur in a number of polymer systems. Its success lies in its ability to describe the temperature dependence of both the initial crystal thickness (L_c^*) and the linear growth rate (ω_r). A large volume of data has been shown to fit the relationship:

$$L_c^* = \frac{C_1}{\Delta T} + C_2 \tag{6.87}$$

and

$$\omega_r = \beta \exp\left(-\frac{K_g}{T_c \Delta T f}\right) \tag{6.88}$$

where C_1, C_2, K_g and f are constants. Plots of $\log(G)/\beta$ as a function of $1/T_c\Delta T$ consist of lines with relatively abrupt changes in the slope coefficients as

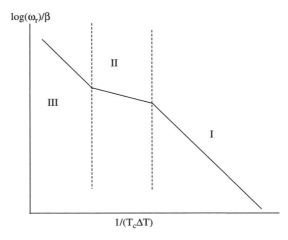

Figure 6.12 Schematic of the predictions of the LH theory for polymer crystal growth.

indicated in Figure 6.12. The growth rate exhibits approximately linear behaviour between the transitions from one regime to another.

The above equations can be simplified if we use

$$\Delta G = \Delta H^0 - T_c \Delta S = \Delta H^0 \left[1 - \frac{T_c}{T_m^0}\right]$$
$$= \frac{\Delta H^0}{T_m^0}[T_m^0 - T_c] = \frac{\Delta H^0 \Delta T}{T_m^0} \quad (6.89)$$

and insertion of eqn (6.88) into the rate equations (6.74), (6.83) and (6.86) allows calculation of various constants for the crystallization process in the expression for K_g shown in Table 6.3.

The transition from regime I to regime II crystallization was reported[25] as being sharp and accompanied by a distinct change in supermolecular structure from axialitic to spherulitic. The LH theory predicts[24] that the ratio of the slopes (K_g) in regimes I and II growth regions is 2:1. The transition from regime I to regime II can be predicted by the so-called Z test.[28] The quantity Z is defined by

$$Z = \frac{\omega_i L^2}{4\omega_s} \quad (6.90)$$

where L is the length of the substrate, and estimated using eqn (6.90) has been ascribed values of the order of 0.1 μm. Regime I occurs when $Z < 0.1$ and regime II when $Z > 1$. Although there is no doubt that these transitions occur, whether the simple growth processes described by theory are always applicable is a matter of debate. Diffusion is an important factor in the crystallization process and to some extent is allowed for by the incorporation of the reptation

Table 6.3 Growth rate equations and growth rate data for linear polyethylene.[4,24]

	Regime I	Regime II	Regime III
$K_g{}^a$	$\dfrac{4b\sigma\sigma_L}{\Delta H^0 k}$	$\dfrac{2b\sigma\sigma_L}{\Delta H^0 k}$	$\dfrac{4b\sigma\sigma_L}{\Delta H^0 k}$
ΔT (K)	<17	17–23	>23
ω_s/ω_i	≫ 1	< 1	≪ 1
Supermolecular structure	Axialitic	Spherulitic	Spherulitic

^{a} $\omega_r = \beta\exp(-K_g/T_c\Delta Tf)$; units of K_g are those of ΔH^0 (J m^{-3}).

theory in the above description. The most contentious issue is whether the folds are tight or random.[29,30] As we will see later, the possibility of rearrangement involving the *stems* changing their conformation on the surface has to be a possibility and aids the formation of tight folds. Determination of the value of L has also been a point of uncertainty. Using decoration techniques the persistence length, which is the average distance between adjacent secondary nucleation sites, has been determined to have a maximum value of 200 nm,[31–33] which is a significantly lower value than would be predicted by the LH theory.

The persistence length is determined by the ratio of the growth rate (ω_s) across the surface to the secondary nucleation rate (ω_i). The LH theory predicts a change in the growth rate because ω_s is weakly temperature dependent whereas ω_i is strongly dependent. Alternative explanations for changes in rate have been put forward and include effects such as the temperature dependence of the interfacial energies, viscosity effects and nucleation processes.

6.9.2 Sadler–Gilmer Theory

In Chapter 2 it was pointed out that the shape, or *habit*, of a crystal depended on growth mechanism and required consideration of the frequency of occurrence of various defects being created in the crystal surface. A temperature exists, the roughening temperature (T_r), at which growth occurs from the defects; below this temperature growth occurs at the facets as discussed in Chapter 2. When crystallization of polyethylene is carried out at higher temperatures and in poorer solvents, single crystals with slightly rounded {100} sectors are obtained. Crystallization at very high temperatures leads to leaf-shaped crystals.

In the case of the crystallization of low molecular mass materials at temperatures above T_r, the growth is proportional to ΔT, whereas experiments on crystal growth in the case of polymers indicate that the growth rate is proportional to $1/T_c\Delta T$. Sadler[34] suggested that the difference in temperature dependence above T_r was of an entropic origin. In colloid science entropic barriers as a consequence of a distribution of chain conformations are well known and are discussed in Chapter 11.

The Sadler–Gilmer theory[26] leads to a prediction of the growth rate:

$$\omega_r \propto \exp\left(-\frac{K_g}{T_c \Delta T}\right) \quad \text{and} \quad L_c^* = \frac{C}{\Delta T} + \delta L_c \tag{6.91}$$

which is in agreement with experiment. If the absorption energy of the *stem* onto the surface is large compared with kT then roughening will not occur. If it is assumed that rather than the whole *stem* adding as a single step, only part of the *stem* adds to the surface and the remainder then collapses behind it then roughening becomes possible. A short element of six CH_2 units would have an interaction energy of approximately $0.6kT$ and attachment of such a short section would allow roughening. The attachments can be divided into various forms: 'blind attachment', 'pinning' and 'detachment'.[4] Polymer chains in the melt will have a variety of different conformations depending on the temperature. As the temperature is lowered so linear all-*trans* elements will be favoured; a chain that has a particular conformation in the melt becomes attached to the surface but its conformation is not necessarily suitable for later stages of crystal growth. The *stem* length may lie below the limit of thermodynamic stability allowing the chain to become detached from the site. There is clearly a critical length of the all-*trans* element for stable attachment to occur but this may also incorporate chain folds and/or loops that can prohibit further growth. For growth to continue either this element has to become detached or this unfavourable conformation has to change into a more favourable conformation. The Sadler–Gilmer theory incorporates the possibility of dynamic exchange of the type depicted in Figure 6.13.

The dynamics of this process are encapsulated in the Sadler–Gilman model[26] that considers the balance between the forward and backward reactions in terms of the ratio k_-/k_+:

$$\frac{k_-}{k_+} = \exp\left(\frac{2\varepsilon}{kT_m^0} - \frac{m\varepsilon}{kT_c}\right) \tag{6.92}$$

where ε is the interaction energy, m is the number of neighbouring sites, T_m^0 is the equilibrium melting point and T_c is the crystallization temperature. As with other theories in polymer science, the connectivity of the chain elements influences the allowed possible ways in which attachment and detachment can occur. The statistical mechanical analysis has to be constrained to consider only those chains that exist at the surface. Two different methods have been used to derive kinetic expressions. The first approach sets up the rate equations for the various steps and solves these numerically and provides averaged values of the various constants. The second approach uses Monte Carlo simulation and allows the growth process to be mapped in time[26,35] and is in good agreement with experiment.

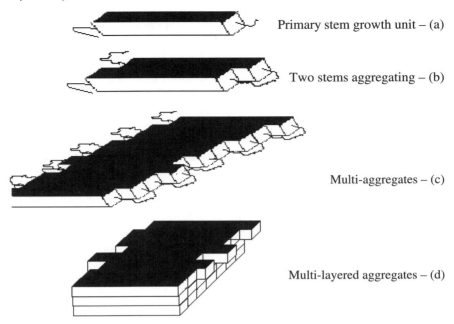

Figure 6.13 The building of a crystal surface; (a) primary stem with disordered cha9ns at each end. There will be a distribution of stem lengths. (b) aggregation of two stems, (c) multli aggregates, and (d) multilayered stems.

6.10 Crystallization via Metastable Phases

In Chapter 4, the plastic crystalline phase of low molar mass hydrocarbons was discussed. Just below the melting point a high degree of rotational freedom is observed whilst restricting the molecules to a defined lattice structure. In polyethylene, studies at elevated pressure have identified the existence of a stable hexagonal structure rather than the usual orthorhombic structure. This hexagonal phase allows rapid crystal thickening and the formation of extended chain crystals without any chain folds. The thickening in the case of the orthorhombic phase crystals is a considerably slower process, but does also occur. In both cases, the folds change to all-*trans* structures and signify the ability for the polymer crystal at elevated temperatures to undergo dynamic changes. The nature of these dynamic changes has been studied using a variety of methods.

Potential impact of rotator phase on crystallization. Molecular dynamics simulations[35] have been carried out to study the conformational defects in the orientationally ordered structures of short-chain molecules and polymer chains. The simulation showed the creation of orientationally ordered structures from a random coil configuration by cooling from the melt. The longer chains exhibit folding with double *gauche* defects with the same sign (... tg(+)g(+)t ...) predominantly located at the chain ends and the kink defects (... tg(+)tg(−)t ...), which do not cause serious deformation of chain molecules, can exist even at the chain interiors. On the other hand, in the case of a single polymer chain, several types of the conformational defects which give rise to large deformation of a polymer chain, such as the double *gauche* defects with opposite sign (. . . tg(+)g(−)t . . .), can be observed in the fold surfaces. A number of similar calculations have been reported,[36–45] all demonstrating that it is possible to incorporate in the simulations subtle effects which allow better comparison between experiment and theory. Changes in the force field allow a level of disorder to be include which is consistent with the slight deviations in lattice parameters observed experimentally.[36] Calorimetric studies of homogeneous nucleation of crystallizing *n*-alkanes (CH_3–$(CH_2)_{n-2}$–CH_3), with values of n which lie between 17 and 60 and low molar mass polyethylene fractions have indicated that the surface energy varies for *n*-alkanes as low as $n = 25$ which is below the value that is normally considered to be the lower limit for chain folding. This behaviour is consistent with the entropic model, in which the surface energy increases when the 'cilia' dangling from the bundle nucleus exceed a length where their entropic cost begins to raise significantly the surface energy. At low n, homogeneous nucleation occurs into the metastable rotator phase that plays a significant role in the nucleation process.[38] A number of studies have examined the potential influence of the rotator phase and conclude that it plays a significant role in the growth process for medium to high molecular weight polymers and may also be evident in low molecular mass materials.[39–45]

Behaviour of other polymer systems. The important of metaphases in the crystallization of *trans*-poly(1,4-butadiene) has a been observed.[46] The polymer can exist in two forms, a monoclinic phase stable at low temperatures and a hexagonal phase that is stable at high temperatures and similar to the hexagonal phase of polyethylene. In heating *trans*-poly(1,4 butadiene) at normal pressure, a transition from monoclinic to hexagonal structure occurs and rapid crystal thickening is observed. Keller *et al.*[47] have suggested that this behaviour may be common to other polymer systems. Crystallization proceeds through the mobile hexagonal phase that permits rapid crystal thickening, with the orthorhombic phase being formed when the crystals reach a certain critical thickness and applies to melt crystallization of polyethylene at normal pressure.

6.11 Molecular Fractionation

Separation of small molecules into different phases is a well-known phenomenon.[48] Polymers that have a different molar mass have sufficient free energy difference to have a tendency to separate and differences in their melting points will allow fractionation into different crystalline species. Lower molar mass materials will crystallize at low temperatures into separate lamellae often located between the dominant lamellae or at the spherulite boundaries.[19,49] At each crystallization temperature there will exist a critical molar mass (M_{crit}) such that the molecules of molar mass greater than M_{crit} are able to crystallize at this temperature whereas molecules of molar mass less than M_{crit} are unable to crystallize. Fractionation has been demonstrated for a range of polymer systems.[50,51] The equilibrium melting point of a given polymer is dependent not only on its molar mass but also on the molar masses of the other species present in the melt:[19]

$$\frac{1}{T_m} - \frac{1}{T_m^0(M)} = \frac{R}{\Delta H}\left[-\ln\nu_p + (x-1) \times (1-\vartheta_p) - x\chi(1-\nu_p)^2\right]$$

(6.93)

where T_m is the melt temperature, $T_m^0(M)$ is the equilibrium melting crystallization temperature of the pure species of molar mass M, R is the gas constant, ν_p is the volume fraction in the melt of the crystallizing species, ΔH is the molar heat of fusion, χ is the interaction parameter and $x = \sum v_i x_i / \sum v_i$ is the volume fraction of the crystallizing species with respect to all other polymers in the melt. The validity of this expression has been shown experimentally. Mehta and Wunderlich[19] have suggested that a particular species nucleates separately and it is the size of this nucleus that governs the growth. Polymers with branched chain structure and differences in tacticity exhibit not only molar mass segregation but also segregation according to structural type. The crystallization temperature is shifted towards lower temperatures with increasing degree of chain branching.

Blends of linear and branched polyethylene normally crystallize in two stages. The components crystallize separately provided that they are of similar molar mass. Linear polyethylene will crystallizes at the highest temperatures, forming regular shaped crystal lamellae. Branched polymers crystallize at lower temperatures in finer, S-shaped lamellae located between the stacks of the dominant lamellae. Although linear and branched polyethylenes are chemically very similar they can phase separate in the molten state.[52] A characteristic of phase separated behaviour is the observation of a dominant lamella structure (Figure 6.14).[53–55]

Metallocene polymers. Since the mid-1990s vinyl polymers synthesized using metallocene catalysts have been commercially available. These polymers have 'purer' structures than those previously available produced using Zeigler–Natta polymerization methods. The characteristics of metallocene polymers are that they can have a very low density of chain defects (side chains) and very narrow molecular mass distributions. Such materials conform very closely to ideal linear polymers and they exhibit many of the classic characteristics expected of such polymers. Metallocene polymers will often have incorporated small amounts of a monomer with a side chain to control the degree of crystallinity. The distribution of these side chains is often much more regular than in the case of the Zeigler–Natta materials and this is reflected in the type of crystalline structure generated. Because the materials are stereochemically and molar mass well defined they exhibit fractionation on slow cooling. This is demonstrated by the DSC analysis of the cooling behaviour. Two distinct peaks are observed for polymers that have molar masses below the critical value for entanglement and reflect the creation of different types of lamellar structures (Figure 6.15). The highest temperature peak at 116 °C reflects the lamellae formed from the

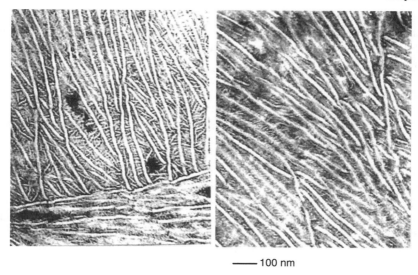

Figure 6.14 Transmission electron micrographs of a polyethylene blend.[57]

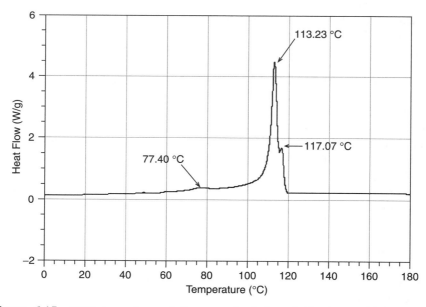

Figure 6.15 DSC trace for metallocene polyethylene indicating multiple melting points.

dominant molar mass species and this forms the main lamellar structure. The secondary structure is associated with the lower melting peak at 112.5 °C. Lower mass fractions form a broad low melting peak at ~77 °C and corresponds to the small crystallites that infill between the lamellae.

Various electron microscopy studies[56] of these materials have shown that very large lamellar structures are observed and that axialitic growth is dominant. The chains fold and add to the lamellae in an ordered fashion to create these large structures. The lamellae have very few bridging *tie* chains and this is evident in the poor environmental stress crack resistance of the materials. If the molar mass is increased above M_c the growth becomes more spherulitic and the lamellae are connected by an increasing number of tie molecules and the environmental stress crack resistance is dramatically improved. There is a very close connection between the morphology generated in the solid state and the bulk mechanical properties.[57]

6.12 Orientation-Induced Crystallization

Many polymers when extruded or in some other way subjected to an external force, which has an impact on the material alignment, exhibit an enhanced degree of crystallinity. The external forces can shear the polymer chains leading to elongation in the direction of the flow. The extent to which alignment is achieved will depend on the rate of oscillation or shear rate to which the melt is subjected. A balance between the natural Brownian motion and the applied forces that attempt to achieve alignment determines the configuration that the polymer chain adopts. The relaxation of the polymer in the melt is well understood and is describe by a combination of Rouse and reptation motion.[58] Reptation motion is only observed for polymers with molecular mass above the critical value for entanglement (M_c) and is several orders of magnitude slower than for the collective motions of the chain described as the Rouse modes. Subjecting a high molar mass polymer can induce alignment and elongation in the flow direction that will place the chain elements in a more favourable position thermodynamically to be able to crystallize. Rubber elasticity theory similarly allows the thermodynamic state of the molecule to be related to the degree of extension in the melt.[59-61] The process which occurs is closely related to that of strain hardening observed when melts are subjected to uniaxial elongational flow.[62] In rubber elasticity theory the degree of extension is connected to the free energy through the entropy. Using this approach the equilibrium melting point is defined by

$$T_m^0 = \frac{H_m - H_c}{S_m - S_c} = \frac{\Delta H^0}{S_m^0 - (R/2)[\lambda^2 + (2/\lambda) - 3] - S_c^0}$$

$$= \frac{\Delta H^0}{\Delta S^0 - (R/2)[\lambda^2 + (2/\lambda) - 3]} \quad (6.94)$$

where λ is the molecular 'draw ratio', which is the ratio of the size of the chain in the flow field to that in the absence of the flow field, R is the gas constant and ΔH^0 and ΔS^0 are, respectively, the enthalpy and entropy changes between the melt and the crystalline state. The second term in the denominator in eqn (6.94) comes from the statistical mechanical theory of rubber elasticity. The effective degree of supercooling at a given crystallization temperature (T_c) increases due to orientation:

$$\Delta T = \Delta T_m - \Delta T_c = \frac{\Delta H^0}{\Delta S^0 - \frac{R}{2}\left(\lambda^2 - \frac{2}{\lambda} - 3\right)} - T_c$$

$$= \frac{\Delta H^0}{\Delta S^0}\left[\frac{\Delta S^0}{\Delta S^0 - \frac{R}{2}\left(\lambda^2 + \frac{2}{\lambda} - 3\right)}\right] - T_c$$

$$T_m^0(\lambda = 1)\left[\frac{1}{1 - \frac{(R/2)[\lambda^2 - (2/\lambda) - 3]}{\Delta S^0}}\right] - T_c \tag{6.95}$$

The kinetics of orientation-induced crystallization has been studied by a number of researchers who have combined the above with the Avrami equation and shown that this approach provides an acceptable description of the processes that occur. Molecular simulations[63] of uniaxial ordering and creation of a ribbon-like polymer structure depend on the effective energy of interaction between the polymer and the flow field. The transition temperature depends on the polymer chain length, decreasing linearly with the reciprocal of the length for chains containing more than 100 monomer units. On the other hand, the change of the entropy is independent of the polymer length for molecules containing more than 30 monomers. This prediction and also calculations based on the Doi–Edwards[64] model for reptation provide good agreement with experimental data.[65] The influence of entanglement and the creation of the amorphous domain can be predicted by consideration of the frequency of the occurrence of entanglements and the associated relaxation rates for disentanglement. The predictions are close to experiment.[66]

Recommended Reading

D.C. Bassett, *Principles of Polymer Morphology*, Cambridge Solid State Series, Cambridge University Press, Cambridge, 1981.
U.W. Gedde, *Polymer Physics*, Chapman and Hall, Glasgow, 1995.
A. Keller and G. Goldbeck-Wood, in *Comprehensive Polymer Science, Supplement 2*, ed. G. Allan, Pergamon Press, Oxford, 1996.
A.S. Vaughan and D.C. Bassett, in *Comprehensive Polymer Science*, ed. G. Allen, Pergamon Press, Oxford, 1989, vol. 2.

References

1. P.J. Flory, *Statistical Mechanics of Chain Molecules*, Interscience, New York, 1969.
2. A. Keller, *Philos. Mag.*, 1957, **2**, 1171.

3. P.J. Barham, R. Chivers, A. Keller, J. Matinez-Salazar and S. Organ, *J. Mater. Sci.*, 1985, **20**, 1625.
4. U.W. Gedde, *Polymer Physics*, Chapman & Hall, 1995, ch. 8.
5. J.I. Lauritzen Jr and J.D. Hoffman, *J. Res. Natl Bur. Std.*, 1960, **64A**, 73.
6. L. Mandelkern, *Physical Properties of Polymers*, American Chemical Society, Washington, DC, 1984, p. 155.
7. G. Ungar, J. Stejny, A. Keller, I. Bidd and M.C. Whiting, *Science*, 1985, **229**, 386.
8. P. Spegt, *Makromol. Chem.*, 1970, **139**, 139.
9. P.H. Geil, F.R. Anderson, B. Wunderlich and T. Arakawa, *J. Polym. Sci. A*, 1964, **2**, 3707.
10. D.C. Bassett, S. Block and G.J. Piermarini, *J. Appl. Phys.*, 1974, **45**, 4146.
11. P.J. Flory, *J. Am. Chem. Soc.*, 1962, **84**, 2857.
12. D. Sadler, in *Crystalline Polymers*, ed. I. H. Hall, Elsevier, Amsterdam, 1984.
13. J.S. Higgins and H.C. Benoit, *Polymers and Neutron Scattering*, Clarendon Press, Oxford, 1994, p. 268.
14. S.J. Spells, A. Keller and D.M. Sadler, *Polymer*, 1984, **25**, 749.
15. B. Wunderlich, *Macromolecular Physics*, Academic Press, New York/London, 1973, vol. 1.
16. D.C. Bassett, *Principles of Polymer Morphology*, Cambridge University Press, Cambridge, 1981.
17. M.L. Williams, R.F. Landel and J.D. Ferry, *J. Am. Chem. Soc.*, 1955, **77**, 3701.
18. B. Wunderlich, *Macromolecular Physics: 3. Crystal Melting*, Academic Press, New York, 1980. pp. 72–73.
19. A. Mehta and B. Wunderlich, *Colloid Polym. Sci.*, 1975, **253**, 193.
20. B. Wunderlich, *Macromolecular, Physics: 2, Crystal Nucleation Growth Annealing*, Academic Press, New York/London, 1978.
21. B. Wunderlich and A. Mehta A, *J. Polym. Sci., Polym. Phys. Ed.*, 1974, **12**, 255.
22. A.J. Kovacs, *Ric. Sci. (Suppl.)*, 1955, **25**, 668.
23. E. Ergoz, J.G. Fatou and L. Mandelkern, *Macromolecules*, 1972, **5**, 147.
24. J.D. Hoffman and R.L. Miller, *Macromolecules*, 1988, **21**, 3038.
25. J.D. Hoffman, L.J. Frolen, G.S. Ross and J.I. Lauritzen Jr, *J. Res. Natl Bur. Std. A. Phys. Chem.*, 1975, **79A**, 671.
26. D.M. Sadler and G.H. Gilmer, *Polymer*, 1984, **25**, 1446.
27. P.G. De Gennes, *J. Chem. Phys.*, 1971, **54**, 5143; *J. Chem. Phys.*, 1971, **55**, 572.
28. J.I. Lauritzen Jr, *J. Appl. Phys.*, 1973, **44**, 4353.
29. J.D. Hoffman, C.M. Guttman and E.A. DiMarzio, *Discuss. Faraday Soc.*, 1979, **68**, 177.
30. D.Y. Yoon and P.J. Flory, *Polymer*, 1977, **19**, 509.
31. J.J. Point, M.C. Colet and M. Dosiere, *J. Polym. Sci. Polym. Phys. Ed.*, 1986, **24**, 357.
32. J.J. Point and M. Dosiere, *Polymer*, 1989, **30**, 2292.
33. J.J. Point and D. Villars, *Polymer*, 1992, **33**, 2263.

34. D.M. Sadler, *Polymer*, 1983, **24**, 1401.
35. G. Goldbeck-Wood and D.M. Sadler, *Mol. Simul.*, 1989, **4**, 15.
36. S. Fujiwara and T. Sato, *Prog. Theor. Phys. Suppl.*, 2000, **138**, 342.
37. T.L. Phillips and S. Hanna, *Polymer*, 2005, **46**, 11003.
38. H. Kraack, M. Deutsch and E.B. Sirota, *Macromolecules*, 2000, **33**, 6174.
39. T.L. Phillips and S. Hanna, *Polymer*, 2005, **46**, 11035.
40. M.J. Nowak and S.J. Severtson, *J. Mater. Sci.*, 2001, **36**, 4159.
41. P.K. Mukherjee and M. Deutsch, *Phys. Rev. B*, 1999, **60**, 3154.
42. K. Nozaki and M. Hikosaka, *J. Mater. Sci.*, 2000, **35**, 1239.
43. E.B. Sirota and A.B. Herhold, *Polymer*, 2000, **41**, 8781.
44. S. Fujiwara and T. Sato, *J. Chem. Phys.*, 2001, **114**, 6455.
45. I.E. Mavrantza, D. Prentzas, V.G. Mavrantzas and C. Galiotis, *J. Chem. Phys.*, 2001, **115**, 3937.
46. S. Rastogi and G. Ungar, *Macromolecules*, 1992, **25**, 1445.
47. A. Keller, M. Hikosaka, S. Rastogi, A. Toda, P.J. Barham and G. Goldbeck-Wood, *J. Mater. Sci.*, 1994, **29**, 2579.
48. R. Koningsveld, W.H. Stockmayer and E. Nies, *Polymer Phase Diagrams: A Text Book*, Oxford University Press, Oxford, 2001.
49. M.I. Bank and S. Krimm, J. Polym. Sci., *Polym. Lett.*, 1970, **8**, 143.
50. X.H. Guo, B.A. Pethica, J.S. Huang and R.K. Prud'homme, *Macromolecules*, 2004, **37**, 5638.
51. X.H. Guo, B.A. Pethica, J.S. Huang, R.K. Prud'homme, D.H. Adamson and L.J. Fetters, *Energy Fuels*, 2004, **18**, 930.
52. P.J. Barham, M.J. Hill, A. Keller and C.C.A. Rosney, *J. Mater. Sci. Lett.*, 1988, **7**, 1271.
53. M. Conde Brana and U.W. Gedde, *Polymer*, 1992, **33**, 3123.
54. U.W. Gedde, *Prog. Colloid Polym. Sci.*, 1992, **87**, 8.
55. A.E. Woodward, *Atlas of Polymer Morphology*, Hanser, Munich/Vienna/New York, 1989.
56. A.J. Muller, M.L. Arnal, A.L. Spinelli, E. Canizales, C.C. Puig, H. Wang and C.C. Han, *Macromol. Chem. Phys.*, 2003, **2004**, 1497.
57. J.H. Magill, *J. Mater. Sci.*, 2001, **36**, 3143.
58. R.T. Bailey, A.M. North and R.A. Pethrick, *Molecular Motion in High Polymers*, Oxford University Press, 1981.
59. H. Janeschitz-Kriegl, *Prog. Colloid Polym. Sci.*, 1992, **87**, 117.
60. G. Eder, H. Janeschitz-Kriegl and S. Liedauer, *Prog. Colloid Polym. Sci.*, 1992, **87**, 129.
61. J.M. Haudin and N. Billon, *Prog. Colloid Polym. Sci.*, 1992, **87**, 132.
62. C. Gabrial and H. Munstedt, *J. Rheol.*, 2003, **47**(3), 619.
63. A.E. Arinstein, *Phys. Rev. E*, 2005, **72**, 51806.
64. M. Doi and S.F. Edwards, *The Theory of Polymer Dynamics*, Clarendon Press, Oxford, 1986.
65. S. Coppola, N. Grizzuti and P.L. Maffettone, *Macromolecules*, 2001, **34**, 5030.
66. K. Iwata, *Polymer*, 2002, **43**, 6609.

CHAPTER 7
Glasses and Amorphous Material

7.1 Introduction

In the previous chapters our attention has been focused on materials that achieve their physical characteristics through order in the solid state. Another very important class of materials are those that produce solids which are characterized by a complete lack of order: these are the glasses. Window glass is the amorphous form of SiO_2 which also occurs as a variety of naturally occurring crystalline minerals: quartz, feldspar micas, amphiboles and pyroxens. Glasses are formed from B_2O_3, P_2O_5, As_2S_3, *etc*. Some small molecules form glasses, such as phenyl ethers, 2-methylpentane and glycerine, as do a number of polymers, such as poly(methyl methacrylate) (PMMA) and polystyrene:

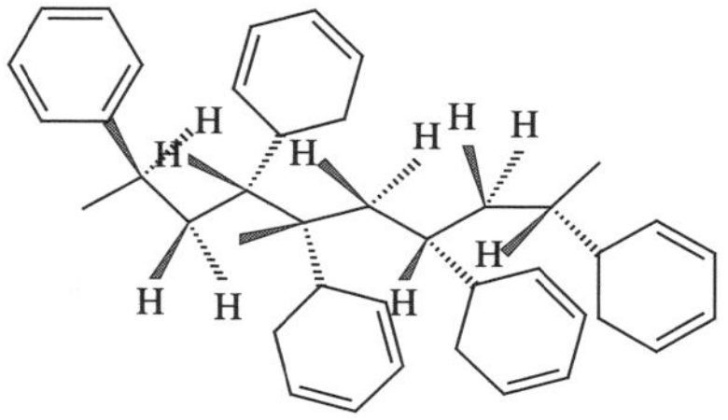

The carbon atom next to the phenyl is potentially a chiral centre, and polystyrene produced by the usual method of radical-initiated polymerization produces an atactic polymer that does not form an ordered solid. However, the use of an ionic route can produce an isotactic form which can lead to a crystalline form of the polymer.[1] The polymer is able to form a helical structure that can pack to form a crystalline solid.

It will be evident that whilst polymers are able to form glasses they are not unique and that the basic characteristics of the glassy state are common to a number of organic and inorganic materials. Some materials, such as *ortho*-terphenyl, form a glass if cooled quickly and subsequently change to a crystalline form.[2]

7.2 Phenomenology of the Glass Transition

The common characteristic exhibited by all glass-forming solids is a similarity in their thermal expansion behaviour. Most crystalline solids exhibit a linear expansion coefficient up to their melting point, whereupon they undergo a discontinuous increase in volume over a small temperature range. Glasses, in contrast, exhibit at very low temperatures a linear expansion coefficient similar to that of a crystalline solid, but at some temperature there is a change in the slope of the expansion coefficient marking the solid taking on a deformable characteristic. The X-ray scattering from glassy solids is usually rather broad and indicates that there is present in the solid a broad spectrum of scattering lengths and usually no clear indication of a primary unit cell structure. High molar mass polymeric materials have the added characteristics of forming a rubbery state which gives the transition from solid to rubber its name: T_g, glass to rubber transition. One of the simplest methods of determining T_g is to measure the change in volume of a polymer with temperature using a dilatometer.[3] The typical plot for a crystallizable glass-forming material such as o-terphenyl (Figure 7.1a) shows certain characteristic features. Fast cooling of the melt allows supercooling to a glass that may or may not be stable. At T_g the temperature dependence becomes parallel to that of the crystalline form. In the case of o-terphenyl the crystalline form is created after storing the glass close to its T_g value for a long time. On heating the crystalline form the expansion follows that of a classic ordered material showing an abrupt change in volume at the melt temperature.

A further facet of glass-forming solids is that the volume of the glassy state depends on the cooling rate (Figure 7.1b). Rapid cooling creates a solid with a higher apparent volume than if the material is slowly cooled. The inflection point depends on the rate of the cooling of the material, indicating that the value of T_g, unlike T_m, is not a fixed quantity and indicates that this transition is defining a metastable state of matter. Despite the apparent flexibility of the

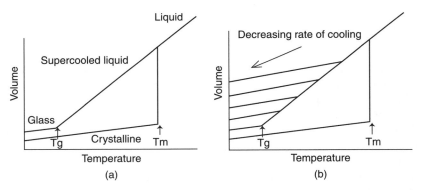

Figure 7.1 Schematic of the variation in the volume with temperature of glassy and crystalline *ortho*-terphenyl: (a) cooling–heating; (b) variation with cooling rate.

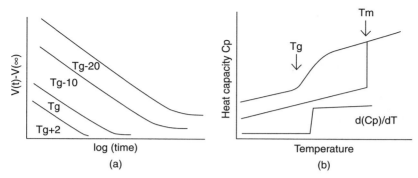

Figure 7.2 (a) Variation of the isothermal volume contraction. $V(t)$ is the volume at time t minus the infinite volume of the solid, crystalline, value as a function of log(time). (b) Variation of the heat capacity C_p and its derivative in the region of T_m and T_g.

T_g value, it is a fairly well-behaved property that has been very well studied and is now fairly well understood.[5] Storage of the glass below T_g allows observation of the variation of the volume with time (Figure 7.2). The exact temperature dependence can change slightly with the form of the glass but most materials exhibit a variation that is approximately logarithmic with time.

The glass undergoes densification on ageing and this process is the basis of *physical ageing*, a phenomenon that has been extensively investigated by Stuick.[5]

An alternative and popular method of studying the glass–rubber transition is through differential scanning calorimetry[3] and the measurement of the heat capacity, C_p. For a crystalline solid the heat capacity will exhibit a discontinuous change at the melting point T_m, and this is designated a *first*-order phase transition (Figure 7.2b). In contrast, the first derivative of the heat capacity undergoes a discontinuous change at T_g and this is therefore designated a *second*-order transition. Since enthalpy exhibits phenomena very similar to those of the volume, it is not surprising that time-dependent *relaxation* processes are observed when the samples are stored below T_g. The relaxation of the enthalpy or the volume is termed *physical ageing*.[5]

7.2.1 Dynamic Mechanical Thermal Analysis

Another characteristics feature of the glass transition is the associated change in the modulus. The stress, *elongation*, is related to the strain, the force applied to a material by the *modulus*. Conventionally there are two approaches to the measurement of the modulus: static and dynamic. The static method involves measurement of the stress–strain profile and from the slope of the curve the elastic modulus can be determined. The dynamic method subjects the sample to a periodic oscillation and explores the variation of the amplitude and phase of the response of the sample as a function of temperature. A small sample of the test material is subjected to displacement as shown in Figure 7.3.

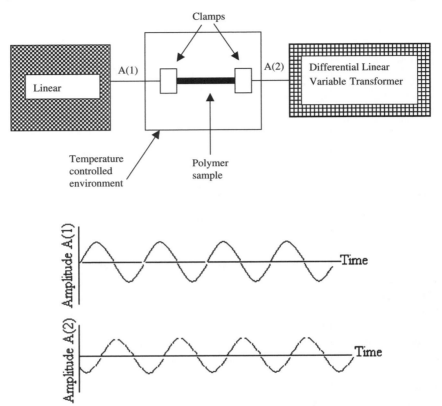

Figure 7.3 Schematic diagram of the dynamic mechanical thermal analysis experiment (top) and a comparison of the amplitudes of the oscillation at points (1) and (2) for a semi-flexible material (bottom).

The oscillation is provided by a linear motor that is able to provide a constant phase and amplitude that is independent of temperature and the forces that are applied to the sample. The variation of the amplitude at point (1) is designated $A(1)$ and that at point (2) is designated $A(2)$. Depending on the stiffness of the material, the amplitude $A(2)$ will vary with temperature and also the frequency of the oscillation. The amplitude $A(1)$ represents the stress which is applied to the sample and $A(2)$ is the corresponding strain that is generated. The stress and the strain are related through the complex modulus E^*; the shift along the time axis of the oscillations is the phase angle ϕ. For a rigid material the phase angle will be approximately zero. As the material softens the phase angle will increase reaching a maximum value at T_g. Softening of the material will increase the difference in the amplitudes until in the limit $A(2)$ would tend to zero as the material becomes infinitely flexible. Comparison of complex functions is best carried out using an Argand diagram in which the stress (σ) and strain (ε) are represented as complex quantities and resolved into real and imaginary components. The σ vector can be resolved into two components: σ' in phase and σ'' out of phase with ε. Two moduli can be defined, G' and G'' representing the in- and out-of-phase components of the complex modulus G^*. The in-phase modulus is defined as $G' = \sigma'/\varepsilon$ and the out-of-phase components as $G'' = \sigma''/\varepsilon$. Since $\sigma' = \sigma \cos \delta$ and $\sigma'' = \sin \delta$, then $G' = \sigma'/\varepsilon = (\sigma \cos \delta)$, and similarly $G'' = G^* \sin \delta$. Using the notation of complex numbers, $\sigma = \sigma' + i\sigma''$ since σ is the vector sum of σ' and σ''. Thus $G^* = \sigma/\varepsilon = (\sigma' + i\sigma)/\varepsilon = G' + iG''$. The ratio $G''/G' = \tan \delta$ is referred to as the *loss factor*. The out-of-phase or loss modulus represents the energy that is not recovered on deformation and is dissipated as heat. If we take a rubber band and subject it to rapid oscillation it will heat up and this is a measure of the energy that has been supplied to the molecules in the form of elongation of the chains that is not recovered when the force is removed. The loss modulus is obtained from the phase information obtained from the dynamic measurements and tells us directly about the ability of the polymer chains in the material to move. The technique and analysis of data are described elsewhere.[4,6]

Most glassy polymers will exhibit a modulus at low temperature which is of the order of 10^{10}–10^9 Pa. At the glass transition the modulus will have dropped to a value of the order of 10^6 Pa and the material will now have rubbery characteristics. This drop in modulus is accompanied by an increase in the loss modulus that will peak at T_g as shown in Figure 7.4.

7.2.2 Dielectric Relaxation Spectroscopy (DRS)

A number of studies have been carried out of the glass transition process using dielectric relaxation spectroscopy. With the advent of computer-assisted measurements[7] the DRS technique has increased in popularity and is now routinely used for the study of molecular mobility in polymeric materials. Chemical bonds may possess dipole moments as a consequence of the differences in

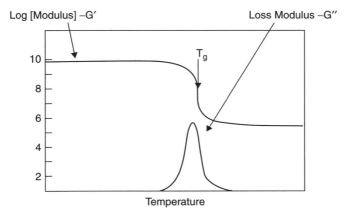

Figure 7.4 Schematic of a dynamic mechanical thermal analysis plot of a glass-forming polymer.

electron density between the atoms forming the chemical bonds. However, whether or not a polymer has a net dipole moment will depend on the symmetry of the polymer and whether the vectorial components of the dipoles are able to cancel. A polymer such as polytetrafluoroethylene has a high degree of regularity and symmetry and consequently a very small dipole moment despite the bond dipole being high. If a molecule is placed in an electric field the molecule will attempt to move to achieve a minimum energy situation that corresponds to the dipole being aligned with the field. If the field instead of being static is allowed to oscillate the ability of the dipoles to follow the applied field depends on their intrinsic mobility. A characteristic frequency exists at which the frequency of oscillation and the natural relaxation will be equal. This is the so-called *relaxation frequency*. The dependence of the electronic polarizability (α) on the frequency of oscillation (ω) depends on physical characteristics of the system. An electric field induces a distortion of the electronic clouds associated with the chemical bonds and is responsible for the refractive index (n) of the material measured at optical frequencies, 10^{15} Hz, and is designated the *electronic polarization* (Figure 7.5). According to the Clausius–Mossotti equation:

$$\frac{n^2 - 1}{n^2 + 2} = \frac{4\pi \rho N_A}{3M} \alpha_e \qquad (7.1)$$

where N_A is Avogadro's number, M is the molar mass and ρ is the density. On lowering the frequency into the optical–infrared region (10^{13} Hz) additional contributions are observed which correspond to the vibrations, rocking and twisting motions that are characteristic of the molecule.

These are resonant processes and conventional infrared spectroscopy investigates the dielectric loss processes associated with these resonances and are designated *atomic polarization* and *bond polarization*. In Figure 7.5 the vertical lines denote the resonance features that are characteristics of a particular molecular structure and occur in the infrared. These resonances are used by

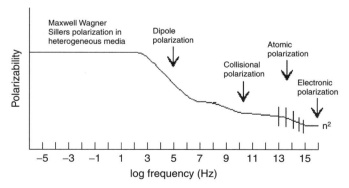

Figure 7.5 Schematic of the frequency dependence of the polarizability as a function of frequency for a polar system.

chemists to identify the nature of the molecular structure. At lower frequencies, molecules can exhibit an additional polarization that is due to the distortions in the electron cloud when molecules undergo inelastic collisions and energy is transferred into excited vibrational states. This contribution is designated *collision polarization*. However the main contribution to the polarizability arises from the alignment of the permanent dipoles and their motion is controlled by the local force field in which they find themselves. In the case of a small molecule, the viscosity of the fluid determines the rate at which the dipole can align. In the case of a polymer solid, the rate of alignment depends on how rigidly the dipole is attached to the polymer chains. The relaxation process can occur anywhere in the accessible frequency range that is currently between 10^{-5} Hz and 10^{11} Hz.

The dielectric relaxation experiment can be visualized in terms of a simple capacitor in which the dipoles are aligned by an applied electric field (E). The magnitude of the charge–displacement field that is observed on the plates depends on the polarizability of the media between the plates of the capacitor. The greater the polarizability the larger the charge that can be carried by the plates. The polarization (P) induced per unit volume for a material placed between parallel plates in a capacitor can be related to the susceptibility χ of the material to be polarized:

$$P = \chi \varepsilon_0 E \tag{7.2}$$

where ε_0 is the permittivity of free space. The capacitance (C) of the capacitor is defined as the amount of charge it can store per unit voltage and the dielectric permittivity is defined by

$$\varepsilon = \frac{C}{C_0} \tag{7.3}$$

where C_0 is the capacitance when the capacitor is in vacuum. The dielectric permittivity ε is defined as

$$\varepsilon = (1 + \chi) \tag{7.4}$$

Combining eqn (7.2) with eqn (7.4) gives

$$P = (1 + \chi)\varepsilon_0 E = \varepsilon \varepsilon_0 E \tag{7.5}$$

In a simple experiment removal of the applied field allows the polarization in the capacitor to decay as a consequence of the randomization of the dipoles. The decay will be determined by a *relaxation time* τ. The decay of the polarization in the capacitor can be described by

$$P(t) = P_0 \exp\left(-\frac{t}{\tau}\right) \tag{7.6}$$

As in the case of the dynamic mechanical experiment we are dealing with mathematical complex quantities. The dielectric permittivity should be expressed as a complex quantity ε^*:

$$\varepsilon^* = \varepsilon' - i\varepsilon'' \tag{7.7}$$

where ε' is the dielectric permittivity and ε'' is the dielectric loss. Normally experiments are performed in the frequency rather than the time domain.[6,7] The time domain can be transformed into the frequency domain using the Laplace transform:

$$P^*(\omega) = L\left(-\frac{\partial [P(t)]}{\partial t}\right) \tag{7.8}$$

The angular frequency ω dependence of the complex permittivity can be described by

$$\varepsilon^* = \varepsilon_\infty + \frac{(\varepsilon_0 - \varepsilon_\infty)}{(1 + i\omega\tau)} \tag{7.9}$$

where ε_∞ is the high-frequency limiting value of the of the permittivity and ε_0 is the low-frequency limiting value or static polarizability. In practice, this formulation is only used to describe dipolar processes and an alternative approach is used to describe the collision, atomic and electronic polarizations that for convenience are lumped together as the ε_∞ value. The complex permittivity can be separated into its real and imaginary parts as follows:

$$\varepsilon' - i\varepsilon'' = \varepsilon_\infty + \frac{(\varepsilon_0 - \varepsilon_\infty)(1 - i\omega\tau)}{(1 + i\omega\tau)(1 - i\omega\tau)} \tag{7.10}$$

$$= \varepsilon_\infty + \frac{(\varepsilon_0 - \varepsilon_\infty) - i\omega\tau(\varepsilon_0 - \varepsilon_\infty)}{(1 + \omega^2\tau^2)} \tag{7.11}$$

Separating the variables gives

$$\varepsilon' = \varepsilon_\infty + \frac{(\varepsilon_0 - \varepsilon_\infty)}{(1 + \omega^2\tau^2)} \tag{7.12}$$

$$\varepsilon'' = \frac{(\varepsilon_0 - \varepsilon_\infty)\omega\tau}{(1 + \omega^2\tau^2)} \tag{7.13}$$

Equations (7.12) and (7.13) describe the ideal relaxation process in which it is assumed that all the dipoles are in similar environments and that they all relax with the same time constant. A number of semi-empirical relationships have been proposed to help with the description of the experimental curves in terms of appropriate equations. The most general is that due to Havriliak and Negami:[8]

$$\varepsilon^* = \varepsilon_\infty + \frac{(\varepsilon_0 - \varepsilon_\infty)}{\left[1 + (i\omega\tau)^\beta\right]^\alpha} \tag{7.14}$$

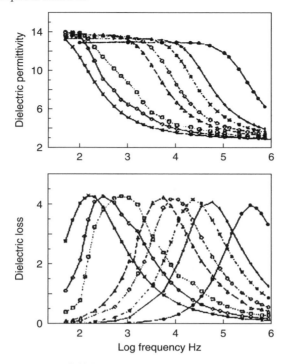

Figure 7.6 A typical set of dielectric relaxation curves of a simple dipole relaxation process.[8]

where α and β are distribution parameters. These parameters are an indication of the breadth of the distribution (β) and the possibility of there being more than one process contributing to the overall relaxation (α). A more complete discussion of the interpretation of these parameters can be found elsewhere.[8] An example of a typical dielectric relaxation is shown in Figure 7.6. As the temperature is lowered the peak in the dielectric loss moves to lower frequency reflecting the slowing down of the relaxation process. The temperature dependence of the relaxation time is determined by the kinetics of the reorientation process and is a reflection of the force field acting on the dipole.

Relaxations tend to divide into two types: those that obey a simple Arrhenius temperature dependence and those that do not. For simple thermally activated processes Arrhenius behaviour is observed. The probability of the dipole reorientating depends directly on the thermal energy distribution. The relaxation time is related to the frequency of maximum dielectric loss:

$$\tau = \frac{1}{2\pi f_{max}} \quad (7.15)$$

and

$$f_{max} = A_1 \exp\left(-\frac{\Delta E^+}{RT}\right) \quad (7.16)$$

where A_1 is the pre-exponential factor and is a function of the activation mechanism, ΔE^+ is the activation governing the reorientation process, R is the gas constant and T is the temperature. Equation (7.16) indicates that $\log(f_{\max})$ plotted against $1/T$ should be linear with a slope of $-\Delta E/RT$. This type of behaviour is observed for dipoles that are able to move independently of one another, such as in a simple liquid, or for a dipole pendant to the main chain such as in the case of the methacrylate group in PMMA and is designated the β process:

The arrow indicates that the carbonyl dipole is rotating relative to the main chain, which is fixed in space. This type of behaviour is observed at low temperature, below T_g of PMMA.

Studies of the dielectric relaxation process in the region of T_g show that the frequency–temperature dependence is not of Arrrhenius type (Figure 7.7). The process being observed corresponds to the rotation of the ester group around the polymer backbone and is designated the α process:

This simplistic representation of the process does not adequately describe the fact that for one ester group to move requires the co-operative motion of the

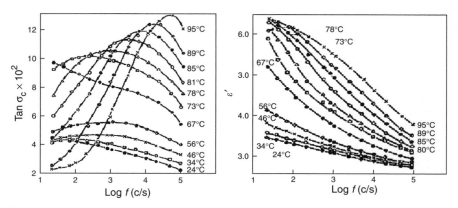

Figure 7.7 The dielectric relaxation of isotactic poly(methyl methacrylate) measured at a number of temperatures.[9]

groups around it. This co-operative process is the basis of the glass transition process and gives rise to the deviation from linearity of the temperature dependence of the relaxation process.

The dielectric relaxation for PMMA is complex at high temperatures and has been shown to be a merge of the relaxation of the side chain β process with the α process. This observation illustrates that the relaxation process reflects the motion of the dipole and this is controlled by a potential energy surface of the type discussed in Chapter 1. At higher temperatures the chain motion has more energy and the possibility of more complex, *coupled*, motions becomes possible. In the case of PMMA the side chains and backbone motions become coupled and it becomes difficult to distinguish unambiguously between the side chain and backbone motions.

7.2.3 Positron Annihilation Lifetime Spectroscopy (PALS)

Further insight into the nature of the glass transition has been obtained using the PALS technique. The natural radioactive decay of ^{22}Na produces a positron (positive electron, e^+) with energy of ~ 1.54 MeV and an accompanying γ-ray. This high-energy particle entering organic matter will be slowed down by collisions and the ionization of atoms in its path. There are three processes which can occur: (1) the positron can decay by collision; and (2) the positron can combine with an electron liberated from one of the atoms by inelastic collision and form a para spin positronium p-Ps (Ps = e^- plus e^+ pair); or (3) form an ortho spin positronium (o-Ps). The p-Ps will decay naturally after a period of ~ 140 ps to two γ-rays. The o-Ps, however, is spin forbidden and has a predicted lifetime of ~ 60 ns. Both p-Ps and o-Ps can only be formed if they can achieve energies comparable to kT and find atomic vacancies within the solid or liquid in which they are formed. It is possible to analyse the decay data and abstract from the total decay a component that is associated with the o-Ps decay and ascribe this to *pick off annihilation*. The o-Ps which is spin forbidden can exchange its electron with an electron on an atom of a molecule which forms the walls of the cavity in which the o-Ps is formed. This exchange process has been explored and it has been shown that the intensity of the decay is related to the number density of the voids in the material, and the lifetime to the size.[10] A number of studies of the glass transition have been carried out, as typified by the behaviour of oligomeric phenyl ethers[8] (Figure 7.8).

Below the glass transition temperature, 230 K, the lifetime is weakly temperature dependent. Above T_g, the lifetime increases to a point T_r whereupon the rate of increase slows down. The intensity of the o-Ps component grows in an approximately linear fashion and does not perceptibly change at T_r. The observation of T_r is an intrinsic feature of T_g. As we have seen from the dielectric relaxation studies, the T_g process is associated with reorientation of dipoles about the backbone of the chain. The o-Ps will reside in a cavity formed from the surrounding polymer chains for a time that is dictated by its exchange lifetime. If, however, during this lifetime the polymer chains which form the

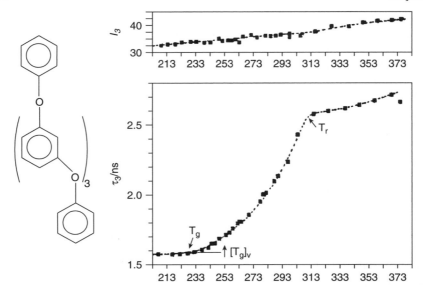

Figure 7.8 Variation of the o-Ps lifetime τ_3 and intensity I_3 as a function of temperature for an oligomeric *m*-phenoxy ether.[11,12]

walls of the cavity are able to move and effectively reduce the size of the cavity, then the effect that this will have on the o-Ps lifetime is to reduce the observed value from that which would be expected in terms of the temperature variation of the void size. This reduction in the apparent void size as a consequence of the rotational motion of the molecules is identified as T_r. A comparison of the dielectric relaxation and o-Ps lifetime behaviour (Figure 7.9) illustrates this feature of T_g.

It is apparent from comparison of the results of the above observations that the glass transition process is associated with the collective motion of elements of the polymer backbone about the polymer axis. For this motion to be able to occur there must exist in the neighbourhood of the moving segment a lack of material, voids, or as it is usually termed, *free volume*. The free volume is therefore the amount of volume required for co-operative motion about the backbone to occur.

7.3 Free Volume and the Williams–Landel–Ferry Equation[13]

The free volume theory is based on the concept introduced by Doolittle which describes the nonlinear behaviour of the viscosity of a liquid as T_g is approached:

$$\eta = A \exp\left(\frac{BV_o}{V_f}\right) \tag{7.17}$$

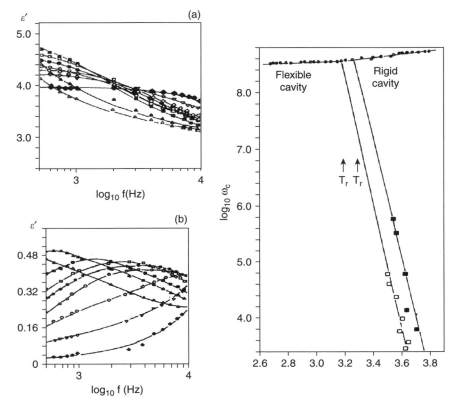

Figure 7.9 Temperature dependence of the dielectric relaxation in the oligomeric phenyl ether shown in Figure 7.8, and a comparison of the effective lifetimes indicating rigid and flexible cavity annihilation behaviour for o-Ps.[11,12]

where η is the viscosity, V_o and V_f are, respectively, the occupied and free volumes and A and B are constants for a particular system. Taking the logarithm of eqn (7.17) one obtains:

$$\ln(\eta) = \ln(A) + \frac{BV_o}{V_f} \tag{7.18}$$

We can now define a parameter f:

$$f = \frac{V_f}{(V_o - V_f)} \cong \frac{V_f}{V_o} \quad \text{since} \quad V_o \gg V_f \tag{7.19}$$

One can rewrite eqn (7.18) as

$$\ln(\eta) = \ln(A) + B\frac{1}{f} \tag{7.20}$$

The value of the parameter f can be defined as f_g at the glass transition temperature and allowed to increase with a value α_f which is equal to the liquid expansion minus that of the glass:

$$f_t = f_g + \alpha_f(T - T_g) \tag{7.21}$$

The viscosity at any temperature T divided by the value at the glass transition temperature is then expressed by

$$\ln\left(\frac{\eta_T}{\eta_{T_g}}\right) = \ln(a_T) = B\left(\frac{1}{f_T} - \frac{1}{f_g}\right) \tag{7.22}$$

where a_T is the so-called *shift factor*. Inserting eqn (7.21) in eqn (7.22) leads to

$$\begin{aligned}\log(a_T) &= B\left(\frac{1}{f_g + \alpha_f(T - T_g)} - \frac{1}{f_g}\right) \\ &= \frac{B}{f_g}\left[\frac{f_g - f_g - \alpha_f(T - T_g)}{f_g + \alpha_f(T - T_g)}\right] \\ &= -\frac{B}{f_g}\frac{T - T_g}{(f_g/\alpha_f) + (T - T_g)} \\ &= -\frac{B'}{2.3 f_g}\frac{T - T_g}{(f_g/\alpha_f) + (T - T_g)}\end{aligned} \tag{7.23}$$

The Williams–Landel–Ferry (WLF) equation has the form

$$\log(a_T) = -17.4\frac{(T - T_g)}{51.6 + (T - T_g)} \tag{7.24}$$

The constants 17.4 and 51.6 are nearly universal and can be used to describe the behaviour of a wide range of materials. The value of 17.4 for the first constant implies that the fractional free volume at T_g is 0.025, *i.e.* of the order of 2.5% for most materials. The second constant 51.6 is the ratio f_g/α_f and this would arise if the value of α_f were equal to 4.8×10^{-4} K^{-1}. There are several other theories that exist, but despite its simplicity the WLF theory has been found to be very successful in describing a significant volume of data.

The WLF theory describes the shifts in the dielectric data as a function of temperature and explains the observed nonlinear dependence on temperature. The α process is therefore not controlled by thermal activation but is a function of the *free volume*. For the segment of the chain containing the dipole to move, it must have sufficient *free volume* to execute the motion.

7.4 How Big is the Element That Moves in the T_g Process?

Whilst the size of the element that moves depends on the detailed stereochemistry of a particular polymer, it is appropriate to consider how big the element might be. Direct experiment evidence can be obtained by examining the

Glasses and Amorphous Material

^{13}C NMR and ultrasonic relaxation behaviour of a simple α-methylstyrene–alkane copolymer system:[13]

The ^{13}C NMR relaxation allows differentiation between the motion of the phenyl and methyl groups and the backbone alkane chain. By studying the temperature dependence of the relaxation times for the various groups it is possible to explore the way in which the activation energy changes with the value of n, the number of CH_2 groups joining the styrene dimers (Figure 7.10).

The relaxation with $n = 0$ approximates to the behaviour of polystyrene in solution. Once the value of n is greater than ~5 the motions have become decoupled and the relaxation of the alkane block is almost independent of that of the styrene dimer. This implies that the size of the group in solution that is required for co-operative motion is approximately 5–6 carbon atoms. Similar analysis has been carried out on solid-state relaxations and it is generally found that the α process involves motion of between 6 and 10 bonds depending on the polymer system. If we consider the co-operative motion of such an element it is possible to envisage that rotation about the backbone can occur without the

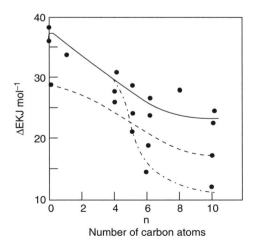

Figure 7.10 Variation of the activation energy for the various groups with the number of carbon atoms in the alkane block: – – –, methyl motion; - · - · -, styrene motion; —, alkane motion.

requirement for there to be significant translational motion of the main chain. Motion of a smaller element would require translational motion of the chain and is only possible close to chain ends:

Clearly for this type of motion to occur there will have to be *free volume* next to the elements of the chain that moves. This is consistent with the concepts of the WLF theory.

7.5 Physical Characteristics of T_g

7.5.1 Factors Influencing the Value of T_g

There are a variety of different factors that will influence T_g of a material. As indicated above, it is not an individual barrier to internal rotation that determines the glass transition or the α relaxation process. The value of $\sim 80\text{–}100$ kJ mol^{-1} for polystyrene is several times the value which would be predicted for rotation about an isolated bond. This observation is consistent with the rotational process involving the co-operative motion of a number of individual bonds. As indicated above, the process involves the solid expanding to create the *free volume* that is necessary for the rotational process to occur. The chains will interact through a range of forces and the overall packing density will be controlled by the *cohesive energy density (CED)* that is a measure of the chain–chain interactions. The CED[14] is equal to the vaporization energy divided by the molar volume. Polymers cannot be vaporized without degradation and the CED is determined by swelling experiments. The CED is equal to the vaporization energy of the low molecular mass liquid that swells the polymer to the greatest extent.

7.5.2 Molar Mass Effects

The molar mass of the polymer is an important factor in consideration of T_g. Since *free volume* is a rate-controlling factor, it may be anticipated that the ends of the chain will be less restricted than the central portion in terms of their ability to move. It is found empirically that the molar mass dependence for a large number of polymers can be described by the following simple relationship:

$$T_g(M) = T_g(\infty) - \frac{K}{M} \qquad (7.25)$$

where $T_g(\infty)$ is the limiting high molar mass value of T_g and K is a constant which for many polymer systems has a value of $\sim 10^5$. A plot of $T_g(M)$ against

Glasses and Amorphous Material

$1/M$ will be linear with a slope of $-K$. PMMA[15] has a value of 2.1×10^5 °C mol g^{-1} whereas polystyrene[16] has a value of 1.7×10^5 °C mol g^{-1}.

7.5.3 Plasticization Effect

The addition of a diluent will increase the *free volume* and consequently can lower the value of T_g. Assuming that the polymer and diluent can both be described by eqn (7.21), and using the subscripts p and d to designate, respectively, polymer and diluent, the free volume of a mixture can be expressed by

$$f_T = 0.025 + \alpha_{fp}(T - T_{gp})V_p + \alpha_{fd}(T - T_{gd})V_d \qquad (7.26)$$

where V_p and V_d are, respectively, the volume fractions of polymer and diluent. At the glass transition temperature f_T becomes equal to 0.025 and T becomes equal to T_g. Rearranging eqn (7.26) gives:

$$T_g = \frac{\alpha_{fp} V_p T_{gp} + \alpha_{fd}(1 - V_p)T_{gd}}{\alpha_{fp} V_p + \alpha_{fd}(1 - V_p)} \qquad (7.27)$$

In cases where the values of α_{fd} and T_{gd} are known, a very good fit to the above equation is often found. Deviations are sometimes observed and reflect specific interactions between the diluent and the polymer.

A PALS study of plasticization of PMMA with dicyclohexylphthalate illustrates the changes that occur when plasticization of a polymer occurs (Figure 7.11).

The incorporation of a low level of plasticizer fills the available *free volume*, the lifetime of the o-Ps is decreased and the temperature to which the solid has to be raised before polymer backbone motion can occur is increased. The phenomenon of the addition of a diluent raising T_g is called *antiplaticization* and is commonly observed for low levels of plasticizer. Further increase in the plasticizer levels leads to a decrease in the temperature at which the lifetime plots change slope indicative of plasticization of the polymer by the diluent.

7.5.4 Incorporation of Comonomer and Blends

In order to increase the range of physical properties available it is desirable to create polymers in which two different monomers are randomly incorporated into the polymer chain. The random copolymer will have a glass transition which is intermediate between those of the homopolymer and can be described by the simple relationship

$$\frac{1}{T_g} = \frac{W_1}{T_{g_1}} + \frac{W_2}{T_{g_2}} \qquad (7.28)$$

where W_1 and W_2 are, respectively, the weight fractions of monomer (1) and (2) in the copolymer that have T_g values $_{g_1}$ and $_{g_2}$. This general mixing law is found to apply for a wide range of materials.

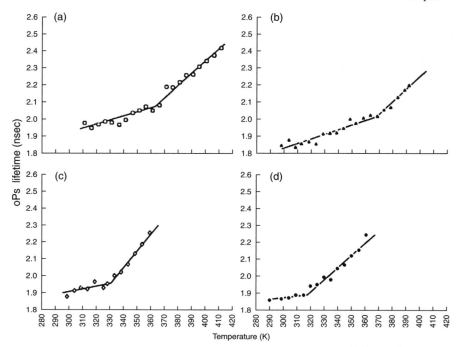

Figure 7.11 Influence of dicyclohexylphthalate on PMMA; measurement of the o-Ps lifetimes: (a) 100% PMMA, (b) 95% PMMA, (c) 90% PMMA and (d) 80% PMMA.

7.5.5 Effects of Chemical Structure

The primary barrier to internal rotational about the backbone will be determined by local interactions between neighbouring groups attached to the bonds about which rotation takes place. Values of T_g for a range of polymers are summarized in Table 7.1.

An extensive tabulation of the T_g data exists in the *Polymer Handbook*. Inspection of the literature indicates that there are a number of values quoted for the same material. The general confusion that appears to exist in relation to T_g is that it is a dynamic entity and hence is subject to the method of measurement. Hence change in the heating rate for the DSC measurement will give a different value as a consequence of the different heating rate representing a different effective frequency of observation. All the values of T_g will be interrelated through the WLF or an equivalent equation. This problem is discussed in the next section.

The mobility of the polymer chains is primarily affected by the barrier to rotation around the chain backbone. The influence of change in structure can be illustrated by considering the effect of substitution in the structure $-(CH_2-CHX)_n-$. If $X = H$, T_g is approximately $-60\,°C$. Changing X to a CH_3 group increases T_g to $-10\,°C$ and introduction of a phenyl group raises T_g further to

Glasses and Amorphous Material

Table 7.1 Glass transition temperatures for a range of polymeric materials.[17,18]

Monomer	T_g (K)	Monomer	T_g (K)
Ethylene	195	Cyanomethyl acrylate	296
Propylene (isostatic)	272	Chloroprene (1,4 *trans*)	238
Propylene (syndiotactic)	265	Vinyl chloride (syndiotactic)	371
Propylene (atactic)	242	Vinyl acetate	307
But-1-ene (isotactic)	223	Amylose triacetate	440
Isobutylene	202	Amylose tributyrate	365
4-Methylpentane (isotactic)	373	Amylose triproprionate	406
Styrene	373	Vinyl ether ether	230
α-Methylstyrene	375	Dimethyl siloxane	150
Butadiene (1,4 *cis*)	218	Vinyl methyl ether	242
Butadiene (1,4 *trans*)	215	Acrylic acid	378
Methyl acrylate (isotactic)	311	Vinyl fluoride	313
Methyl acrylate (atactic)	378	Vinylidene fluoride	333
Methyl acrylate (syndiotactic)	378	Chlorotrifluoroethane	373
Ethyl methacrylate	338	Tetrafluoroethane	390
Butyl methacrylate	293	Ethylene terephthalate	338
Propyl methacrylate	308	Ethylene oxide	232
Hexyl methacrylate	268	Cellulose triacetate	473

100 °C. Substitution in the phenyl ring at the α-position raises T_g to 115 °C. T_g in the case of a naphthalene substitution has a value of 135 °C. T_g for biphenyl is 145 °C. T_g of the more sterically hindered poly(α-methylstyrene) is 175 °C and that of polyacenaphthalene is 265 °C. A further illustration of the influence of subtle changes on T_g can be seen in the case of poly(butyl methacrylate)s. The normal butyl methacrylate has a T_g value of −56 °C, that of secondary butyl methacrylate is −22 °C and that of isobutyl methyacrylate is 43 °C. Once more, the greater the steric interaction the higher the value of T_g.

In general terms:

- Increasing the steric hindrance for rotation about the backbone will lead to an increase in the observed value of T_g.
- Long non-polar side chains will effectively plasticize the structure and will lead to a lowing of T_g.
- Regular structures appear to have higher values of T_g than irregular structures, provided that the former do not represent a more sterically hindered situation. This situation is exemplified by a comparison of the isotactic, syndiotactic and atactic form of PMMA. The isotactic form is sterically hindered and has the lowest value of T_g. The syndiotactic and atactic forms are less sterically hindered and have higher barriers to rotation and higher values of T_g. Calculation of the rotational isomeric potentials for rotation provides a good indication of the T_g values.
- Incorporation of phenyl groups in the backbone will increase the value of T_g through conjugation effects.
- Introduction of heteroatoms, oxygen, sulfur, *etc.*, will lower T_g by increasing the bond angles and reducing the steric interactions.

Poly(dimethyl siloxane) has one of the lowest values of T_g, the O–Si–O bond angle being larger and the bond lengths greater than in a carbon backbone polymer.
- The value of T_g depends on the molecular mass of the polymer, being lower for the lower molar mass materials and reaching a limiting value for molar masses typically above 10^5 Da.
- Inclusion of features in the polymer backbone that disrupt the ability for neighbouring chains to interact will in general lower T_g.

7.6 Kauzmann Paradox

Most discussions of the glass transition raise the question as to whether the process is truly a thermodynamic (second-order) transition or whether it is a kinetic phenomenon which saves the thermodynamic 'catastrophe'. There continues to be much debate about this point which illustrates the lack of the ability of a number of researchers to accept the metastable nature of the glassy state. One way of looking at a glass is to consider it having frozen-in disorder which can relax, but if T_g is sufficiently high, $\sim 200\,°C$ or greater, we are talking about geological times scales for the process. Kauzmann examined the thermodynamic behaviour of supercooled glass-forming liquids by extrapolating their equilibrium properties to low temperature. He found that not very far below the glass-forming temperature but still above 0 K the extrapolated entropy and several other properties of the liquid become less than that of the crystalline solid and highlight the fact that the liquid properties become less than the crystal state properties above 0 K.

7.6.1 Pressure Dependence of the Glass Transition

Typical PVT data for a glass-forming polymer are shown in Figure 7.12. It is apparent that as the pressure is increased so the specific volume is reduced and there is a corresponding increase in the value of T_g. This type of behaviour is generally found for most polymers. As we will see later, the exact behaviour is dictated by the cooling regime that is adopted.

7.6.2 Physical Ageing

Recognizing that T_g has a kinetic component because it is associated with cooperative motion of the elements of the lattice structure moving to create *free volume* for the molecules or polymer chains to move, leads to the idea that T_g can change with storage time. Figure 7.13 indicates the isothermal contraction of glucose after quenching from $T_0 = 40\,°C$ to different temperatures.

Glucose, like all glass-forming liquids, exhibits *physical ageing*. Depending on the extent to which cooling takes place, the rate of the physical ageing will vary. The common feature of all glasses is that this behaviour is nonlinear and does not follow a simple Arrhenius type of behaviour. The process involves

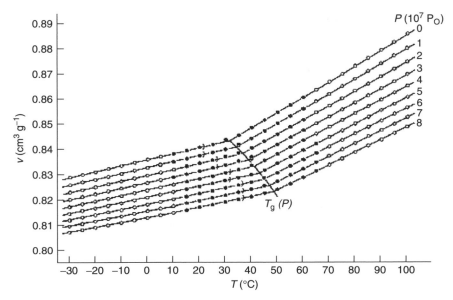

Figure 7.12 The specific volume of poly(vinyl acetate) measured as a function of temperature for pressures between 1×10^7 and 8×10^7 Pa (from reference 19).

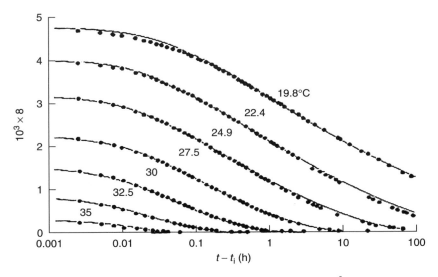

Figure 7.13 Quenching behaviour of glucose cooled from 40 °C.[5]

rearrangement of the organization of the species in the glass and as such will involve a redistribution of *free volume* in the system and hence one would expect it to be controlled by a WLF or similar type of relationship. Similar behaviour can be observed in respect to creep. If a sample is subjected to a load,

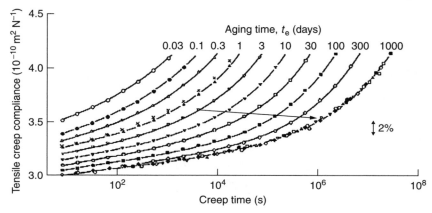

Figure 7.14 Creep behaviour for poly(vinyl chloride).[5]

then after a period of time the sample will deform. The behaviour for poly(vinyl chloride) is shown in Figure 7.14.

All the physical ageing phenomena can be understood as the slow relaxation of the configurational entropy that is frozen into the glass on quenching below T_g. In the case of the simple system *ortho*-terphenyl, the ageing process goes to completion and crystals are formed. With most polymeric materials the development of order is limited and densification is the best description of the process. An extensive review of this topic is to be found in Struick's book.[5]

7.7 Distribution of Free Volume in a Glass

In order to understand many of the properties of the glassy state it is important to consider the nature of *free volume*. The WLF theory is based in the concept of free volume V_f (eqn (7.19)) and leads to a universally accepted and successful method of describing the behaviour of glasses. The question must be asked as to whether *free volume* is a single-valued function at a particular temperature and pressure or whether it is a distribution of values and the WLF parameter is strictly speaking an averaged parameter. Some insight into this problem can be obtained by studying the rate of ring closure of indolinobenzospirans when dispersed in a polymer matrix:

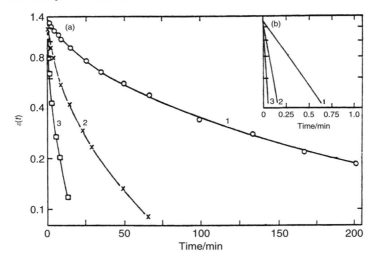

Figure 7.15 Kinetic plots for the ring closure reaction for 8-bromo-2,2'-dimethyl-6-nitro-1'-phenyl-(2H-[1]benzospiropyran-2,2'-indoline) in PMMA at (1) 17 °C, (2) 30.2 °C and (3) 42.0 °C.

Exposure of the 8-bromo-2,2'-dimethyl-6-nitro-1'-phenyl-(2H-[1]benzospiropyran-2,2'-indoline) to light causes it to isomerize into the coloured form. In a solvent it rapidly converts to the bleached form and exhibits first-order kinetics.[20] The process is thermally activated and will require volume for the relative rotation of the two parts of the molecule to effect ring closure. However, if the same process is carried out with the dye in a polymer matrix the process now depends on there being *free volume* available for the rotation of the molecule required for ring closure to occur. Studies of the kinetics (Figure 7.15) indicate that simple kinetics is no longer followed and that a more complex analysis is required. Good fits of the data can be obtained if it is assumed that the process is split into two processes. The initial fast decay is essentially the same as that seen in a liquid and indicates that the ring closure process is only subject to thermal control. This implies that these molecules have sufficient *free volume* available to execute the process.

The longer time kinetics has to allow for the *free volume* being created next to the molecule to allow for the ring closure. To describe the process it is necessary to allow for there to be a distribution of *free volume* in the system and this is best described by a distribution function $Z(p)$ which has the form

$$Z(p) = \exp(\beta\sqrt{B}) \exp(-Bp)\psi(\beta, p) \qquad (7.29)$$

where B describes the kinetics of the ring closure process and p is the probability of the occurrence of the chromophore in a matrix region of sufficient free volume for ring closure to occur. Analysis of a number of chromophores indicates that this type of function allows for the *free volume*

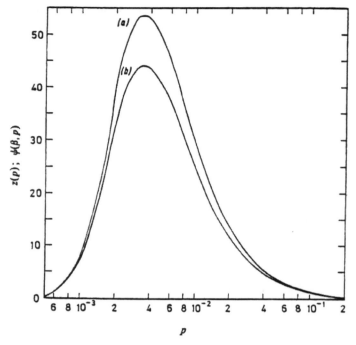

Figure 7.16 Functions (a) $Z(p)$ and (b) $\psi(\beta, p)$ using $B = 2$, $\beta = 0.146$, the values of these parameters having been derived from kinetic data obtained at 17 °C for the system discussed in the text.

distribution and has the form shown in Figure 7.16. The parameter β is the distribution parameter. This experiment illustrates that rather than being single valued, *free volume* is more correctly described by a distribution of values. In a matrix there will be voids which are smaller and larger than a mean value, the latter being the value which is usually described by the WLF equation. If we accept that *free volume* is a distribution then this has profound implications for the way we look at other features of the glassy phase in materials.

If *free volume* is a distribution rather than being single valued then the dynamic, time-dependent, behaviour would conform to a single relaxation process and have an ideal form. However, if there is a distribution of sites at which relaxation can occur with different values of the *free volume*, then there will be a distribution of relaxation times. This is what is observed experimentally and produces the form of the curves shown in Figure 7.13 and 7.14.

7.8 Fragility

It has been observed that for all glasses it is possible to define a *relaxation time* associated with the reorganization of the liquid–glass structure. The relaxation time, which has a theoretical value of infinity at T_K, a temperature just below T_g, decreases with increasing temperature to the value of 200 s at the 'normal' accepted glass transition temperature,[21] and then continues to fall as $T > T_g$ in a manner which is usually predicted from the Vogel–Fulcher–Tammann (VFT) equation:

$$\tau = \tau_0 \exp\left(\frac{DT_0}{T - T_0}\right) \quad (7.30)$$

where τ_0, D and T_0 are constants; $T_0 \sim T_K$ ($<T_g$) is the divergence temperature. A plot of the reduced temperature dependence of relaxation times for several weak and strong forming glasses is shown in Figure 7.17. Deviations from Arrhenius behaviour indicated by the dotted line are an indication of the strength of the interactions present in the liquid. In liquids like SiO_2, BeF_2 and P_2O_5 deviations are barely detectable and these are classed as 'strong' liquids. In other systems, deviations are observed, and the apparent slope of the temperature dependence is many times the vaporization energy or in the case of polymers the bond dissociation energy; these are fragile liquids. Thermodynamically strong liquids like SiO_2 and GeO_2 show small differences in heat

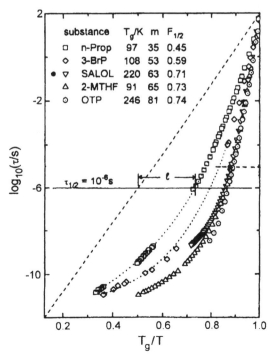

Figure 7.17 T_g scaled Arrhenius plot of the dielectric relaxation times for several glass-forming molecular liquids. The fragility $F_{1/2}$ is defined as $2(T_g/T_{1/2})$. The liquids are from top to bottom n-propanol, 3-bromopentane, salol (phenylsalicylate), 2-methyltetrahydrofuran and *ortho*-terphenyl.

capacity, ΔC_p, between liquid and glass and the entropies approach the crystal values only slowly. Fragile liquids have such high heat capacities relative to their crystals so that a natural extrapolation of the liquid entropy below the melting point indicates disappearance of the excess over the crystal at a temperature far below 0 K, the Kauzmann entropy catastrophe discussed above. Most polymer systems show behaviour that resembles *ortho*-terphenyl and are strong glasses.

The proximity of the divergence temperature T_0 to the 'normal' T_g ($0 < T_0/T_g < 1.0$) provides one measure of the fragility.[22] Using either the VFT or the WLF equations creates problems because these do not accommodate the 'fragility' of the material in their derivations.

A parameter m can be defined as follows:

$$m = (T_g)^{-1} \frac{\partial(\ln \tau)}{\partial(1/T)} \qquad (7.31)$$

where m is the *'fragility'* or *steepness index*. Angell and co-workers[22] have proposed that the fragility can be defined in terms of a length parameter l determined as the deviation from Arrhenius behaviour at an effective relaxation time of 10^{-6} s. There is no fundamental reason for this choice; however, in the case of polymers it does correspond to a length scale which is comparable with collective motions executed by the polymer chain. As indicated above, different techniques give apparently different values for T_g. Part of the reason for this apparent discrepancy is the fact that different methods are looking at different length scales of motion of the system. A dynamic mechanical observation will obviously be expected to see a different collection of motions from those observed by dielectric or nuclear magnetic resonance studies. A more detailed discussion of this topic is presented elsewhere.[6] The origins and definitions of fragility have been extensively discussed in the literature. The main property which emerges is that fragility is an indication of the extent to which the motions are associated with T_g and co-operative in nature.

7.9 Theories of T_g

A considerable amount of research has been reported into the modelling of the glassy state and in particular attempts to predict the distribution of the relaxation processes which take place close to or just above T_g. The comparison between dynamic and thermodynamic approaches has been discussed by Wales and Doye.[23] They have used a model which is similar to that used to model crystallization. Each of the possible configurations of the species involved in glass formation are depicted in terms of a series of interconnected energy states; movement from one potential energy surface (PES) to another state is controlled by an energy barrier. It is assumed that the entities in the liquid attempting to organize themselves into a crystalline state are able to execute two types of motion: vibrations about a local minimum and less frequent jumps over a significant barrier to form a new more stable state. Using this approach it is possible to set up a series of partition functions which allow calculation of thermodynamic properties of the system. The advantage of this approach is that it naturally predicts a distribution of relaxation times and by adjusting the energy values can accommodate both strong and weak fluids. The principal problem is that no theory helps with the visualization of the glass state, which is best assumed to be a disordered array of interacting species which interact to varying degrees and inhibit the creation of the order that would lead to crystal growth. An extensive review of the various approaches to the modelling of the relaxation behaviour of glasses has been published previously[24] and a detailed discussion of this topic is beyond the scope of the present text.

Recommended Reading

R.T. Bailey, A.M. North and R.A. Pethrick, *Molecular Motion in High Polymers*, Oxford University Press, 1982.

G.B.N. McKenna, in *Comprehensive Polymer Science*, ed. G. Allan and J. C. Bevington, Pergamon Press, 1989, p. 311.

L.C.E. Stuick, *Physical Ageing in Amorphous Polymers and Other Materials*, Elsevier, Amsterdam, 1978.

References

1. B. Wandelt, D.J.S. Birch, R.E. Imhof and R.A. Pethrick, *Polymer*, 1992, **33**, 3558.
2. R.A. Pethrick and B.D. Malholtra, *J. Chem. Soc., Faraday Trans. 2*, 1982, **78**, 95.
3. D. Campbell, R.A. Pethrick and J.R. White, *Polymer Characterization*, Stanley Thornes, Cheltenham, UK, 2000.
4. A. Eisenberg, *Physical Properties of Polymers*, American Chemical Society, Washington, DC, 1984, p. 55.
5. L.C.E. Stuick, *Physical Ageing in Amorphous Polymers and Other Materials*, Elsevier, Amsterdam, 1978.
6. R.T. Bailey, A.M. North and R.A. Pethrick, *Molecular Motion in High Polymers*, Oxford University Press, 1982.
7. R.A. Pethrick and D. Hayward, *Prog. Polym. Sci.*, 2002, **27**, 1983.
8. S. Havriliak and S.J. Negami, *Dielectric and Mechanical Relaxation in Materials*, Hanser, Munich, 1997.
9. G.P. Mikhailov and T.I. Borisova, *Polym. Sci. USSR*, 1961, **2**, 387.
10. R.A. Pethrick, *Prog. Polym. Sci.*, 1997, **22**, 1–47.
11. R.A. Pethrick, F.M. Jacobsen, O.A. Mogensen and M. Eldrup, *J. Chem. Soc., Faraday Trans. 2*, 1980, **76**, 225.
12. R.A. Pethrick and B.D. Malholtra, *Phys. Rev. B*, 1983, **22**, 1256.

13. A.V. Cunliffe and R.A. Pethrick, *Polymer*, 1980, **21**, 1025.
14. R.A. Hayes, *J. Appl. Polym. Sci.*, 1961, **5**, 318.
15. R.B. Beevers and E.F.T. White, *Trans. Faraday Soc.*, 1960, **56**, 744.
16. F. Buche, *Physical Properties of Polymers*, Wiley Interscience, New York, 1962.
17. R.J. Andrews and E.A. Grulke, in *Polymer Handbook*, ed. J. Brandrup, E.H. Immergut and E.A. Grulke, Wiley Intercsience, New York, 1999.
18. W. Kauzmann, *Chem Rev.*, 1948, **43**, 219.
19. J.E. MacKinney and M. Goldstein, *J. Res. Natl Bur. Stand. Sect. A*, 1978, **78A**, 331.
20. M. Kryzewski, B. Nadolski, A.M. North and R.A. Pethrick, *J. Chem. Soc., Faraday Trans. 2*, 1980, **76**, 351.
21. C.T. Moyniham, P.B. Macedo, C.J. Montrose, P.K. Gupta, M.A. DeBolt, J.F. Dill, P.W. Drake, A.J. Easteal, P.B. Elterman, R.P. Moeller, H.A. Sasabe and J.A. Wilder, *Ann. N. Y. Acad. Sci.*, 1976, **279**, 15.
22. J.L. Green, K. Ito, K. Xu and C.A. Angell, *J. Phys. Chem. B*, 1999, **103**, 3991.
23. D.J. Wales and J.P.K. Doye, *Phys. Rev. B.*, 2001, **63**, 214204.
24. G.B.N. McKenna in *Comprehensive Polymer Science*, ed. G. Allan and J.C. Bevington, Pergamon Press, 1999.

CHAPTER 8
Polymer Blends and Phase Separation

8.1 Introduction

In the preceding chapters, the systems considered have been essentially single components and the physical properties exhibited by these systems are determined by the chemical structure and molar mass of the molecules. Early in the commercial use of polymers, it was recognized that it was possible by mixing various monomers in the reaction to obtain polymers with different properties from those of the homopolymers. Some of these copolymers had properties that were simple averages of the properties of the homopolymers, others had different characteristics.[1,2] Rather than having to make different copolymers the possibility of generation of blends of the homopolymers was investigated. With certain pairs of polymers, homogeneous mixtures were created that had properties that were an average of those to the homopolymers; others did not form blends. As a result there has been considerable effort devoted to the study of the blending processes and the types of morphology that are created from mixing either monomers or polymers. There are a variety of different ways of describing blends. The term *compatible* is often used to describe mechanically processable blends that resist gross phase separation and/or give desirable properties. Blends that are homogeneous at some temperature may under other conditions phase separate and these are referred to as partially or nearly miscible blends. According to the above definition polymer blends can be divided into three basic groups:[3]

Group 1. Miscible blends:
- polystyrene–poly(2,6-dimethyl 1,4-phenylene oxide)
- poly(methyl methacrylate)–poly(vinylidene fluoride)

Group 2. Partially miscible blends:
- polystyrene–poly(vinyl methyl ether)
- poly(ethylene oxide)–poly(ether sulfone)
- phenoxy resin–poly(ether sulfone)

Group 3. Immiscible blends:
- polyethylene–poly(methyl methacrylate)
- polystyrene–poly(methyl methacrylate).

Some systems are made compatible by the addition of a third component that is termed a *compatibilizer* or *emulsifier*. A considerable number of studies have been carried out and it is desirable to be able to rationalize the behaviour of these systems in terms of a theoretical framework.[4]

8.2 Thermodynamics of Phase Separation

Various attempts have been made to rationalize the behaviour of multicomponent systems.[1,2] Using the classical framework of Gibbs, it is possible to understand the behaviour of many polymer blends simply by considering the change in the free energy with composition. If

$$\left(\frac{\partial^2 G}{\partial V^2}\right)_T = -\left(\frac{\partial P}{\partial V}\right)_T = 0 \tag{8.1}$$

where G is the free energy, V is the volume, P is the pressure and T is the temperature, then this defines the spinodal condition and represents a boundary between a stable single phase and one in which two phases are stable. In a homogeneous binary mixture, we can use the usual thermodynamic conventions of additivity to create the properties of the mixtures. If the system is miscible then the free energy change G with composition can be depicted by the curves shown in Figure 8.1a.

Lowing the temperature above some critical value T_c may lead to a variation as depicted in Figure 8.1b. Since the laws of dilute solutions require $G_m(x_2)$ to have an infinite slope at both ends of the x_2 axis, and negative at $x_2 = 0$ and positive at $x_2 = 1$, there must be two positively curved portions of the $G_m(x_2)$ curve surrounding a negatively curved portion, separated by two points of inflection at the compositions x_{2s1} and x_{2s2}. Between these two compositions a

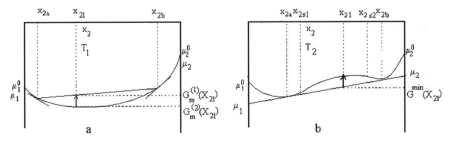

Figure 8.1 Isobaric $G_m(x_2)$ curves for a partially miscible binary mixture: (a) above the critical temperature; (b) below the critical temperature.

single-phase system will separate spontaneously because any concentration fluctuation, however small, will reduce G_m and so trigger further separation until the minimum value is reached at G_m^{min} (x_{21}). Such a separation is called a spinodal decomposition, the two points of inflection being known as spinodal points and defined by

$$\left(\frac{\partial^2 G_m}{\partial x_2^2}\right)_{P,T} = 0 \quad \text{(spinodal)} \qquad (8.2)$$

and the condition for *instability* is

$$\left(\frac{\partial^2 G_m}{\partial x_2^2}\right)_{P,T} < 0 \quad \text{(unstable)} \qquad (8.3)$$

The ranges $(0, x_{2a})$ and $(x_{2b}, 1)$ are stable regions and the remaining ranges (x_{2a}, x_{2s1}) and (x_{2s2}, x_{2b}) are called metastable because, although as homogenous systems they have a larger G_m value than the two phases value represented by the line $\mu_1\mu_2$, they are still stable with respect to immediately neighbouring concentrations in the sense of Figure 8.1a. Metastable systems require some form of nucleation to reach their stable two-phase state. Unstable systems can do so without nucleation.

8.2.1 Thermodynamics of Polymer–Polymer Miscibility

The theoretical modelling of the behaviour of simple liquids and polymers can be approached from consideration of the systems as behaving like gases and as solids. The first group of theories are commonly referred to as *equation of state* theories and assume that the gas laws can be adapted to describe the behaviour of liquids. Flory and co-workers proposed an equation of state theory for the thermodynamics of polymer melts in which they blended the two approaches.[1] Each of the pure components is characterized by three state parameters, V^*, T^* and P^*, that can then be used using conventional thermodynamic expression to obtain density, thermal expansion coefficients and thermal pressure coefficients. In polymer mixtures, two additional terms were introduced, X_{12} and Q_{12}, associated with the enthalpy and entropy of the mixture and describe specific interactions between the two components. Mixing two components can give rise to additional interactions not present in either of the components. For instance, poly(vinyl alcohol) will undergo hydrogen bonding interactions with poly(methyl acrylate) and an additional term is required to account for the additional interactions. It is normal to describe a set of reduced thermodynamic variables:

$$\tilde{V} = V/V^*; \quad \tilde{T} = T/T^* = kT/z\varepsilon; \quad \tilde{P} = P/P^* = PV^*/z\varepsilon \qquad (8.4)$$

which are respectively the volume, temperature and pressure, z is the coordination number of the equivalent lattice and k is the Boltzmann constant. The subscripts 1 and 2 are used in the following derivation to designate the two components of the mixture. The resulting equation of state is

$$\frac{\tilde{P}\tilde{V}}{\tilde{T}} = \frac{\tilde{V}^{1/3}}{(\tilde{V}^{1/3} - 1)} - \frac{1}{\tilde{T}\tilde{V}} \qquad (8.5)$$

At atmospheric pressure where $\tilde{P} \sim 0$ this equation becomes

$$\tilde{T} = \frac{(\tilde{V}^{1/3} - 1)}{\tilde{V}^{4/3}} \qquad (8.6)$$

The $\tilde{V}$ of the mixture can be related to the thermal expansion α by

$$\alpha = \left(\frac{1}{V}\right)\left(\frac{\partial V}{\partial T}\right) \qquad (8.7)$$

or

$$\alpha T = \left(\frac{\tilde{T}}{\tilde{V}}\right)\left(\frac{\partial \tilde{V}}{\partial \tilde{T}}\right)_P \qquad (8.8)$$

The two terms in this equation are obtained from eqn (8.6) to give

$$\alpha T = \frac{3(\tilde{V}^{1/3} - 1)}{(4 - 3\tilde{V}^{1/3})} \qquad (8.9)$$

or

$$\tilde{V}^{1/3} = \frac{(3 + 4\alpha T)}{(3 + 3\alpha T)} \qquad (8.10)$$

The thermal pressure coefficient γ can be obtained as follows:

$$\gamma = \left(\frac{\partial P}{\partial T}\right)_V = \frac{P^*}{T^*}\left(\frac{\partial \tilde{P}}{\partial \tilde{T}}\right)_{\tilde{V}} \qquad (8.11)$$

The parameter $(\partial \tilde{P}/\partial \tilde{T})_{\tilde{V}}$ is evaluated by differentiating eqn (8.5) with respect to $\tilde{T}$ at constant $\tilde{V}$. Substitution in eqn (8.11) and combining with eqn (8.6) gives

$$P^* = \gamma T \tilde{V}^2 \qquad (8.12)$$

The volume fractions of the constituent polymers in the mixture are given by

$$\phi_2 = \frac{m_2 V^*_{sp2}}{(m_2 V^*_{sp2} + m_1 V^*_{sp1})} \qquad (8.13)$$

and

$$\phi_1 = 1 - \phi_2 \tag{8.14}$$

where m_i is the mass of component i and V^*_{spi} is its hard core volume per gram. In addition the volume fractions, site fractions, are defined as

$$\theta_2 = \left(\frac{S_2}{S_1}\right) \frac{\phi_2}{[\phi_1 + (S_2/S_1)\phi_2]} \tag{8.15}$$

and

$$\theta_1 = 1 - \theta_2 \tag{8.16}$$

where S_i is the surface area per unit volume ratio for component i which is usually calculated using Bondi's tabulations of group surface areas[4] and volumes.[2] P^* of the mixture is related to the contact interaction term X_{12} via

$$P^* = \phi_1 P_1^* + \phi_2 P_2^* - \phi_1 \phi_2 X_{12} \tag{8.17}$$

The X_{12} term has energy per unit volume dimensions and is concentration independent, unlike the classic Flory–Huggins interaction parameter that is dimensionless and concentration dependent. Both parameters are still temperature and pressure dependent. T^* is related to P^* through

$$\frac{P^*}{T^*} = \frac{\phi_1 P_1^*}{T_1^*} + \frac{\phi_2 P_2^*}{T_2^*} \tag{8.18}$$

These equations indicate the way in which the thermodynamic properties of the system change with the composition. In a blend of polymers, new interactions can be created which are not present in the homopolymer system. For instance, polar groups can induce dipoles or quadrupoles that will increase the interaction energy between the polymer chains and can change the entropy of the system. Such interactions explain the unusual properties obtained when fluorine atoms are present in a polymer system.

8.2.2 Enthalpy and Entropy Changes on Mixing

The enthalpy change on mixing, ΔH_M, is the difference in energy between the mixture E_{012} and the pure components E_{01} and E_{02} and is given by

$$\Delta H_M = E_{012} - (E_{01} + E_{02}) \tag{8.19}$$

Equation (8.19) can be rewritten using the above equations in the form[3]

$$\Delta H_M = (m_1 V^*_{sp1} + m_2 V^*_{sp2}) \left(\frac{\phi_1 P_1^*}{\tilde{V}_1} + \frac{\phi_2 P_2^*}{\tilde{V}_2} - \frac{P^*}{\tilde{V}}\right) \tag{8.20}$$

The enthalpy interaction parameter χ_H is obtained from the partial molar heat of mixing. For component 1 this has the form

$$\Delta \tilde{H}_1 = \tilde{H}_1 - \tilde{H}_0 = \left(\frac{\partial \Delta H_M}{\partial N_1}\right)_{T,P,N_2}$$
$$= \left(\frac{\partial \Delta H_M}{\partial N_1}\right)_{N_2,T,V} + \left(\frac{\partial \Delta H_M}{\partial \tilde{V}}\right)_{N_2 T_1 N_1} \left(\frac{\partial \tilde{V}}{\partial N_1}\right)_{N_2,T,P} \quad (8.21)$$

and is obtained using the usual thermodynamic arguments for the expansion of standard functions. Substitution of the above equations yields

$$\Delta \tilde{H}_1 = P_1^* V_1^* (\tilde{V}_1^{-1} - \tilde{V}^{-1}) + \left(\frac{\alpha T}{\tilde{V}}\right)\left(\frac{\tilde{T}_1 - \tilde{T}}{\tilde{T}}\right) + \frac{(V_1^* X_{12})}{\tilde{V}}(1 + \alpha T)\theta_2^2$$
$$= RT\chi_H \phi_2^2 \quad (8.22)$$

Similarly the entropic interaction parameter, χ_S, is obtained from the partial molar excess entropy changes on mixing and is given by

$$T\Delta \tilde{S}(\text{excess}) = -P_1^* V_1^* \left(\frac{3\tilde{T}_1 \ln(\tilde{V}_1^{1/3} - 1)}{(\tilde{V}^{1/3} - 1)} - \frac{\alpha T(\tilde{T}_1 - \tilde{T})}{\tilde{T}\tilde{V}}\right)$$
$$+ \frac{V_1^* \theta_2^2 (\alpha T X_{12} + T\tilde{V} Q_{12})}{\tilde{V}} = RT\chi_S \theta_2^2 \quad (8.23)$$

Combining eqn (8.22) and (8.23) gives the excess chemical potential of component 1 in the mixture:

$$\Delta G_1(\text{excess}) = \Delta H_1 - T\Delta S_1(\text{excess}) = \Delta \mu_1(\text{excess})$$
$$= P_1^* V_1^* \left(\frac{3\tilde{T}_1 \ln(\tilde{V}^{1/3} - 1)}{(\tilde{V}^{1/3} - 1)} + (\tilde{V}_1^{-1} - \tilde{V}^{-1})\right) + V_1^* \theta_2^2 \frac{(X_{12} - T\tilde{V} Q_{12})}{\tilde{V}}$$
$$= RT\chi_e \theta_2^2$$
$$(8.24)$$

Therefore we can deduce that

$$\chi_e = \chi_H + \chi_S \quad (8.25)$$

For a polymer system such as ethylene–vinyl acetate with 45% acetate mixed with chlorinated polyethylene with a 52% chlorine[1] content, the value of X_{12} would be -4.9 J cm^{-3} and Q_{12} has a value of -0.0108 J cm^{-3} °C^{-1}. Since χ_H is negative and χ_S is positive then χ_e is small and negative, which indicates that the mixture will be homogeneous. Experimentally it is observed that this mixture at 83.5 °C is homogeneous and becomes heterogeneous at 90 °C.[3]

8.3 Phase Separation Phenomena

Mixing two polymer systems will create a range of new interactions. In general, specific interactions are exothermic, causing negative changes in heat of mixing which in turn favours miscibility but involves a penalty in entropy. The entropy change on mixing consists of combinatorial and residual parts. The latter arise because of the specific interactions whereas the former are due to statistical mixing of the two polymers involved. The change in free energy of the system is therefore expressed as

$$\Delta G_M = \Delta H_M - T(\Delta S_C + \Delta S_R) \tag{8.26}$$

A negative free energy change is necessary but not a sufficient condition for homogeneity between two polymers. More appropriately the shape of ΔG as a function of concentration of one of its constituents at a temperature T describes homogeneity or heterogeneity of the mixture as discussed in Section 8.2.

8.3.1 The Phase Diagram for Nearly Miscible Blends

Studies of polymer solutions and binary mixtures indicate that there exist two critical temperatures: the upper critical solution temperature (UCST) and the lower critical solution temperature (LCST). The LCST is the lowest temperature for which two phases can be observed and the USCT is the highest temperature at which two phases can be observed for a mixture. The bimodal condition is described by

$$\Delta \mu_i = \Delta \mu_i(\text{combinational}) + \Delta \mu_i(\text{excess}) \tag{8.27}$$

or

$$\begin{aligned}\Delta \mu_i =& RT\left[\ln \phi_1 + \left(1 - \frac{r_1}{r_2}\right)\phi_2\right] \\ &+ P_1^* V_1^* \left[\frac{3\tilde{T}_1 \ln\left(\tilde{V}_1^{1/3} - 1\right)}{(\tilde{V}^{1/3} - 1)}(\tilde{V}_1^{-1} - \tilde{V}^{-1}) + P_1(\tilde{V} - \tilde{V}_1)\right] \\ &+ \frac{V_1^* \theta_2^2 (X_{12} - T\tilde{V} Q_{12})}{\tilde{V}}\end{aligned} \tag{8.28}$$

and r_1 and r_2 are the lattice sites occupied by the components 1 and 2. Similarly the chemical potential of component 2 is

$$\Delta\mu_2 = RT\left[\ln\phi_2 + \left(1 - \frac{r_1}{r_2}\right)\phi_1\right]$$
$$+ P_2^* V_2^* \left[\frac{3\tilde{T}_2 \ln\left(\tilde{V}_2^{1/3} - 1\right)}{\left(\tilde{V}^{1/3} - 1\right)}\left(\tilde{V}_2^{-1} - \tilde{V}^{-1}\right)\tilde{P}_2\left(\tilde{V} - \tilde{V}_2\right)\right] \quad (8.29)$$
$$+ \frac{V_2^* \theta_1^2 (X_{12} - T\tilde{V}Q_{12})S_2}{S_1 \tilde{V}}$$

The spinodal is obtained by applying the stability condition. Note that two terms have been added to the equations that introduce the pressure dependence of the functions:

$$\frac{\partial^2 \Delta G_M}{\partial \phi_2} = \frac{\partial \Delta \mu_i}{\partial \phi_2} = 0 \quad (8.30)$$

Differentiating eqn (8.28) with respect to ϕ_2 gives the spinodal condition:

$$-\frac{1}{\phi_1} + \left(1 - \frac{r_1}{r_2}\right) - \frac{P_1^* V_1^*}{RT_1^*} \frac{A}{(\tilde{V} - \tilde{V}^{2/3})} + \frac{P_1^* V_1^*}{RT}\left(\frac{1}{\tilde{V}^2} + P_1\right)A$$
$$+ \frac{V_1^* X_{12}}{RT} \frac{2\theta_2}{\tilde{V}} \frac{\theta_1 \theta_2}{\phi_1 \phi_2} - \frac{V_1^* X_{12}}{RT} \frac{\theta_2^2}{\tilde{V}^2} A \quad (8.31)$$
$$- \frac{V_1^* Q_{12}}{R}\left(\frac{2\theta_2}{\tilde{V}} \frac{\theta_1 \theta_2}{\phi_1 \phi_2}\right) = 0$$

where

$$A = \frac{\partial \tilde{V}}{\partial \phi_2} = -\frac{\partial \tilde{V}}{\partial \phi_1} \quad \text{or} \quad \frac{\partial \tilde{V}}{\partial \phi_2} = \frac{B - C\left(\frac{\tilde{P}}{\tilde{T}} + \frac{1}{\tilde{T}\tilde{V}^2}\right)}{\frac{2}{\tilde{V}^3} - \frac{\tilde{T}}{3\tilde{V}^{5/3}}\frac{(3\tilde{V}^{1/3} - 2)}{(\tilde{V}^{1/3} - 1)^2}} \quad (8.32)$$

and

$$B = \frac{\partial \tilde{P}}{\partial \phi_2} = \frac{\tilde{P}}{P^*}\left[P_1^* - P_2^* - \theta_2 X_{12}\left(1 - \frac{\theta_1}{\phi_2}\right)\right] \quad (8.33)$$

$$C = \frac{\partial \tilde{T}}{\partial \phi_2} = \frac{\tilde{T}}{\tilde{P}} B + \frac{T}{P^*}\left(\frac{P_2^*}{T_2^*} - \frac{P_1^*}{T_1^*}\right) \quad (8.34)$$

These equations can be used to simulate the behaviour of various polymer mixtures.[5] For the simulations the contact entropy Q_{12} is the only adjustable parameter. X_{12} and Q_{12} are empirically interrelated:

$$Q_{12} \approx 2 \times 10^{-3} \times X_{12} \; (\text{K}^{-1}) \quad (8.35)$$

8.4 Parameters Influencing Miscibility

As can be seen from the above equation, X_{12} and Q_{12} are the parameters that control miscibility. In the absence of strong interactions such as hydrogen bonding, dipole-induced dipole or strong dipolar interactions, physical parameters or the equations of state parameters of the homopolymer become important. Clearly the equations of state parameters are also determined by polymer–polymer interactions but now the important factor is the relative balance between these interactions. Detailed theories have been produced which look in more detail at the effects of the various interactions.[6]

8.4.1 Molar Mass Dependence of Phase Diagrams[1]

One of the most important variables in a polymer system is the molar mass of the components. Figure 8.2 illustrates the phase diagrams for a variety of systems.

For strictly binary systems, appropriate differentiation of the free enthalpy of mixing leads to

$$\frac{\Delta\mu_1/RT}{m_1} = \frac{\ln\phi_1}{m_1} + \frac{1}{m_1} - \frac{\phi_1}{m_1} - \frac{\phi_2}{m_2} + \left(X_{12} - \phi_1\frac{\partial X_{12}}{\partial \phi_2}\right)\phi_2^2 \qquad (8.36)$$

$$\frac{\Delta\mu_2/RT}{m_2} = \frac{\ln\phi_2}{m_2} + \frac{1}{m_2} - \frac{\phi_1}{m_1} - \frac{\phi_2}{m_2} + \left(X_{12} - \phi_2\frac{\partial X_{12}}{\partial \phi_1}\right)\phi_1^2 \qquad (8.37)$$

For such systems the spinodal and critical conditions yield

$$\text{Spinodal}: \frac{1}{m_1\phi_1} + \frac{1}{m_2\phi_2} - 2X_{12} + 2(1-2\varphi_2)\frac{\partial X_{12}}{\partial \phi_2} + \phi_2(1-\phi_2)\frac{\partial^2 X_{12}}{\partial \phi_2^2} = 0 \qquad (8.38)$$

$$\text{Critical point}: \frac{1}{m_1\phi_1^2} - \frac{1}{m_2\phi_2^2} - \frac{6\partial X_{12}}{\partial \phi_2} + 3(1-2\phi_2)\frac{\partial^2 X_{12}}{\partial \phi_2^2} + \phi_2(1-\phi_2)\frac{\partial^2 X_{12}}{\partial \phi_2^3} = 0 \qquad (8.39)$$

The last three terms in eqn (8.38) reduce to $-2X_{12}$ and in eqn (8.39) to zero. If X_{12} does not depend on concentration then the critical points can be identified:

$$\phi_{2c} = \frac{m_1^{1/2}}{m_1^2 + m_2^2} \qquad (8.40)$$

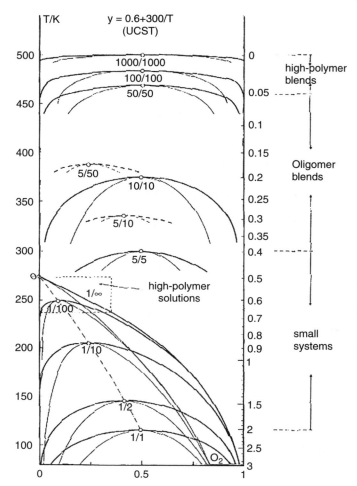

Figure 8.2 Binodals (heavy curves) as described by eqn (8.36) and (8.37) and spinodals (light curves) according to eqn (8.38) showing the effect of molar mass size (m_1/m_2). Open circles are critical points specified by eqn (8.40) and (8.41) and connected by the dash-dot curve.[1]

and

$$X_{12} = \frac{1}{2}(m_1^{1/2} + m_2^{1/2})^2 \qquad (8.41)$$

For small molecules $m_1 = m_2 = 1$ but for high polymer solutions $m_1 = 1$, $m_2 > 1$, the critical value of X_{12} approaches $\frac{1}{2}$ for very large m_2 and the critical volume fraction of polymer becomes very small. When both m_1 and m_2 are very large X_{12c} is very small.

In general, as the molecular mass of the polymer is increased the minimum temperature at which phase separation starts is reduced. For an

80-fold increase in the molar mass the cloud point temperature reduces by about 60 °C.

8.4.2 Effect of Pressure on Miscibility

Mixtures with UCST behaviour usually become less miscible on applying pressure. Typically the pressure dependence is $\partial T/\partial P$ and is of the order of 0.01–0.05 °C atm^{-1}.

8.4.3 Addition of Block Copolymers

The addition of block copolymers to a binary polymer mixture can aid compatibilization of the components of the blend by reduction of the interfacial energy between the two phase separated domains. Copolymers can depress the build-up of the interfacial energy between the phases and/or make a favourable contribution to the entropy changes on mixing at the homogeneous/heterogeneous phase separation temperature with the result that phase separation is delayed.

8.4.4 Refinements of Theory

The theory presented above does not allow for lattice compressibility. This has been incorporated in the theory by modification of the interaction parameter X_{12} according to the theory of Sanchez and Lacombe:[7]

$$\chi_1 = R\rho_1 \left\{ X_{12} + \frac{1}{2}\psi^2 \tilde{T}_1 P_1^* \beta_1 \right\} \tag{8.42}$$

where ψ is a dimensionless function and β_1 is the isothermal compressibility of component 1. The first term is the energetic contribution and the second one is an entropic contribution. According to this theory the following inequality must hold:

$$\frac{\partial(\Delta\mu_1/kT)}{\partial\phi_1} = \frac{1}{2}\left(\frac{1}{r_1\phi_1} + \frac{1}{r_2\phi_2}\right) - \tilde{\rho}\left(X_{12} + \frac{1}{2}\psi^2 \tilde{T} P^* \beta\right) \succ 0 \tag{8.43}$$

where the first term is the combinatorial contribution, $\tilde{\rho}X_{12}$ is an energetic contribution and $\frac{1}{2}\tilde{\rho}\psi^2\tilde{T}P^*\beta$ is an entropic contribution from the equation of state. The combinatorial entropy makes a larger contribution on both sides of the spinodal curve. To relax this effect a correction entropy factor has been incorporated in the theory.

8.5 Kinetics of Phase Separation: The Spinodal Decomposition

The spinodal is the limit of metastability beyond which a homogeneous phase can no longer exist. Spinodal decomposition that differs from nucleation and growth is the mechanism by which a homogeneous blend starts to phase separate at the temperatures inside the spinodal region. The process is essentially the growth in the variation of concentration with time, a diffusion process driven by the free energy. The flux of material is against the concentration gradient resulting in a negative diffusion coefficient. In addition, impurities play a small or zero role in this mechanism in contrast to nucleation and growth.

At the early stage of spindonal decomposition infinitesimal fluctuations in the concentration start to grow, the phase sizes are very small and co-continuous/interconnected with decomposition occurring spontaneously.

A theoretical model based on a diffusion equation was developed by Cahn and Hilliard[8] and introduces a diffusion equation that relates the interdiffusional flux J of the two phases ($\tilde{J} = \tilde{J}_1 = -\tilde{J}_2$) to the gradient of chemical potential differences:

$$-\tilde{J} = M\Delta(\mu_1 - \mu_2) \tag{8.44}$$

where M is the diffusion mobility and is simply the ratio of diffusional flux to the local chemical potential. Thermodynamic considerations show that M is always positive.

The change in chemical potential for a heterogeneous mixture at the early stage of phase separation is given by

$$\mu_1 - \mu_2 = \frac{\partial \Delta G}{\partial \phi} - 3K\nabla^2 \phi \tag{8.45}$$

where K is the gradient energy coefficient and is usually determined experimentally. However, it can be estimated independently from the dimensions of the polymers:

$$K = \frac{1}{6}\left[\frac{M_1}{M_{w1}\phi_1} R_{g1}^2 + \frac{M_2}{M_{w2}\phi_2} R_{g2}^2\right] \tag{8.46}$$

where M is the monomer mass, M_w is the molar mass of the polymer and R_g is the radius of gyration of the corresponding components. Substituting these results in eqn (8.44) gives

$$-\tilde{J} = M\frac{\partial^2 \Delta G}{\partial \phi^2}\nabla\phi - 2MK\nabla^2 \phi \tag{8.47}$$

By taking the divergence of the above equation the general Cahn–Hilliard diffusion equation is obtained and this forms the basis of the interpretation of

Polymer Blends and Phase Separation

the spinodal decomposition phenomena:

$$\frac{\partial \phi}{\partial t} = M \frac{\partial^2 \Delta G}{\partial \phi^2} \nabla^2 \phi - 2MK\nabla^4 \phi + \text{nonlinear terms} \quad (8.48)$$

The solution to eqn (8.47) is usually written in terms of the growth in amplitude of the concentration fluctuation with growth rate of $R(q)$ the scattering parameter which reflects phase separation:

$$R(q) = -M \frac{\partial^2 \Delta G}{\partial \phi^2} q^2 - 2MKq^4 \quad (8.49)$$

where q is the wavenumber, $q = (4\pi/\lambda) \sin(\theta/2)$, λ is the wavelength and θ is the scattering angle. From this equation it is apparent that the sign of $R(q)$ is governed by $\partial^2 \Delta G/\partial \phi^2$, since M, K and q are positive quantities. Thus $R(q)$ is negative in the homogeneous and metastable regions and only becomes positive within the spinodal phase region. Hence the phase separation should proceed spontaneously within the spinodal region at a scale governed by the values of q that yield a positive growth rate. The maximum growth rate appears at

$$q_m^2 = -\frac{1}{4K} \frac{\partial^2 \Delta G}{\partial \phi^2} \quad (8.50)$$

which corresponds to the most rapidly growing wavelength of

$$\lambda_m = \frac{2\pi}{q_m} \quad (8.51)$$

The scale of phase separation for polymer blends at the early stage of spinodal decomposition is very small and is of the order of a few hundreds or thousands of angstroms and is typical of the morphology found in many blended systems.

8.6 Specific Examples of Phase-Separated Systems

In the above discussion the theory has considered primarily the case of the mixing of two polymeric species. In Figure 8.1 it can be seen that similar behaviour is observed for a polymer dispersed in a solvent. Several technologically interesting examples of phase separation demonstrate the effect of transition from low molar mass to high molar mass on the phase behaviour of the mixture.

8.6.1 High-Impact Polystyrene[9]

Polystyrene is extensively used in a range of applications, such as car headlamp lenses, where its mechanical properties are important. Polystyrene has a T_g value of 100 °C and can withstand a reasonable impact. However, improvement of its impact properties is commercially desirable and requires incorporation of a

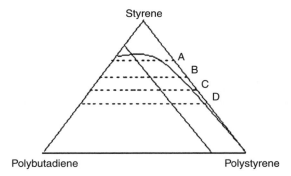

Figure 8.3 Ternary phase diagram for mixtures of styrene–polybutadiene–polystyrene.

mechanism for energy dissipation below T_g. High-impact polystyrene (HIPS) is a dispersion of polybutadiene within a matrix of polystyrene. The tertiary phase diagram for styrene–polybutadiene–polystyrene is shown in Figure 8.3. Polybutadiene will dissolve in styrene monomer to form a homogeneous single phase.

Initially droplets of a polystyrene-rich phase are formed within a polybutadiene-rich phase, designated in Figure 8.3 as tie line A. As the polymerization proceeds the ratio of the two phases will change until between B and C the volume fraction of the polystyrene-rich phase exceeds the volume fraction of the polybutadiene-rich phase and the mixture will phase invert so that the polystyrene phase becomes continuous (Figure 8.3).

The actual points of phase inversion depend on the viscosities of the phases and hence the molar mass of the polymer and on the rate of stirring within the reactor. The resultant morphology of a commercial material is shown in Figure 8.4. The polystyrene phases are trapped within the butadiene phase during the phase inversion and the complicated morphology formed is responsible for many of the advantageous physical properties of these blends.

8.6.2 Rubber Toughened Epoxy Resins[10]

Amine-cured epoxy resins are extensively used as structural matrices in composite manufacture and as such the improvement of their mechanical properties is very desirable. As in the case of polystyrene the addition of a rubber phase provides an energy dissipation mechanism below T_g and this increases the toughness of the overall material. Carboxy-terminated butadiene acrylonitrile (CTBN) is initially soluble in the reaction mixture of digylcidyl ether of bisphenol A (DGEBA) and the low-temperature cure system triethylenetetra-amine (TETA):

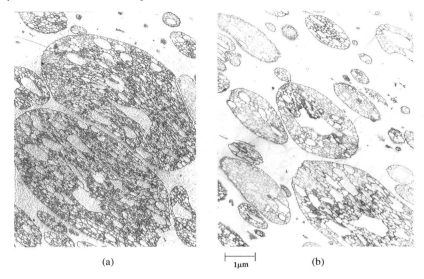

Figure 8.4 Transmission electron micrographs of HIPS showing the styrene dispersed within the butadiene phase that is in turn dispersed within the styrene phase. The isoprene phase is stained black.

Carboxy-terminated butadiene acrylonitrile random copolymer (CTBN)

As in the case of the polystyrene system, increasing the molar mass of the epoxy resin polymer dispersed in the CTBN-rich phase will lead to phase separation.

Two chemically dissimilar polymers will naturally attempt to phase separate and the structure that is formed will reflect the way in which this process occurs and the driving forces associated with the process. Phase separation is used to achieve rubber toughening in thermoset resin systems. Low molar mass CTBN copolymer is soluble in the simple mixtures of monomers used to create amine-cured epoxy resins systems. However, as the molecular mass of the epoxy resin increases so the balance of entropy and enthalpy of mixing of these components changes and a driving force for phase separation is created.

Since the process of phase separation is both a kinetic and thermodynamic driven process the features that are created depend on the temperature and speed of the cure process. Rubber toughening is improved if the particles of the CTBN are small and evenly dispersed. The size of the particles formed depends on the temperature of the cure and the rate at which it occurs (Figure 8.5). In this example, contrast can be imparted to the sample by staining with osmium tetroxide. The osmium tetroxide can add to the double bonds of the butadiene. From the point of view of electron microscopy these regions have now a much

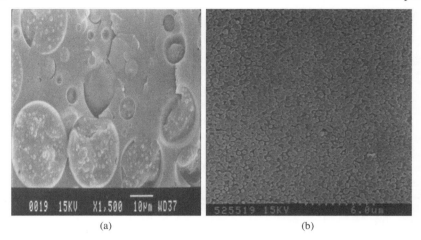

Figure 8.5 Electron micrographs of CTBN particles dispersed in an amine-cured epoxy resin. The morphology changes with the concentration of CTBN: (a) 3% CTBN; (b) 8% CTBN.

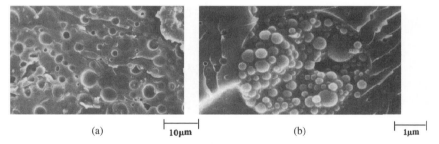

Figure 8.6 Morphology of TETA–DGEBA–CTBN systems: (a) low-resolution image showing the distribution of CTBN phases in the matrix; (b) enlargement of one of the CTBN phases showing the included epoxy nodules within the CTBN phase.

greater electron scattering efficiency than they had previously and contrast with the surrounding unstained areas of the epoxy resin.

Further increase in the concentration of the epoxy resin polymer will lead to a phase inversion within the CTBN phase. The nodules in the scanning electron microscopy images are of the phase separated epoxy phase (Figure 8.6). This behaviour has striking similarities to the HIPS system and the phase diagrams for this relatively polar system are strikingly similar to those of HIPS.

8.6.3 Thermoplastic Toughened Epoxy Resins

For structural application epoxy resins are cured at high temperatures and use an aromatic amine hardener rather than the aliphatic compounds indicated

Polymer Blends and Phase Separation 223

above. A typical aerospace resin uses 4,4′-diaminodiphenylsulfone (DDS) together with DGEBA and a thermoplastic, poly(ether sulfone) (PES), as the toughening agent:[11]

4,4'-diaminodiphenylsulfone (DDS)   poly(ether sulfone) (PES)

As with the other examples discussed above, the PES is initially soluble in the mixture of DDS and DGEBA forming a homogeneous solution. As the cure process proceeds phase separation occurs and, depending on the concentration of the PES, the dominant morphology changes.[12] Between 0 and 2.5% of PES a homogeneous material is obtained. Above this concentration a particulate morphology is observed, with PES being the dominant component of the particulate phase. At approximately 22.5% the morphology is of a co-continuous structure and this is observed up to about 30%. Above 30% of PES in the system phase inversion occurs and the continuous matrix is now PES. These changes are illustrated in the electron micrographs in Figure 8.7.

In all the above examples the mechanical property enhancement is a consequence of spinodal decomposition of an initially homogeneous mixture to form a co-continuous phase separated structure.

8.6.4 Epoxy Resins

The synthesis of epoxy resins involves an addition process and usually involves the reaction of a primary aromatic or aliphatic amine with a di- or higher function epoxy compound. The initial reaction will create a linear molecule that contains secondary amine functions. These secondary amine functions are less reactive than the primary amines and the result of the initial reaction is to grow

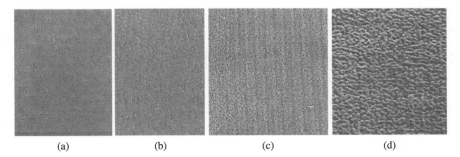

Figure 8.7 Electron micrographs showing the effects of an increase in PES concentration on the phase structure of an amine-cured epoxy resin: (a) 10% PES, (b) 20% PES, (c) 30% PES and (d) 35% PES.

linear polymers. As the reaction proceeds, the possibility of the creation of branched chain structures and ultimately cyclic ring structures increases:

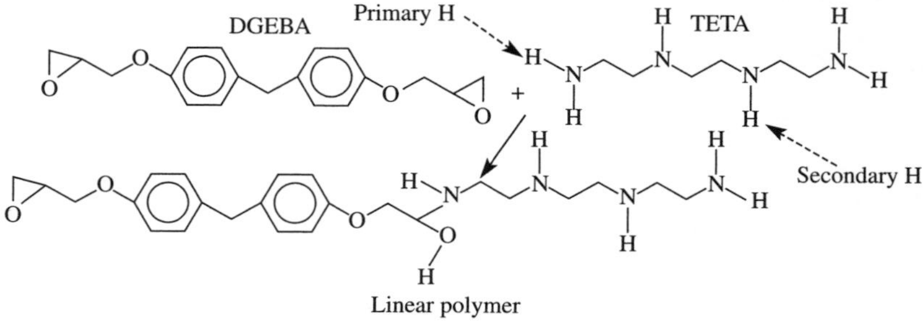

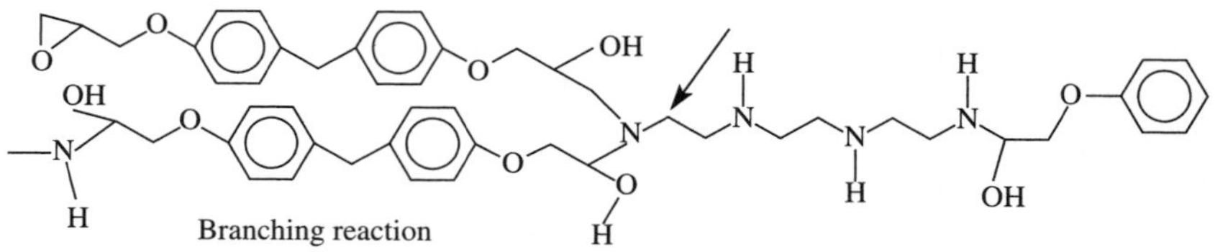

[Chemical structure diagrams showing DGEBA + TETA reaction forming Linear polymer, and Branching reaction]

The final thermoset system is ideally a three-dimensional linked cyclic structure which extends through all space. However, in general, this process is never completed because not all the reactive functions can be consumed. In certain areas the formation of cyclic structures can be completed at an early stage in the reaction and these entities are now thermodynamically less compatible than the linear polymers with the monomers. As a consequence these crosslinked entities have higher values of T_g than the less highly crosslinked matrix in which they are dispersed. There is some evidence that these entities can phase separate during the course of the reaction and will influence the physical properties of the matrix that is formed. For instance when water enters the epoxy resin matrix it can plasticize the more open regions lowering T_g by about 10 °C for every 1% of water absorbed; however, these more highly crosslinked regions are less susceptible to moisture uptake and their T_g is often hardly depressed by exposure to water.

8.7 Block Copolymers: Polystyrene-*block*-Polybutadiene-*block*-Polystyrene (SBS) Block Copolymer

In the search for new materials, the idea of block copolymers emerged. Block copolymers are materials produced by the careful control of the synthesis a polymer and contain regular sequences of more than one type of monomer. In 1965, Shell brought to the market a number of polystyrene-*block*-polybutadiene-*block*-polystyrene and polystyrene-*block*-polyisoprene-*block*-polystyrene

Polymer Blends and Phase Separation

materials. These materials show high strength and have a rubber elasticity that is comparable to that of rubber. This system has been extensively studied and is reviewed by Price and co-workers.[13] The behaviour reported for this system is typical of that found for all block copolymers, but can depend on the thermal history of the sample.

8.7.1 General Characteristics

The two polymers, polystyrene and polybutadiene, are immiscible (Figure 8.8). This has already been illustrated in the case of HIPS. Thermodynamically the blocks of the polymer will be incompatible and will attempt to phase separate.

Electron microscopy of stained samples of SBS copolymers (Figure 8.10) has shown that phase separation occurs in a controlled manner and produces highly regular structures in the solid state. Depending on the length of the block, typical values of m_1 and m_2 being of the order of 2000–3000 with n having a value of the order of 1000, different physical characteristics are observed. Polymers have also been produced in which the ratio of m_1 to m_2 is ~ 1 but varied in relation to n give differences in the morphology. Figure 8.9 represents these differences schematically for an AB or an ABA block copolymer system. As the concentration of the component A in the AB block is varied so the morphology changes.

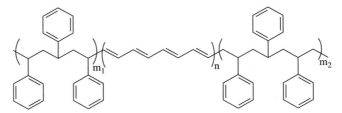

Figure 8.8 The idealized structure of the polystyrene-*block*-polybutadiene-*block*-polystyrene block copolymer; m_1 and m_2 denote the number of monomers in the styrene block and n denotes the number of monomers in the butadiene block.

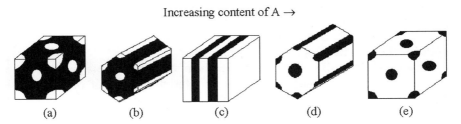

Figure 8.9 Effect of composition on block copolymer morphology: (a) spheres of A in matrix of B; (b) cylinders of A in matrix of B; (c) alternating A and B lamellae; (d) cylinders of B in matrix of A; (e) spheres of B in matrix of A.[13]

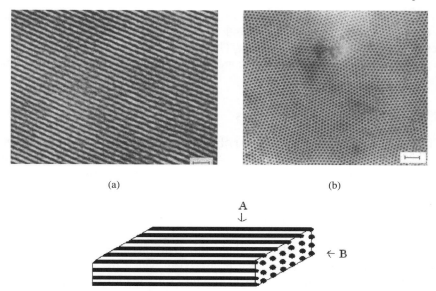

Figure 8.10 Osmium tetroxide stained electron micrographs of polystyrene-*block*-polybutadiene-*block*-polystyrene block copolymer produced by extrusion. (a) Micrograph demonstrating the orientation of the styrene cylinders in the direction of the extrusion (direction A in the schematic) and (b) shows a close-packed structure transverse to the extrusion direction (direction B in the schematic). Scale bar = 1 μm.

In practice, it is found that the degree of regularity of the domain structure depends very much on the physical treatment to which a sample has been subjected (Figure 8.10). Imposition of an external stress, as in the case of extrusion of the polymer through a die, will induce alignment of the phase-separating morphology and create the type of structures shown in Figure 8.10. The size of the features observed in the electron micrographs is dictated by the molar mass of the blocks in the polymer chain. The radius of the cylinders is determined by the radius of gyration of the blocks in the copolymer. The perfection of the morphological structure will depend on a combination of thermodynamic and kinetic factors. The thermodynamics will drive the system to phase separation and the kinetics will control the rate at which the phase separation is achieved. Similar structures are observed in diblock copolymers. Hence SBS diblock copolymer will exhibit a similar morphology but its mechanical characteristics are very different.

In the case of the SBS block copolymer the matrix structure is pinned by the styrene domains and whilst mobility is introduced once T_g of the butadiene phase has been exceeded, the material will not flow until the temperature has been raised above T_g of polystyrene. In contrast, whilst the low-temperature mechanical characteristics are similar with the material exhibiting elastic properties once T_g has been reached, the subsequent characteristics are less predictable since flow is possible below T_g of the polystyrene phase that is now unable to

Polymer Blends and Phase Separation 227

anchor the material. Clearly there is an interface between the domains and there has been much discussion as to the thickness of the interfacial regions.

The most important factor deciding the type of equilibrium domain structure adopted by a system is the weight fraction of the components. The molar mass appears to play a secondary role in this respect, but strongly influences the location of the boundary defining the transition from homogeneous to microphase separated states. For SBS the body-centred cubic (bcc) polystyrene spheres are stable up to approximately 20 wt%, hexagonally packed polystyrene cylinders between approximately 20 and 40%, lamellae from 40 to 60%, polybutadeine cylinders from 64 to 82% and bcc polybutadiene spheres above 84%. Between 82 and 84% an orthorhombic structure has been reported which may be a metastable form. The phase diagram does, however, show some molar mass dependence, as shown in Figure 8.11.

The microdomain size D_i and lattice repeat distance d of periodically organized microdomains have the following approximate molecular mass dependence:

$$D_i \sim (M_i)_n^{2/3}; \quad d \sim (M)_n^{2/3} \tag{8.52}$$

where $(M_i)_n$ and $(M)_n$ are the number average molar masses of the blocks forming the domains and the overall number average molar mass of the copolymers, respectively. Because of the way block copolymers are prepared their molar mass distribution is often very narrow, typically 1.01.

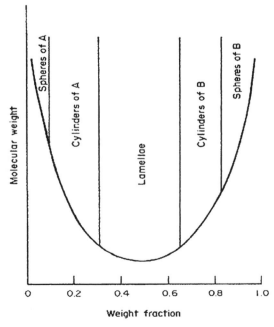

Figure 8.11 Phase diagram for the geometry, stability and microdomains of an AB-type block copolymer.[13]

8.7.2 Thickness of the Domain Interface

The thickness of the interfacial region is donated by δ and the phase is assumed to be strongly segregated so that $\delta \ll D_A$, D_B (Figure 8.12). Away from these interfacial regions the domains consist either of pure A or pure B at the density of the homopolymer melt, $\rho_{0,A}$ and $\rho_{0,B}$ respectively.

The enthalpy of mixing ΔH_{mix} may be estimated using

$$\Delta H_{mix} = \gamma \xi - V\phi_A \phi_B \chi_{AB} kT \qquad (8.53)$$

where ξ is the total surface area of the domains, γ is the surface tension between the phases, V is the total volume of the system, ϕ_A and ϕ_B are the volume fractions of A and B, respectively, and χ_{AB} is the Flory–Huggins interaction parameter for the system. The second term, which really drives the transition so that unfavourable high-energy A–B contacts are reduced, gives the enthalpy change on going from a randomly mixed state to the domains assuming no volume change, whilst the first term is the surface term. From space filling considerations we have

$$\xi D_A \rho_{0,A} = \xi D_B \rho_{0,B} = N_P \qquad (8.54)$$

where N_P is the total number of copolymer chains. Hence

$$\Delta H_{mix} = \gamma N_P (D_A \rho_{0,A})^{-1} - V\phi_A \phi_B \chi_{AB} kT \qquad (8.55)$$

and γ depends on the interactions between the blocks in the interfacial region and may be estimated for example in terms of χ_{AB} and δ. The entropy of mixing ΔS_j arises from the entropy change in placing the A–B junctions within an interfacial volume (V_j) and is estimated as

$$\Delta S_j = N_P k \ln\left(\frac{V_j}{V}\right) \approx N_P k \ln\left(\frac{\delta_{0,A}}{[D_A(\rho_{0,A} + \rho_{0,B})]}\right) \qquad (8.56)$$

Finally, $\Delta S_{conformation}$ arises from the constraints that the blocks of the copolymer are restricted to domains. Because chains in the polymer melt behave as if they were ideal, *i.e.* as if there were no excluded volume effects, then it is frequently argued that Gaussian statistics should be applied to these domains of pure A and B and provided the inequality $L_k N_k^{1/2} < D_k < L_k N_k$ holds good

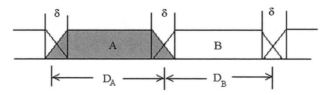

Figure 8.12 Density distribution of A and B segments in a block copolymer exhibiting microphase separation.

Polymer Blends and Phase Separation

(k represents A or B) this leads to the result

$$\Delta S_{\text{conformation}} = k_A D_A^2 (N_A L_A^2)^{-1} + k_B D_B^2 (N_B L_B^2)^{-1} \qquad (8.57)$$

where k_A and k_B are constants which may be estimated from the solution of the equation of motion for a random flight chain contained within the domain. If the above results are combined together the scaling law is obtained:

$$D_A \sim (N_A + N_B)^{2/3} \qquad (8.58)$$

Meier[16] has estimated the thickness of the boundary layer to be given by

$$\gamma = \frac{kT\chi_{AB}\delta}{2} \qquad (8.59)$$

There has been considerable interest in the estimation of the interfacial thickness but experiment has shown it to be of the order of a few monomer units.

8.8 Polyurethanes[14,15]

Polyurethanes are a very important class of materials that are readily synthesized by the reaction of an isocyanate with either a hydroxyl-containing polyester or polyether to form a polyurethane or reaction with an amine to form a urea. The urethane containing one NH and a carbonyl per bond has the capability of forming stable hydrogen bonds and in a typical chain-extended polyurethane phase separation is promoted to give a structure shown schematically in Figure 8.13. The 'rooftop' structure of the MDI promotes favourable packing and this is further stabilized by the π–π interactions between neighbouring phenyl groups and the favourable alignment of the urethane groups to achieve hydrogen bonding. This 'hard' block structure has a melting point of approximately 158 °C.

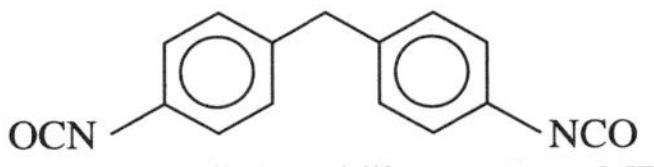

Methylenediphenyldiisocyanate - MDI

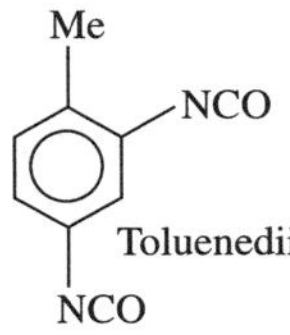

Toluenediisocyanate-TDI

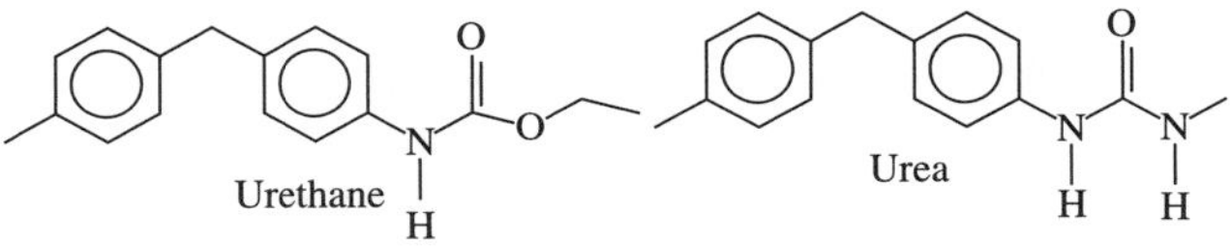

Urethane

Urea

The urethane linkages can also hydrogen bond to the polyester and hence, unlike the SBS block copolymers, the interface between the hard block and the 'soft' predominantly ester or ether phase is diffuse rather than being sharp. The

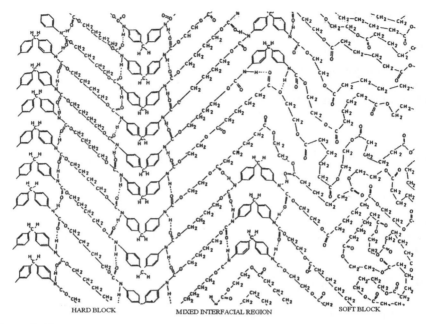

Figure 8.13 Schematic of the structure of a polyurethane.

ester or ether phases have values of T_g which are in the range $-20\,°\text{C}$ to $20\,°\text{C}$ depending on molar mass. This 'soft' phase imparts flexibility to the material and allows it to develop rubbery characteristics. Polyurethanes are extensively used in footwear and other applications where flexibility combined with structural integrity is required.

Evidence for phase separation can be obtained from a variety of different methods including dynamic mechanical thermal analysis; however, microscopy does not give clear images in the way that is observed for SBS.

There is current interest in the use of block copolymers to help create structures that have potential sensor applications,[17] e.g. block copolymers of polystyrene–poly(methyl methacrylate) (PS-b-PMMA). The nature of the organization that is created in thin films is influenced by the factors influencing phase separation of the polymers and very importantly the surface energy of the substrate on which they are deposited (Figure 8.14). If a substrate is patterned and then certain areas chemically modified, a substrate is created with variation in the surface energy across the surface. This is discussed in more detail in Chapter 9. The differences in surface energy will influence the morphology created.

On the left of Figure 8.14b, the surface is chemically homogeneous and the lamellae form a fingerprint morphology that lacks long-range order. On the right the surface is chemically stripped with a periodicity L_s that matches the block copolymer periodicity L_o and induces the lamellae to form perfectly ordered structures over large areas, similar to that observed for SBS

Polymer Blends and Phase Separation

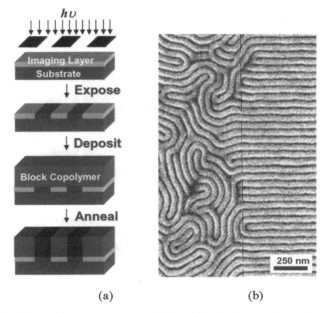

(a)   (b)

Figure 8.14   The substrate is patterned before the block copolymer is deposited (a). The resulting structure is shown in (b) and illustrates the effects of differences in surface energy on the morphology created. The dashed line indicates the boundary between the modified and unmodified areas.[17]

copolymers discussed above. It is possible using these patterns to deposit various inorganic materials and create functional ordered structures which can be used to fabricate sensors. The creation of these devices is only possible because of the principles of thermodynamics which govern the formation of these morphologies.

Recommended Reading

R.A. Brown, A.J. Masters, C. Price and X.F. Yuan, in *Comprehensive Polymer Science*, ed. G. Allan, Pergamon, 1989, vol. 2, p. 155.
S. Rostami, in *Multicomponent Polymer Systems*, ed. I.S. Miles and S. Rostami, Longman Scientific & Technical, Harlow, UK, 1992, p. 63.

References

1. R. Koningveld, W.H. Stockmayer and E. Neis, *Polymer Phase Diagrams: A Textbook*, Oxford University Press, 2001.
2. J.S. Rowlinson and F.L. Swinton, *Liquid and Liquid Mixtures*, Butterworths, 1982.

3. S. Rostami, in *Multicomponent Polymer Systems*, ed. I.S. Miles and S. Rostami, Longman Scientific & Technical, Harlow, UK, 1992, p. 63.
4. A. Bondi, *J. Phys. Chem.*, 1964, **68**, 441.
5. S. Rostami and D.J. Walsh, *Macromolecules*, 1984, **17**, 315.
6. S.I. Kuchanov and S.V. Panyukov, in *Comprehensive Polymer Science, Second Supplement*, ed. G. Allan, Pergamon, 1996, p. 441.
7. I.C. Sanchez and R.H. Lacombe, *Macromolecules*, 1978, **11**, 1145.
8. W.G. Cahn and J.E. Hilliard, *J. Chem. Phys.*, 1958, **28**, 258.
9. D.J. Walsh, in *Comprehensive Polymer Science*, ed. G. Allan, Pergamon, 1989, vol. 2, p. 135.
10. C. Delides, D. Hayward, R.A. Pethrick and A.S. Vatalis, *J. Appl. Polym. Sci.*, 1993, **47**, 2037.
11. A.J. MacKinnon, S.D. Jenkins, P.T. McGrail and R.A. Pethrick, *Macromolecules*, 1992, **25**, 3492.
12. A.J. MacKinnon, S.D. Jenkins, P.T. McGrail and R.A. Pethrick, *Polymer*, 1993, **34**, 3252.
13. R.A. Brown, A.J. Masters, C. Price and X.F. Yuan, in *Comprehensive Polymer Science*, ed. G. Allan, Pergamon, 1989, vol. 2, p. 155.
14. R.A. Pethrick and C. Delides, *Eur. Polym. J.*, 1981, **17**, 675.
15. R.A. Pethrick, C. Delides, A.V. Cunliffe and P.G. Klein, *Polymer*, 1981, **22**, 1205.
16. D.J. Meier, *J. Polym. Sci., Part C*, 1969, **26**, 81.
17. M.P. Stoykovich and P. F. Nealey, *Mater. Today*, 2006, **9**(9), 20–29.

CHAPTER 9
Molecular Surfaces

9.1 Introduction

In the case of a simple single crystal such as sodium chloride (Chapter 1), the surface can be pictured as being a plane of uniformly spaced atoms. This is a very simplistic view of a single crystal surface and in reality it will be decorated with defects. The morphology of a solid is dictated by a series of dynamic and thermodynamic factors and can be influenced by the specific interactions that are associated with the surface. Gibbs recognized that the free energy of the interface can be significantly different from that of the bulk. The energy of the interface reflects the imbalance of forces acting on the molecules located at the boundary. Technologically the nature of the surface can have profound effects on physical properties such as wear, adhesion and friction and subjective properties such as gloss and appearance. In order to understand the properties of the interface it is necessary to consider the balance of forces that exist in the region of the interface. There are three types of interface that need to be considered: solid–liquid, solid–air and solid–solid. The latter, solid–solid, was introduced when considering copolymer blends in Chapter 8. The design of the interface between a polymer and another material can be critical to the use of that polymer in a specific application, in particular in adhesion science.

9.2 Gibbs Approach to Surface Energy

In the case of the polymer–air interface, the ability to stick to the surface, resist wear or have a certain aesthetic characteristic can be dependent on the way the polymer molecules are organized at the interface. A number of physical phenomena are related to our understanding of the properties of an interface: spreading of oil on a surface, adhesion of two bodies in contact (e.g. chewing gum sticking to a pavement), interaction of fluids with biological materials (e.g. blood in arteries), *etc.*

At a molecular scale an interface is an imbalanced system of interactions and will have an associated imbalance of chemical potential and free energy. The molecules at the air–solid surface are attracted by the molecules in the bulk and are unable to escape. As discussed in Chapter 1, a simple molecule in the 'bulk' will be surrounded by a number of molecules that reflect the crystal packing geometry. In a typical face-centred cubic (fcc) arrangement we might have 14 nearest neighbouring molecules. To create the surface, half of the unit cell is removed so that now the number of interacting nearest neighbours is reduced to 11. The balance of the forces on the molecule will be to attract the molecule into the bulk. Molecules will prefer to be 'in the bulk' since they would be thermodynamically more stable if they are completely surrounded by other molecules.

Similar imbalances of forces will exist at the liquid–solid and the solid–solid interfaces. In the case of the liquid–liquid or liquid–solid interfaces when the energies of the two phases approach one another then mixing can occur. If the two phases have essentially the same energy then mutual diffusion will occur and a homogeneous mixture can be created. Liquid will mix with other liquid if their energies a similar and the surface tension is low.

9.2.1 Contact Between a Liquid and a Surface

Many of the important features of the interface can be understood by considering the problem of a liquid drop in contact with a surface (Figure 9.1). Gibbs defined a parameter, the *surface tension*, designated as γ, to represent the imbalance of forces at the interface.

Surface tension is defined in terms of the energy change that is required to produce a new surface. The work is equal to Fdx, where dx is the movement in the direction perpendicular to the surface and F is the force created by the molecules in the bulk. Therefore $Fdx = \gamma dA$, where dA is the change in area of the surface and γ is the surface tension which has dimensions of force per unit distance, mN m^{-1}. Typical values of the surface tension of some common liquids are summarized in Table 9.1.

9.2.2 Derivation of Young's Equation and Definition of Contact Angle

With regard to the schematic in Figure 9.1, consider the following interfaces: the liquid and solid, the liquid and gas, and the solid and gas. At the point a the three phases connect and the forces must balance if the system is at equilibrium. The forces can be resolved into components parallel to the substrate and

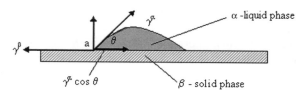

Figure 9.1 Schematic of a liquid droplet in contact with a solid surface.

Molecular Surfaces

Table 9.1 Values of surface tension of some common liquids and an indication of the dominant interactions.

Liquid (type of molecule)	γ (mN m^{-1})	Force involved in surface tension
Mercury (Hg)	476	Metallic
Water (H$_2$O)	72.75	Hydrogen bonds
Octane (C$_8$H$_{18}$)	21.69	van der Waals
Benzene (C$_6$H$_6$)	28.88	van der Waals+π–π induced dipole interactions
Carbon tetrachloride (CCl$_4$)	26.77	van der Waals+collisional dipole interactions

perpendicular to the substrate. The labels are respectively for the surface tension of the liquid–air interface, γ^α; solid–air interface, γ^β; and solid–liquid interface, $\gamma^{\alpha\beta}$. When the droplet is at equilibrium, *Young's equation* applies:

$$\gamma^\beta = \gamma^{\alpha.\beta} + \gamma^\alpha \cos\theta \tag{9.1}$$

The measurement of the force $\gamma^{\alpha\beta}$ is not practical and hence an alternative approach is usually adopted. The work of adhesion that involves removal of the liquid from the solid can be shown to be equal to the work required to create a new surface from a column of a fluid:

$$W = \gamma^\alpha + \gamma^\beta - \gamma^{\alpha\beta} \tag{9.2}$$

This work of adhesion can be combined with the Young equation:

$$W = \gamma^\alpha + \gamma^\beta - \gamma^{\alpha\beta} + \gamma^\alpha \cos\theta - \gamma^{\alpha\beta} \tag{9.3}$$

where θ is the angle between the liquid surface and the substrate. Thence

$$W = \gamma^\alpha - \gamma^\alpha \cos\theta \tag{9.4}$$

which is the work associated with the sticking of the liquid to the surface. The parameter θ has values that lie between 0° and 180°, and defines whether or not a liquid will wet a substrate:

(a) spreading film
$\theta = 0°$, cos $\theta = 1$

(b) wetting droplet
$\theta = 90°$, cos $\theta = 0$

(c) non-wetting droplet
$\theta = 180°$, cos $\theta = -1$

$W^{\alpha\beta} = 2\gamma^\alpha$ (wetting)
$\theta < 90°$

$W^{\alpha\beta} = \gamma^\alpha$
$\theta = 90°$

$W^{\alpha\beta} = 0$ (non-wetting)
$\theta > 90°$

The work of cohesion, the material with itself, is equal to $2\gamma^\alpha$ and has a contact angle of 0°. In general for a simple fluid the work of adhesion and cohesion are equal and for $\theta < 90°$, the liquid is *wetting* the solid and when $\theta > 90°$ the liquid is *not wetting* the solid. In the simplest example, silicone oil that will lower the surface energy could be present as a contaminant.

This situation can be depicted as shown below; the differently shaded bars illustrate the variation in the surface energy:

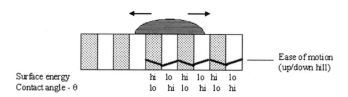

As the size of the droplet increases, the contact angle will vary reflecting the nature of the surface energy immediately at the point of contact between the three phases. If the droplet size is now reduced 'hysteresis' effects can be observed. The liquid having wet the surface may detach leaving a molecular thin layer on the substrate. The effective surface energy is now that of the detaching liquid from the molecular absorbed layer and not that of the solid. The contact angle for the droplet reducing the advancing contact angle θ_a is greater than the retarding contact angle θ_r. The important factor in determining the nature of the process is the relative magnitude of the attraction of the liquid molecules for each other compared to the intermolecular forces across the liquid–solid interface at any given point on the solid surface:

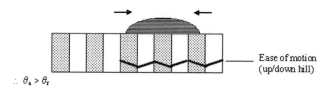

The drop will readily leave the low-energy patch but will not want to move off the high-energy patch; hence θ_r is slightly smaller than θ_a.

Measurements of hexamethylethane, C_8H_{18}, indicate a contact angle of $\theta = 115°$, whereas a C_{16} cycloparaffin has a value of $\theta = 105°$, indicating that in the case of C_8H_{18} the surface is made up of CH_3 groups, whereas in the case of the cycloparaffin the surface is covered with CH_2 groups. The C_{16} cycloparaffin is more readily wetted by water than C_8H_{18}. This technique can be used to determine the difference in the surface tension between different organic compounds.

In the presence of a surface, the total Gibbs free energy includes the normal bulk free energy terms and an additional free energy term which is associated with the creation of the interface:

$$G = H - TS + \gamma A \tag{9.5}$$

where G is the Gibbs free energy, S is the entropy, H is the enthalpy and T is the temperature. The surface tension has the following thermodynamic definition:

$$\gamma = \left(\frac{\partial G}{\partial A}\right)_{T,P} \tag{9.6}$$

The associated thermodynamic properties are as follows for the entropy and enthalpy:

$$\Delta S = -\frac{\partial \gamma}{\partial T} \quad \text{and} \quad \Delta H = \gamma - T\frac{\partial \gamma}{\partial T} \tag{9.7}$$

This classical definition is for any interface. The interfacial energy is a consequence of the interaction between molecules and contains contributions from mutually attractive intermolecular forces due to combined effects of dispersions, dipole, induced dipole and hydrogen bonding interactions. At short distances molecular species are repulsive whereas at longer distances the molecules become attractive. A variety of different interaction potentials can be used; however, the simplest is an inverse square law, in which case the interaction energy has the form

$$W_c = \left(\frac{A}{12\pi}\right)(r_0^{-2} - r^{-2}) \tag{9.8}$$

where r is the separation distance between the entities and A is the Hamaker constant representing the cumulative strength of all types of interaction. If the separation distance is increased to infinity the macroscopic interfacial tension and the strength of the molecular interactions are related by

$$\gamma = \frac{A}{24\pi r_0^2} \tag{9.9}$$

The surface tension of a surface influences the interaction with other phases and in particular solids and liquids. A liquid in contact with a solid or another liquid will attempt to balance the energy at the interface. For a drop of liquid on an interface, the surface energy must balance and the angle it makes with the surface is defined by eqn (9.1). Spontaneous wetting of the surface will occur when $\theta = 0$. In practice, the surface tension, contact energy, is not constant but can vary across the surface as a result of either changes in composition or surface texture. For a heterogeneous surface consisting of two domains the observed contact angle θ_c is defined by

$$\cos\theta_c = f_1 \cos\theta_1 + f_2 \cos\theta_2 \tag{9.10}$$

where f_1 and f_2 are the surface fractions of components 1 and 2, and θ_1 and θ_2 are their contact angles. The differences in contact angle can reflect the presence of different atomic species or surface topography. As we will explore later, surface contamination can have a major effect on the surface energy and contact energy.

9.3 Surface Characterization

There are a variety of different methods available for the inspection of the surface of a material.[1-13] The techniques available can be divided into groups:

(i) Classical surface assessment methods, contact angle measurements.
(ii) Visualization of the surface: optical, electron and atomic force microscopy.
(iii) Spectroscopic assessment of the surface: attenuated total reflection infrared, fluorescence and visible spectroscopy.
(iv) X-Ray and neutron diffraction analysis.
(v) Ion beam analysis: electron recoil and Rutherford backscattering.
(vi) Vacuum techniques: X-ray photoelectron spectroscopy (XPS), secondary ion mass spectroscopy (SIMS), Auger electron spectroscopy (AES).

Each of these techniques can provide information on a particular aspect of the surface and add to our understanding of the factors that influence the way molecules are arranged at a surface.

9.3.1 Classical Surface Assessment Methods, Contact Angle Measurements[4]

The simplest method of assessment of the energy of a surface involves the study of contact angles. The most obvious method is the direct measurement of the tangent between a liquid drop and a solid. A liquid drop in contact with a solid will behave according to eqn (9.4). It is possible to determine the contact angle by measuring the dimensions of a liquid drop.

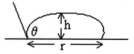

For very small drops of the order of 10^{-4} ml the distorting effect of gravity is negligible and the drop takes the shape of a spherical segment. In this case, the contact angle θ can be calculated using

$$\tan\frac{\theta}{2} = \frac{h}{r} \qquad (9.11)$$

or

$$\sin\theta = \frac{2hr}{h^2 + r^2} \qquad (9.12)$$

Molecular Surfaces

where h is the drop height and r is the radius of the drop base. The drop height is usually smaller than the base radius and more difficult to measure. In such a case, the contact angle may instead be calculated for a spherical drop from the volume of the drop and its radius at its base:

$$\frac{r^3}{V_d} = \frac{3\sin^3\theta}{\pi(2 - 3\cos\theta + \cos^3\theta)} \tag{9.13}$$

where V_d is the drop volume. The contact angle can be determined to $\pm 2\text{--}3°$ but depends very much on the properties of the surface: roughness uniformity, whether it absorbs the solvent, *etc.*

The energy of the surface can be obtained by observing the variation of the contact angle for liquids of different surface tension and extrapolation to a virtual contact angle of 0° (Figure 9.2). At a contact angle of zero the liquid spreads over the polymer surface. This value of the surface tension above which spontaneous spreading does not take place is termed the critical surface tension. The spreading coefficient, S, is defined as $S = \gamma_l(\cos\theta - 1)$ and does not imply that spreading is impossible for $\gamma_l > \gamma_c$; gravitational and other factors may cause the liquid to spread over the surface.

The absolute values of the surfaces tensions for the liquids can be determined independently allowing an unambiguous measurement of the surface tension through the contact angle. Typical values for a range of polymers are summarized in Table 9.2.

Surface roughness is an important factor in consideration of the surface tension. If we consider the surface as depicted in Figure 9.3a we can see that the roughness can effectively vary the contact angle.

If gravity is assumed not to play a role, then the observed contact angle θ_0 is given by

$$\theta_0 = \theta_i + \alpha \tag{9.14}$$

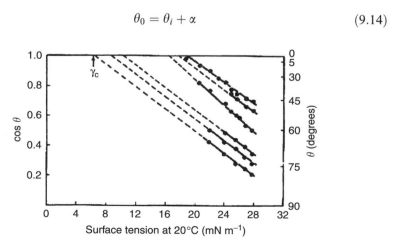

Figure 9.2 Contact angle on various perfluorinated low-energy surfaces of *n*-alkane liquids.[11]

Table 9.2 Critical surface tensions for some typical polymers at 20°C.

Polymer	Critical surface tension, γ_c
Polytetrafluoroethylene	18
Polytrifluoroethylene	22
Poly(vinylidene fluoride)	25
Poly(vinyl fluoride)	28
Polyethylene	31
Polytrifluorochloroethylene	31
Polystyrene	33
Poly(vinyl alcohol)	37
Poly(vinyl chloride)	39
Poly(vinylidene chloride)	40
Poly(ethylene terephthalate)	43
Poly(hexamethylene adipamide)	46

where α is the angle of inclination of the surface at the point of liquid–solid contact. The maximum and minimum observed angles are given by

$$\theta_{\max} = \theta_i + \alpha_{\max} \quad (9.15)$$

and

$$\theta_{\min} = \theta_i - \alpha_{\max} \quad (9.16)$$

where $\alpha_{\max}$ is the maximum inclination of the surface. If the surface is very rough a liquid with a large equilibrium contact angle may not completely wet the surface (Figure 9.3b). The roughness can be related to the contact angle by[15]

$$\cos\theta_w = r\cos\theta \quad (9.17)$$

where θ_w is the equilibrium contact angle on a rough surface, θ is the equilibrium contact angle observed on a smooth surface and r is the surface roughness or the average ratio of the two areas. The effect of high values of surface roughness explains the water repellent characteristics of some materials. Duck feather fibres are about 8 μm in diameter and separated at a similar distance with the effect that the effective contact angle is raised from a value for the material of 95° to a value of 150°, and so water drops just roll off. Hence the expression: 'like water off a duck's back'.

A further factor encountered in practice is the absorption of a liquid by the substrate. The adsorption will match the substrate to the fluid and the contact angle will steadily reduce and wetting will be affected. This is observed with blotting paper where the substrate absorbs the liquid with which it is in contact.

9.3.2 Visualization of the Polymer Surface

Visual or other methods of visualization of the surface can provide a significant amount of information on the nature of the surface.

Molecular Surfaces 241

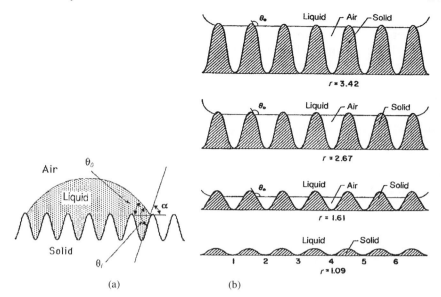

Figure 9.3 Surface roughness and influence of roughness on wetting.[10]

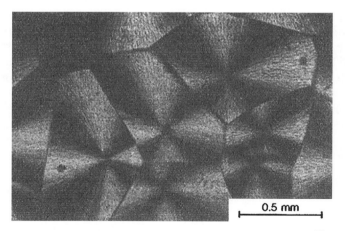

Figure 9.4 Spherulite structure in poly(butane-1) after impingement.[14]

9.3.2.1 Optical Microscopy[14,16]

The simplest and most direct method of examination of a surface is by use of optical microscopy. Surface features of the order of 1 μm or greater are usually detectable. To achieve greater contrast, polarizing microscopy is often used.[14] In the case of liquid crystalline materials (Chapter 3), polarized microscopy, since it explores the alignment of the molecular axis, can reveal both order and alignment in materials. In the case of polymers with a polarizable group either as part of the backbone or aligned with the chain axis, polarizing microscopy can identify regions of order (Figure 9.4).

Optical methods are useful for observation of spherulitic structures (Figure 9.4) and other forms of higher order organization but will not give information on lamellar nanoscale structures. The differences in birefringence between the different orientations adopted by the crystal lamellae as they radiate in all directions and the non-crystalline amorphous regions gives rise to a Maltese cross effect. If the lamellae twist in phase with one another then this gives rise to rings. Birefringence measurements can be made using a compensator and can provide useful information on the nature of the chain alignment in the material. In the case of materials where there is no birefringent species present, visualization of order can be achieved by absorption of a suitable birefringent molecule onto the aligned areas.

9.3.2.2 Electron Microscopy[9,15]

There are two methods of electron microscopy commonly encountered in the study of polymers: scanning electron microscopy (SEM) and scanning transmission electron microscopy (STEM). Both techniques use a focused high-energy beam of electrons to illuminate the sample which is conventionally held in a high vacuum. Recently a variant of SEM, called environmental SEM (ESEM) has been developed which allows the sample to be held at close to ambient pressures.[17] ESEM uses differential pumping to allow the electron beam to be focussed and shaped and it is only in the region of the impingement on the sample that the beam is brought to near atmospheric pressure. This technique has the attraction that it allows examination of specimens that may have their morphology changed by being exposed to high vacuum. It has allowed study of liquid droplets in contact with polymer fibres and examination of highly hydrated biological specimens.

The interaction of the electron beam with the sample can be summarized as follows:

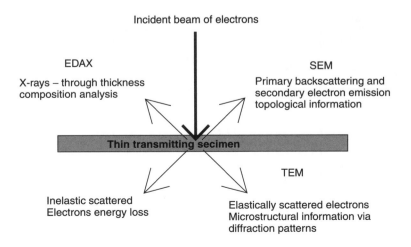

The electrons undergo a combination of inelastic and elastic scattering events. The elastically scattered electrons are collected from the back of thin samples and the diffraction patterns provide information on the crystalline structural order in the irradiated region. Using a beam of the order of 100 kV the region illuminated is $\sim$0.5–70 µm diameter. Special instruments are available which provide high resolution and operate at 500 to 1000 kV. The principal difference between SEM and TEM is the thickness of the sample studied. If the sample is very thin then TEM can be used but requires the section be cut from a liquid nitrogen frozen sample using a diamond-bladed microtome.

9.3.2.3 Scanning Transmission Electron Microscopy (TEM)[9,16]

The type of instrument used for TEM is designed to give atomic resolution. The original instruments used photographic detection of the diffraction images but modern instruments use electronic imaging and allow imaging of very small features.

9.3.2.4 Scanning Electron Microscopy (SEM)[9,16]

In conventional SEM, the sample is placed in a vacuum chamber and exposed to a focused electron beam. The electron and X-ray emission are then analysed to produce a visualization of the polymer structures and atomic composition. To understand how these images are created it is necessary to consider briefly the mechanism of the interaction of the electron beam with the sample. The electron beam will typically have an energy of between 100 kV and 500 kV and is focused down to a spot that will have a cross-section of a few hundred nanometres or less depending on the resolution being sought. This high-energy beam will impinge on the sample surface and as the electrons penetrate the solid they will undergo a range of inelastic and elastic scattering events.

9.3.2.5 Inelastic Electron Scattering

Interaction with the electron cloud of the molecules can lead to a small amount of energy being transferred and the scattered electron has a slightly different energy and a change in direction compared to the incident electron. The primary electron can produce emission of secondary electrons from the atoms with which they interact. These secondary electrons will either be trapped in the solid and produce charging of the substrate or if they have sufficient energy to overcome the surface work function will escape and can be detected. The probability of scattering occurring will depend on the atomic member of the atom producing the scattering; the higher the value of Z, the greater the scattering. The secondary electrons that do not have sufficient energy to escape can be trapped in the surface and lead to surface charging. To avoid this problem, samples for SEM are often coated on a thin metal layer or have a layer of conducting graphite deposited on the surface. Provided that the surface charge is minimized then the scanning of the surface allows the atomic distribution to be determined and images of the surface to be created. It is important to understand that the secondary electrons will come from a layer that is effectively several hundred nanometres or even micrometres deep. Although the image looks like a surface, it is in effect a composite of the events that create the secondary scattering. Some of the electrons will have sufficient energy to kick off one of the core electrons and the result will be that the subsequent relaxation of the electron structure is accompanied by the emission of an X-ray for the atom corresponding to a characteristic of a core electron transition. The X-rays are emitted from a zone that is defined in terms of the region where secondary scattering occurs and are not restricted by the constraints of the escape energy that control the secondary electron emission. As a consequence the depth probed by the X-ray emissions can be several micrometres and will be typically a cylinder of about 0.5 to 1 µm in diameter and a depth 1–2 µm. It will evident that energy dispersive X-ray analysis (EDAX) gives average data over a very significant volume and is not very specific. It is possible in favourable situations to image the surface selectively by selecting an particular X-ray line which is characteristic of that element, this approach being widely used in metals but not very useful for carbon and other low molar mass materials.

9.3.2.6 Elastic Electron Scattering

Some of the electrons are electrically scattered, *i.e.* their direction is changed without changing their energy. These scattering events occur according to Bragg's laws of diffraction and hence analysis of the scattering pattern provides information on the order of the scattering centres in the solid. Just as conventional X-ray scattering provides information on the crystalline order in solids, so electron scattering can in principle provide similar information. Unlike X-rays, the elastically scattered electrons can undergo secondary scattering and then the information on the structure is lost. Electron microscopy provides a very powerful tool for the visualization of polymer structures at the micrometre to sub-micrometre scale and has provided researchers with a wealth of data on the topography of polymer surfaces. Many apparently smooth surfaces are found to be rough when viewed at the nanometre scale (Figure 9.5).

'A picture is worth a thousand words': this is very true when it comes to understanding surface structure. A more detailed discussion of the theory and experimental methods can be found elsewhere.[16–18,9]

9.3.3 Atomic Force Microscopy[19]

An important recent addition to the methods available is atomic force microscopy (AFM).[19] This technique, as its name implies, measures the force between a fine probe that has a tip that is of sub-micrometre dimensions and the surface. There are several variations of the basic method. A simple instrument is shown diagrammatically in Figure 9.6a. The potential energy curves being sensed can be visualized as being similar to that shown in Figure 9.6b.

Molecular Surfaces

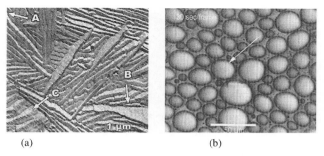

Figure 9.5 Electron micrographs of (a) an etched surface of linear polyethylene showing (A) a group of edge-on lamellae and (B) a basel surface and (C) lamellae side surfaces with striations from pulled-off molecular stems.[18] (b) ESEM of corn oil droplets in a water/oil emulsion.[17]

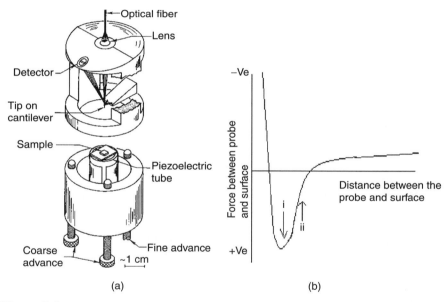

Figure 9.6 Schematic of an atomic force microscope (a) and the force profile being sensed during the scan (b).

A light beam is reflected from the scanning tip onto a quadruple detector. The detector output is able to indicate movement in the z-direction and allows the piezoelectric tube to adjust and maintain a constant distance between the tip and sample (Figure 9.5b). By mapping the variation of the force when the point is scanned across the surface, it is possible to visualize the underlying topography of the surface. The force is measured by detecting the deflection of a spring that supports the sensing element using a laser interferometer. The microfabricated cantilever has a length of only 100 μm. The optical lever ratio

246   Chapter 9

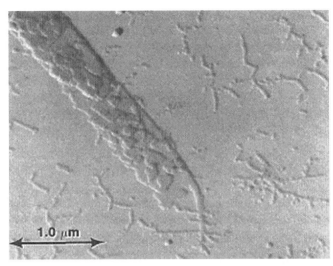

Figure 9.7 AFM image of a deposit of high molecular mass polyethylene deposited on a mica surface.[18]

used is around 800. The force with which the tip pushes against the sample is as low as 2×10^{-9} N. Various modes of scanning have been developed. The simplest is to find the minimum in the force profile in region (i) and attempt to maintain the tip at this distance. The change of force with separation, however, is not very sensitive to change in distance and hence the resolution of the scan of the surface is not very high. A higher resolution can be achieved by moving to the regions where the force profile is steeper (ii); here the movement of the tip will be more sensitive to the change in the profile. Additional sensitivity can be achieved if the tip is modulated and the so-called 'tapping' mode is adopted. Both modes of operation of the instrument allow the surface to be mapped and atomic resolution can be achieved with atomic solids. In the case of polymers the resolution that can be achieved is usually of the order of nanometres[20] (Figure 9.7). The AFM image shows isolated clusters of polymers and also an entangled collection of chains that are attempting to form an ordered structure.

A combination of optical, electron and atomic force microscopy can provide a very useful visualization of the surface of a material. It must be emphasized that each method has its strengths and weaknesses. The optical methods are unable to probe the structure below ~ 1 μm, but have the advantage of readily revealing order through the use of birefringence. Electron microscopy is capable of significantly higher resolution and can readily approach the 10 nm or better level. However, it must always be remembered that the image is the result of complex scattering processes and the height of features is often difficult to determine. AFM is capable of achieving a high resolution in the z-direction, perpendicular to the surface; however, the finite size of the tip limits the resolution of features in the x- and y-directions, although nanometre dimensions are accessible.

Molecular Surfaces 247

Whilst the above methods can provide a visualization of the surface topography, they are unable to address two questions. Firstly, they are unable to provide information on the density distribution relative to the notional surface, and secondly, they are unable to allow identification of the surface elemental, atomic, composition.

9.4 Spectroscopic Assessment of the Surface: Attenuated Total Reflection Infrared, Fluorescence and Visible Spectroscopy[16]

The infrared spectra of thin films and surfaces can be examined using attenuated total reflection (ATR) methods. Unlike conventional infrared spectroscopy the beam impinges on the surface as a consequence of being reflected from a crystal that is in contact with the surface to be examined (Figure 9.8).

In the context of morphology, the principal information that is obtained from such experiments is the conformational distribution of the polymer chains. In recent years, microscopes have been developed which allow the simultaneous observation of the surface and spectroscopic examination of the surface. The illuminating beam is brought down the optical axis and the reflected light is then collected using fibre optics. The spot size is typically several tens of micrometres but can be smaller and does allow characterization of domains or phase structure that has dimensions of this order. The techniques and their application are covered in detail elsewhere.[16]

9.5 X-Ray and Neutron Diffraction Analysis[13]

One of the most interesting questions that one can ask is whether the density changes abruptly or in a gradual manner as the interface is approached.

9.5.1 Neutron and X-ray Reflectivity[21,22]

If we consider a wave impinging on an interface between two materials, refraction will occur reflecting the difference between the two refractive indices. For a strictly planar wave there will also be specular reflection (Figure 9.9).

The reflectivity is defined as the ratio of the intensities of the reflected and incident beams and should be differentiated from the reflectance which is the ratio of the amplitudes of the incident and reflected waves. The reflectance in general is a complex number because there is usually a change in phase of a wave on reflection whereas reflectivity is a real number varying from zero to unity. The specular reflection can provide information on the composition distribution normal to the surface. The reflectivity is a function of both the angle of incidence of the beam to the surface and the refractive index changes of the substrate. The reflectivity is a function of the length scale of interactions of

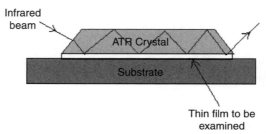

Figure 9.8 Schematic of the ATR experiment. The infrared beam is guided along the surface of the sample. The infrared beam is totally internally reflected within the ATR crystal but probes the surface of the film to a depth that is of the order of several micrometres. This type of experiment can be used to provide important information of the conformational distribution in the case of polymers. Similar experiments can be used to examine the Raman and visible–UV spectra of the surface.

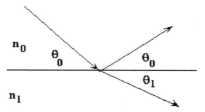

Figure 9.9 Schematic of incident, reflected and refracted beams at an interface between two materials with, respectively, refractive indices n_0 and n_1; $n_1 < n_0$ and $\theta_1 < \theta_0$ which is typically the case for neutrons incident on a material.

the wave with the surface. Visible red light from a helium–neon laser incident at 45° will have a perpendicular component of its wave vector of about 7×10^{-3} nm, so we expect measurements of light reflectivity to be sensitive to interfacial features at a length of ~ 100 nm. To probe length scales of the order of 10–100 Å requires the use of a wave with a perpendicular component of wave vector between 0.1 and 0.01 Å. This can be achieved either using neutrons or X-rays. Neutrons are useful in that there can be selective targeting by doping the polymer to be studied with deuterium.

The method used parallels that for conventional optics.[22,23] Consider a plane wave travelling in medium 0, incident on the smooth surface of medium 1 (Figure 9.10). The associated wave vectors in each medium are k_0 and k_1 and the refractive index at the boundary is given by $n = k_1/k_0$. This refractive index can be written as

$$n = 1 - \lambda^2 A + i\lambda C \qquad (9.18)$$

where the complex term accounts for any absorption in medium 1. For neutron beams

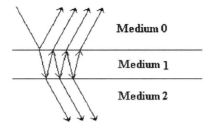

Figure 9.10 Multiple reflection and transmission for a beam incident on a layer of refractive index different from that of the bulk of the substrate.

$$A = \frac{Nb}{2\pi} \quad \text{and} \quad C = \frac{n\sigma_A}{4\pi} \tag{9.19}$$

where N is the atomic number density of medium 1 and b the bound atom coherent scattering length. The terms b and σ vary in an unpredictable manner across the periodic table. For the atoms occurring in most polymers, absorption cross-sections are either zero or negligible and the absorption and can be ignored in eqn (9.18). For polymeric species and solvents of low relative molar mass, Nb can be replaced by the scattering length density of the polymer segment or solvent molecule, ρ, and

$$\rho = \frac{N_A d \sum_i b_i}{m} \tag{9.20}$$

where d is the physical density of the polymer, m is the molar mass of the segment or solvent molecule and $\sum_i b_i$ is the sum of the bound atom coherent scattering length of the atoms making up the interfacial region. For X-rays similar expressions are obtained:

$$\rho = \frac{N_A d}{m} \sum_i z_i r_\rho \tag{9.21}$$

where r_ρ is the electron radius (2.82×10^{-15} m) and $\sum_i z_i$ is the sum of the electrons in the species.

At the interface between two dissimilar materials the grazing angle of incidence, θ_0, is related to the angle of reflection, θ_1, by

$$n_0 \cos \theta_0 = n_1 \cos \theta_1 \tag{9.22}$$

If medium 0 is air, then n_0 is 1 and eqn (9.22) becomes

$$\cos \theta_0 = n_1 \cos \theta_1 \tag{9.23}$$

When n_1 is less than 1 there will exist a critical angle above which the angle of reflection will be real for all incident angles. At this critical angle of incidence,

θ_c, θ_1 is zero and

$$\cos \theta_c = n_1 \tag{9.24}$$

In the case of neutron reflectivity the $\cos \theta_c$ term will be small allowing simplification of eqn (9.24) to give

$$\theta_c = \left(\frac{\lambda^2}{\pi}\rho\right)^{1/2} \tag{9.25}$$

For a smooth surface the components of the incident beam's wave vector normal to the surface are k_{z0} which is equal to $(2\pi/\lambda)\sin\theta_0$ in air and in the polymer k_{z1} is $(k_{z0}^2 - 4\pi\rho)^{1/2}$ which can also be written as $(k_{z0}^2 - k_c^2)^{1/2}$, where k_c is the value of the component of the wave vector normal to the surface at the critical angle. The reflectance of the interface between media 0 and 1 is given by the Fresnel formula:

$$r_{01} = \frac{k_{z0} - k_{z1}}{k_{z0} + k_{z1}} \tag{9.26}$$

The reflectivity R is $r_{01}r_{01}^*$, where r_{01}^* is the complex conjugate of r_{01}; hence for a smooth interface the Fresnel reflectivity is

$$R_F(Q) = \left(\frac{Q - (Q^2 - Q_c^2)^{1/2}}{Q + (Q^2 - Q_c^2)^{1/2}}\right)^2 \tag{9.27}$$

where Q is the momentum transfer normal to the surface defined as $Q = (4\pi/\lambda)\sin\theta$. If we now replace $Q_c = 4\pi^{1/2}\rho^{1/2}$ we find that when $Q \gg Q_c$ we can simplify eqn (9.27) to

$$R_F(Q) = \frac{16\pi^2\rho^2}{Q^4} \tag{9.28}$$

At high values of Q the product $Q^4 R(Q)$ should become constant and have a scattering length which is determined by the bulk density of the material.

The above simplified case assumes that the refractive index/density changes abruptly at the interface. As we will see later, this need not be the case, and in practice there may be a gradual change from the value of air to the bulk values. To describe this situation the substrate may be divided into a series of layers, each of a constant but slightly different refractive index/density. To understand the analysis we shall consider the simple case of a single overlayer (Figure 9.10).

In this more complex case eqn (9.27) has the form

$$R_F(Q) = \left|\frac{r_{01} + r_{12}\exp(2i\beta)}{1 + r_{01}r_{12}\exp(2i\beta)}\right|^2 \tag{9.29}$$

where the r_{ij} terms are the Fresnel reflectances calculated for each interface and β is the phase shift or optical path length in medium 1:

$$\beta = \frac{2\pi}{\lambda} n_1 d \sin\theta_1 \tag{9.30}$$

Careful examination of the equations indicates that the reflected beam in medium 0 will be constructed from a series of beams of different path length and hence the detector will see the result of the interference between the reflected beams leading to a series of maxima and minima. The separation of the minima in Q, ΔQ, is related to the layer thickness by

$$\Delta Q = \frac{2\pi}{d} \tag{9.31}$$

Similar formulae in terms of the Fresnel reflectances may be built up for the situation with a small number of discrete layers. A larger number of layers demands the use of a more general method in which the 'surface' is considered in terms of a series of layers of slightly different characteristics. The models that are used in practice consider the density variation to be a smooth function of the distance from the surface but also may include specifically a different surface layer to account for surface roughness and the possibility of molecular segregation at the surface.

The analysis is discussed in more detail elsewhere,[13] but it takes the form proposed by Born and Wolf[23] for stratified media (Figure 9.11).

The layers in the interface (Figure 9.11) can be represented by a characteristic matrix of the form for an m-layered system:

$$M_m = \begin{bmatrix} \exp(i\beta_{m-1}) & r_m \exp(i\beta_{m-1}) \\ r_m \exp(-i\beta_{m-1}) & \exp(-i\beta_{m-1}) \end{bmatrix} \tag{9.32}$$

where β is the optical path length of layer m as defined before and r_m is the product of the Fresnel reflectance of the mth interface, r_m^f, and a damping term is introduced that takes account of the roughness:

$$r_m = r_m^f \exp(-0.5 Q_{m-1} Q_m \langle \sigma \rangle^2) \tag{9.33}$$

where $\langle\sigma\rangle^2$ is the mean square roughness of the interface and $Q_i = (4\pi/\lambda)\sin\theta_i$. For a total of n layers a resultant 2 by 2 matrix is obtained by multiplying these n characteristic matrices:

$$M_n = M_1 \times M_2 \times M_3 \times \cdots \times M_n \tag{9.34}$$

$$= \begin{bmatrix} M_{11} & M_{21} \\ M_{21} & M_{22} \end{bmatrix} \tag{9.35}$$

The reflectivity of the whole multilayer is then obtained as

$$R(Q) = \frac{M_{21} M_{21}^*}{M_{11} M_{11}^*} \tag{9.36}$$

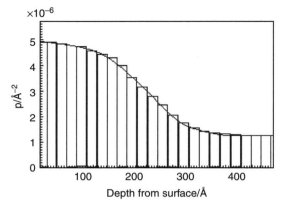

Figure 9.11 The gradual change in scattering length density (continuous line) and its approximation as a series of discrete layers.

As indicated previously, the use of deuterium substitution allows the introduction of contrast between these doped molecules and the rest of the hydrogenated matrix. The details of such experiments are discussed elsewhere.[13]

The neutron and analogous X-ray measurements can provide density contours from the surface that can be used to understand the way in which the conformational distribution of the chains varies form the surface to the bulk. The conformational distribution defines the size of the polymer coil and hence the density of the material.

9.6 Ion Beam Analysis: Electron Recoil and Rutherford Backscattering

Another method of probing the density distribution relative to the surface is Rutherford backscattering. The principle of the method is that of scattering of a heavy particle, such as a helium ion, by the nucleus. Such events are rather infrequent but sufficiently occur to allow study of atomic distributions. Observation of the ions that are scattered back from a sample gives an indication of the distribution of the atomic density relative to the surface.[24] As with other ion probe methods, the ability of a scattered ion to escape from the surface will depend on its energy relative to an escape parameter and hence the information obtained depends on the depth of the scattering event relative to the surface. The typical depth resolution for an organic material is ~ 300 Å. This method has yet to find wide application to polymer systems and is best used to follow the distribution of heavy atomic species in a matrix of lighter material.

Forward recoil spectroscopy. This is a variant of the Rutherford backscattering method. However, rather than detecting the energy of the scattering ions it detects the recoiling target nuclei by using low grazing angles. This method has been studied more extensively for polymers because it is able to differentiate

Molecular Surfaces

between deuterium and hydrogen and can be used to parallel the type of doping experiments that study neutron scattering. The topic of forward recoil spectroscopy has been reviewed by Jones and Kramer[25] and will not be considered further here. The limited availability of the necessary high-energy beam sources has limited the extent to which this method has been used for the study of polymer systems. Segregation of deuterated polymer in a hydrogenated–deuterated polystyrene blend has been reported by Jones and Kramer[25] and illustrates the power of the method.

9.7 Vacuum Techniques: X-ray Photoelectron Spectroscopy (XPS), Secondary Ion Mass Spectroscopy (SIMS), Auger Electron Spectroscopy (AES)

Whilst a number of methods exist for the study of surfaces, three vacuum analysis methods have been most successfully applied to the characterization of polymer surfaces: X-ray photoelectron spectroscopy (XPS), Auger electron spectroscopy (AES) and secondary ion mass spectroscopy (SIMS).

9.7.1 X-Ray Photoelectron Spectroscopy

XPS, or as it is sometimes called electron spectroscopy for chemical analysis (ESCA), allows characterization of the elemental composition of the top 10–30 Å of a material. Its disadvantage is that it is a high-vacuum technique and hence can only be used for the study of materials whose structure is not sensitive to the application of a vacuum.

The technique involves the irradiation of the sample with a fine beam of X-rays. The X-rays ionize the core electrons, which are emitted as photoelectrons and their energy is analysed (Figure 9.12).

The incident X-ray beam has an energy given by $h\nu$, where ν is the frequency. The X-rays are able to ionize an electron from the core K shell which is emitted with an energy given by

$$E_k = h\nu - E_b - e\Phi \tag{9.37}$$

where E_k is the measured kinetic energy of the electron and E_b is the binding energy of the core electron state. The X-ray source will be typically an Mg K_α source with an energy of 1253.6 eV or an Al K_α source with an energy of 1486.5 eV. A correction for the energy involved in the electron escaping from the surface is required: this is designated $e\Phi$, its precise value being dependent on the sample and spectrometer. Since all samples will contain carbon as a contaminant, it is usual to calibrate the energy by adjusting the values to the accepted values of carbon, 285.0 eV.

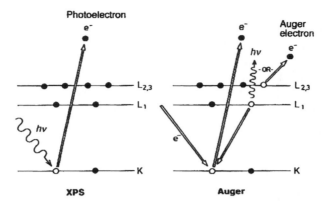

Figure 9.12 Schematic showing X-ray photoelectron and Auger processes.

Table 9.3 Binding energies and shifts for atomic species commonly encountered in polymers.

	Binding energy (eV)					
Element	$1s_{1/2}$	$2s_{1/2}$	$2p_{1/2}$	$2p_{3/2}$	$3s_{1/2}$	$3p_{3/2}$
Hydrogen	14					
Boron	188			5		
Carbon	284			7		
Nitrogen	399			9		
Oxygen	532	24		7		
Fluorine	686	31		9		
Sodium	1072	63		31		
Aluminium		118	74	73		
Silicon		149	100	99	8	3
Phosphorus		189	136	135	16	10
Sulfur		229	165	164	16	8
Chlorine		270	202	200	18	7

Each element has a distinct set of binding energies that are characteristic of that atom (Table 9.3).

Photoelectrons are labelled according to the principal quantum numbers (n) with values 1, 2, 3, 4, 5 or 6 and the angular quantum numbers with values of 0, 1, 2 or 3 commonly designated by the letters s, p, d and f. Doublets are present in the non s level as a result of spin orbital coupling. Two possible states exist and are specified by the quantum number ($j = l \pm s$, where s is the spin quantum number), when l exceeds 0. For example, a 2p level has a doublet designated by $2p_{1/2}$ and $2p_{3/2}$ and a 3d core level has a doublet designated by $3d_{3/2}$ and $3d_{5/2}$. After a photoelectron has been ejected from an inner shell of an atom, the excited atom can relax by one of two mechanisms. The hole created by the ejection of the photoelectron can be filled by an electron from an outer shell, releasing an amount of energy that can be emitted as a quantum of X-ray radiation; or the energy can be given to another electron in the same level or

Molecular Surfaces

lower level (Figure 9.12). In Section 9.3.2.2, these emitted X-rays were discussed as regards EDAX for element identification. If an electron is ejected from the $L_{2,3}$ shell then this Auger electron is then emitted with a kinetic energy approximately given by

$$E_k = E(K) - E(L_1) - E(L_{2,3}) \tag{9.38}$$

where $E(K)$, $E(L_1)$ and $E(L_{2,3})$ are the atomic energy levels. The two processes mentioned above are called X-ray fluorescence and Auger emission, respectively. For low atomic number elements ($Z < 30$), Auger emission tends to be the dominant process. Thus, in an XPS spectrum, Auger electron peaks also appear and occasionally overlap with photoelectron peaks. The kinetic energy of Auger electrons is characteristic of the elemental composition and is independent of the excitation energy, while the kinetic energy of the photoelectrons depends on the X-ray energy.

Clark and co-workers[26] using pure polymers and model compounds have reported a number of values that can be used as reference data. Some typical values for groups encountered in polymer science are listed in Table 9.4.

Subtle interactions between the nucleus and the core levels produce small but measurable *chemical shifts*. The chemical shifts vary from 1 eV for Br, to 1.5 eV for Cl to about 2.9 eV for F. A shift of as much as 8.5 eV has been observed for a $-CF_3$ moiety in a polyimide. The effect of oxygen that is conjugated to the carbon can lead to significant shifts: 3.8 eV in a polyimide. An example of the type of spectrum obtained for a polymeric material is shown in Figure 9.13 for PF6MA.[27]

The deconvolution of the spectrum is carried out on the basis of the peaks having an approximate Gaussian shape and adjusting the peaks to recognized shifts. Because of the large shifts produced by fluorine substitution, polymers either containing fluorine or modified as a result of being exposed to fluorine plasma or reactions have been extensively studied. As expected the incorporation of fluorine lowers the surface energy. The primary effect of nitrogen functionalities varies with the substituent and C 1s shifts of 0.2, 0.6, 1.8 and 1.8 eV are obtained for $-N(CH_3)_2$, $-NH_2$, $-NCO$ and NO_2, respectively. C 1s in $-C{\equiv}N$ exhibits a shift of 1.4 eV.

An example of the use of deconvolution to study poly(vinyl alcohol) (PVA) adsorbed on poly(vinylidene fluoride) (PVDF) is shown in Figure 9.14. PVA is obtained from poly(vinyl acetate) and the hydrolysis process can leave a residual low content of the acetate in the polymer. The spectrum of the virgin PVDF is shown as the dotted curve in Figure 9.14. This spectrum demonstrates that in ideal conditions quantities of the order of 1% can be detected.

Secondary spectral features such as shake up satellites are frequently observed for polymeric materials containing unsaturated hydrocarbons. Theoretical studies have shown that these are associated with $\pi-\pi^*$ transitions in aromatic materials. The shake up peak is clearly visible in the spectra of an oligomeric polyalkylthiophene that is being subjected to attack by O_3[29] and polystyrene[28] (Figure 9.15). It should be noted that the difference between the binding energy for aliphatic carbons in the backbone of polystyrene and the

Table 9.4 Examples of chemical shifts for some polymer entities.

Group	Structure	Lowest peak (eV)
Polyethylene	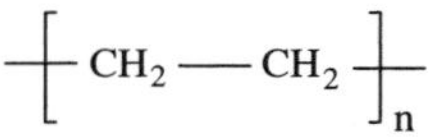 $-[CH_2-CH_2]_n-$	284.8 (C_{1s})
Polypropylene	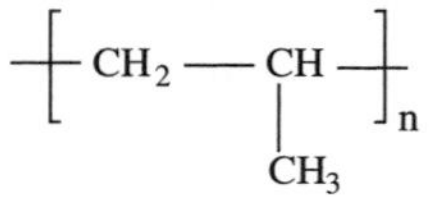 $-[CH_2-CH(CH_3)]_n-$	284.9 (C_{1s})
Poly(vinyl chloride)	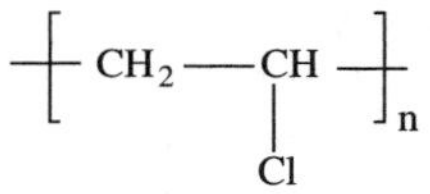 $-[CH_2-CH(Cl)]_n-$	286.5 (C_{1s})
Poly(vinyl bromide)	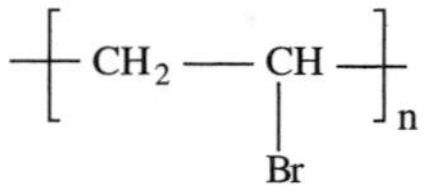 $-[CH_2-CH(Br)]_n-$	286.5 (C_{1s})
Polytetrafluoroethylene	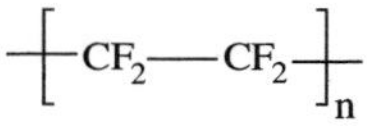 $-[CF_2-CF_2]_n-$	292.0 (C_{1s})
Polystyrene	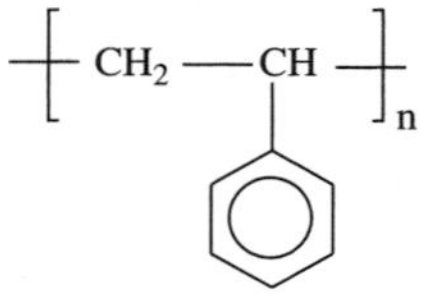 $-[CH_2-CH(C_6H_5)]_n-$	284.2 (C_{1s})
Pyromellitic acid dianhydride	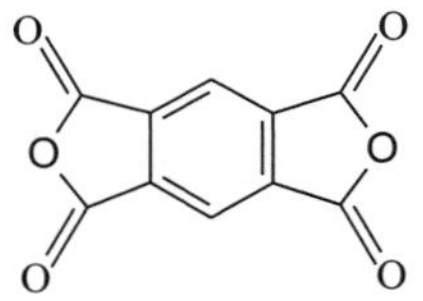 (pyromellitic dianhydride structure)	286.9 (C_{1s})

Table 9.4 (*continued*)

Group	Structure	Lowest peak (eV)
Pyromellitic diimide		285.7 (C_{1s})

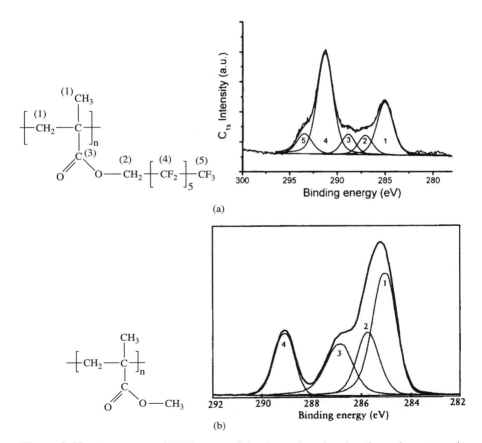

Figure 9.13 Structure and XPS traces of the C_{1s} region showing the carbon atoms in different environments: (a) PF6MA; $-CH_x-C-$ (1), $-CH_2O$ (2), $-CF_2-$ (4), $-CF_3$ (5);[27] (b) poly(methyl methacrylate) obtained using an Al K_α source allowing higher resolution.[28]

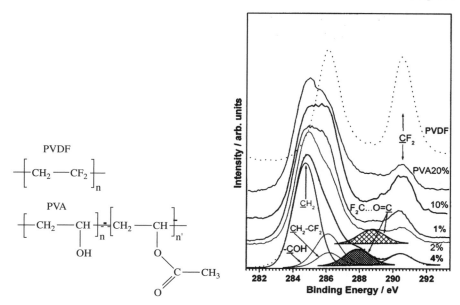

Figure 9.14 Carbon 1s spectra for virgin PVDF and 4% PVA adsorbed on the PVDF surface.[29]

ring carbons is very small making identification of the polymer difficult. The intensity of the satellite as a function of the main photoelectron peak is constant at about 10%, although slight changes occur depending on the structure of the polymer involved. The shake up peak provides a quantitative way in which the surface concentration of phenyl groups following a particular treatment method may be estimated. The peak also provides a means of estimating surface modification brought about by ring opening reactions.

The O 1s and N 1s both vary in a narrow range of about 2 eV at about 533 and 399 eV, respectively. Oxidized nitrogen functions, however, exhibit much higher N 1s binding energies that vary from 405 to 408 eV for a change from –ONO to –ONO$_2$.

9.7.2 Electron Mean Free Path, Attenuation and Escape Depth

In order to be able to analyse the spectra it is important to understand the location of the atoms that are giving rise to the events being observed. For XPS and Auger techniques the electron that is being measured must be detected without it undergoing significant interaction with the polymer material through which it is moving. As it moves through the polymer it is capable of undergoing inelastic interactions leading to ionization or excitation of valence and inner electrons, as well as vibrational excitations. Photoelectrons suffer loss of kinetic energy as a result of inelastic scattering. This limits the non-loss emission to a mean depth of only a few atomic layers below the surface and thus makes this

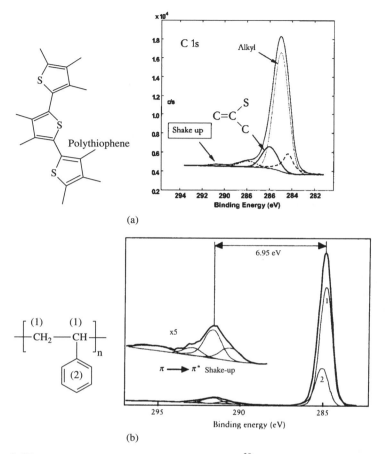

Figure 9.15 Fitted C 1s spectrum of polythiophene[30] (a) and polystyrene (b). In both XPS spectra the shake up peak is visible at ∼6.95 eV above the C 1s peak.

technique surface sensitive. The intensity of the photoelectrons that suffer no loss in kinetic energy after travelling a distance z is found to follow an exponential decay law:

$$I(z) = I(i) \exp\left(\frac{-z}{\lambda_a(E_k)\cos\theta}\right) \quad (9.39)$$

where $I(i)$ is the initial intensity of the photoelectron flux generated at a given point in the solid, $\lambda_a(E_k)$ is the attenuation length of the photoelectron with kinetic energy E_k, θ is the angle between the direction of the emitted photoelectrons to the analyser and the surface normal and z is the distance measured from the event to the surface (Figure 9.16). For a semi-infinite sample, the photoelectron intensity I^∞ can be obtained by integrating eqn (9.39) from $z=0$

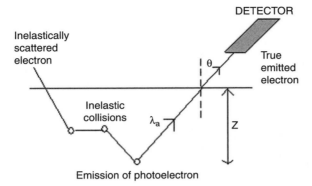

Figure 9.16 Schematic showing the photoelectron trajectories after creation.

to $z = \infty$. The terms *inelastic mean free path*, *attenuation length* and *escape depth* are used. The problem of the mean free path has been extensively investigated and it is found that it follows a 'universal curve'[31] (Figure 9.17). The mean free path reaches a minimum at between 20 and 50 eV.

The variation of the mean free path with electron energy E_k can be described by an equation that has the form

$$\lambda_a(E_k) = \frac{538 a_A}{E_k^2} + 0.41 a_A (a_A E_k)^{0.5} \qquad (9.40)$$

where E_k is the energy of the photoelectron in eV, a_A^3 is the volume of the atom in nm^3 and λ_a is in nm. For many solids the following alternative relationship has been found to be useful:

$$\lambda_a(E_k) = \frac{(49 E_k^{-2} + 0.11 E_k^{0.5})}{\rho} \qquad (9.41)$$

where ρ is the density of the material in g cm^{-3}.

In XPS a method of determining the attenuation length is to use the overlayer method. The variation of the intensity of a species is determined as a function of the thickness of the overlayer. If the substrate is a silicon wafer and the overlayer has a thickness d, then the intensity of the silicon signal will be I_s, which according to eqn (9.39) is given by

$$I_s = I_s^\infty \exp\left(\frac{-d}{\lambda_a^s \cos\theta}\right) \qquad (9.42)$$

where λ_a^s is the attenuation length for the photoelectron generated in the substrate. The photoelectron intensity for the overlayer is given by

Molecular Surfaces

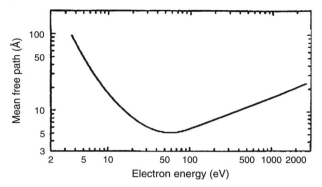

Figure 9.17 The 'universal curve' for the mean free path as a function of the electron kinetic energy according to eqn (9.40)

Table 9.5 Escape depths for some common elements with Mg K_α and Al K_α.

Element	Core level	Kinetic energy (eV) Mg K_α	λ_k (nm)	Kinetic energy (eV) Al K_α	λ_k (nm)
F	1s	568	2.0	3825	5.2
O	1s	722	2.3	3979	5.3
N	1s	855	2.5	4112	5.4
C	1s	970	2.6	4227	5.5
Si	1s	1105	2.8	4362	5.6
F	2s	1223	3.0	4480	5.7
O	2s	1230	3.0	4487	5.7

$$I_0 = I_0^i \int_d^0 \exp\left(\frac{-z}{\lambda_a^0 \cos\theta}\right) \partial z \tag{9.43}$$

$$= I_0^i \lambda_a^0 \cos\theta \left[1 - \exp\left(\frac{-d}{\lambda_a^0 \cos\theta}\right)\right] \tag{9.44}$$

$$= I_0^\infty \left[1 - \exp\left(\frac{-d}{\lambda_a^0 \cos\theta}\right)\right] \tag{9.45}$$

where I_s^∞ and I_s^0 are the photoelectron intensities of the pure substrate and the overlayer at infinite thickness, respectively. Examples of typical values for some common elements are listed in Table 9.5.

In the context of surfaces it is interesting to note that changing from Mg K_α to Al K_α almost doubles the escape depth and hence allows the distribution of elements to be studied deeper into the surface.

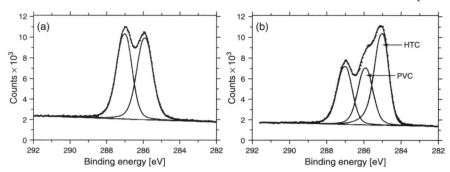

Figure 9.18 C 1s spectra of (a) poly(vinyl chloride) (PVC) and (b) PVC contaminated with hexatriacontane (HTC).[42]

The surface sensitivity of XPS is illustrated by comparison of the spectra of a clean surface and one contaminated with hexatriacontane (Figure 9.18). The presence of the contamination at the polymer surface is easily identified.[42]

9.7.3 XPS Depth Profiling

The normal application of XPS will give information on the atomic composition over the entire sampling depth. It is useful to be able to explore whether surface segregation may be occurring and for this to be achieved it is appropriate to use the escape depth as a tool for the analysis of different depths.

Angular resolution XPS. Inherent in eqn (9.42) is the angle of the emission θ of the photoelectron. The escape depth is defined as the distance normal to the surface at which the probability of an electron escaping without significant energy loss due to the inelastic scattering process drops to e^{-1} of its original value. The escape depth is given by

$$\zeta = \lambda_a \cos \theta \tag{9.46}$$

where ζ is the escape depth, λ_a is the attenuation length and θ is the angle of emission. A specimen recorded at $\theta = 0°$ will have the maximum sample depth, which is twice the sampling depth of a spectrum recorded at a value of $\theta = 60°$. The angular dependence studies are the most frequently used non-destructive approach to depth sampling. The mean depth below the surface from which the photoemissions occur can be described by[32]

$$d = 3\lambda_a \sin \theta_2 \tag{9.47}$$

where d is the depth and θ_2 is the angle between the surface and the take off angle which is the complement of θ as defined in Figure 9.16. The value of λ depends on E_k^x, where typically $0.5 < x < 1$. Some typical values determined using poly(methyl methacrylate) are listed in Table 9.6.

Different X-ray energies. As indicated in Table 9.5, the escape depth is a function of the photoelectron energy. This is illustrated by a study of surface

Molecular Surfaces

Table 9.6 XPS sampling depths as a function of core kinetic energy and take off angle.

	Kinetic energy (eV)		Sampling depth (nm)					
			$Mg\ K_\alpha$			$Al\ K_\alpha$		
Core level	$Mg\ K_\alpha$	$Al\ K_\alpha$	10°	45°	90°	10°	45°	90°
F 1s	568	801	0.8	3.4	4.8	1.1	4.5	6.3
O 1s	723	956	1	4.1	5.8	1.3	5.2	7.3
N 1s	852	1185	1.1	4.7	6.6	1.4	5.7	8
C 1s	967	1200	1.3	5.2	7.3	1.5	6.2	8.7
Si 1s	1152	1385	1.5	5.9	8.4	1.7	6.9	9.7

Table 9.7 Angle and source dependence showing atomic ratios for a study of segmented polyurethanes.

	Ratio				
Anode	C/O	N/C	N/O	Angle of emission (°)	Depth (nm)
Mg	4.0	0.04	0.15	75	2.7
Mg	4.0	0.04	0.17	45	7.3
Mg	4.0	0.05	0.19	0	10.3
Ti	4.0	0.07	0.32	0	21.5

segregation in segmented poly(ether-urethane) block copolymers.[33] The molecular structure of the polymer is similar to that shown in Figure (8.13) has the form

$$\left[\underset{}{\bigcirc} - CH_2 - \underset{}{\bigcirc} - NHCOO \left[CH_2CH_2O \right]_m CONH - \underset{}{\bigcirc} - CH_2 - \underset{}{\bigcirc} \right]_n$$

The urethane linkage is hydrogen bonded with other urethanes and being part of the methylene diphenyl entity forms a phase separated entity which melts at $\sim 158°C$ and is termed the 'hard' block. Examination of the chemical structure indicates that the nitrogen will be incorporated only in the hard block and hence is an indicator of its location in the material. If in addition to looking at different angles, sources with different energies are used, a range of different sample depths can be explored (Table 9.7).

If there were no segregation of the hard segments then the atomic ratios should be independent of the angle and the source used. Segregation of the soft block at the surface will be detected as an increase in the N/C and N/O ratios with an increasing sampling depth. The data in Table 9.7 indicate that the polyether is preferentially segregated at the surface. This segregation is important in certain applications. This type of polyurethane is used in blood handling equipment and the compatibility of the materials with blood is critical to avoid damaging blood cells and inducing thrombosis. The segregated polyether can be easily hydrated by the blood and forms a 'soft' interface that minimizes damage. Table 9.7 illustrates that the angular and energy variation allows exploration of depths between 2 and 22 nm.

Sputtering. For analysis of thicker films destructive methods have to be used. For inorganic materials bombarding the surface with a heavy ion, such as argon, allows the slow removal of the surface and exposure of sublayers. In the case of organic and other covalently bonded materials this approach is not very helpful since the species that are created may not be representative of the original material.

9.7.4 Secondary Ion Mass Spectrometry (SIMS)

In the context of XPS ion bombardment of a polymer was not considered appropriate; however, if carefully controlled and supported with appropriate analytical techniques it is a very useful method for surface analysis. In SIMS two processes have to be considered: sputtering (emission of particles) and ionization of particles (Figure 9.19).

At low incident ion flux single events may occur (Figure 9.19a); however, as the beam flux is increased then multiple events must be considered to become more frequent (Figure 9.19b). The single knock off regime is associated with ion beams of less than 1 keV, the energy transferred to the target being only capable of producing primary recoil events. Atoms from the target are ejected if they have sufficient energy to overcome the surface binding energy. The linear cascade regime occurs at slightly higher energy where now the impact of the incident ion is able not only to produce ejection of ions but is also able to transfer sufficient energy to the surface to induce secondary ionization. The incident ions used are often Ar^+ or Xe^+ ions; for high resolution Ga, In, Sn, Au and Cs ions have been used. A number of theoretical models have been produced which attempt to model the processes occurring; however, these have only a limited success. The important features to recognize are:

- The incident ion can produce the ejection of a neutral particle but also can eject ions. The accumulation of Ar^+ ions will lead to the surface becoming charged.
- If the surface becomes positively charged it will influence the energy that the incoming ion will have and is able to transfer to the substrate.
- If a negative ion is ejected this will increase the positive charge at the surface. Unless this residual charge is neutralized it will affect the surface

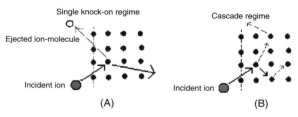

Figure 9.19 Schematic of the SIMS process.

Molecular Surfaces 265

energy and hence the ability of subsequent ions to be ejected. It is common to use an electron gun to flood the surface. This neutralizes the charges that are trapped in the surface.
- Secondary interactions due to multiple collisions will assist the fragmentation of the molecules and as the molar mass is reduced so their ability to leave the surface will increase.
- The ejected ions will have been generated in the topmost layers and rarely will have come from a distance of more than 1 nm from the surface.

The SIMS spectrometer is simply a source of mono-energetic and collimated ions that are targeted at a surface that is contained in a vacuum. The ejected ions are then carefully collected and their molar mass analysed. The ability to differentiate between different species will depend on knowledge of the fragmentation pattern or the ability to identify a unique ion that can be associated with a particular structure.

Briggs and co-workers[34-39] have reported spectra for a range of polymer systems. A typical problem might be the differentiation of polyethylene, polypropylene and polyisobutylene (Figure 9.20). The species that are ejected will be composed of both positive and negative ions. The ions are collected by the use of an accelerating voltage that can be either positive of negative relative to the sample. Depending on the sign of the collecting voltage, the spectrum of the collected species will change. In practice most measurements are made collecting the positive charges; however, in certain cases examination of the negative ion spectrum can significantly help identification of species.

Although there are a number of mass peaks for each polymer, the assignment of particular peaks to a unique species can sometimes be difficult. It is possible that a particular atomic mass unit (amu) can arise from different species. In the spectra in Figure 9.20 the 69 amu peak is assigned to a dimethylcyclopropyllium ion:

$$CH_3-\overset{H}{\underset{}{C}}-\overset{H}{\underset{}{C}}-CH_3$$
$$\underset{\overset{|}{H}}{\overset{\oplus}{C}}$$

which can be formed by a rearrangement of the fragmentation products of the chain. The 69 amu peak is strongest in polypropylene but is also present in the other polymers.

Other fragments that have been identified are:

$$\underset{CH_3}{\overset{H}{>}}C=CH-\overset{\oplus}{C} \quad 69\text{ amu} \qquad \underset{CH_3}{\overset{H}{>}}C=CH-\overset{CH}{\underset{CH_3}{\overset{\oplus}{C}}} \quad 83\text{ amu} \qquad \underset{CH_3}{\overset{H}{>}}C=CH-\overset{CH_3}{\underset{CH_3}{\overset{\oplus}{C}}} \quad 97\text{ amu}$$

The allyl structure is the more likely assignment for the 69 amu ion.

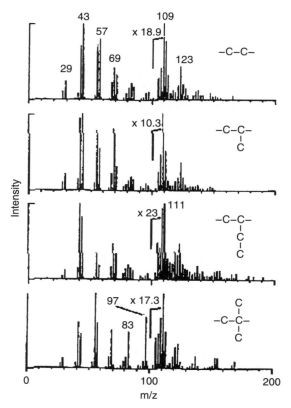

Figure 9.20 Positive ion spectra of (top to bottom) low-density polyethylene, polypropylene, poly(but-1-ene) and polyisobutylene.

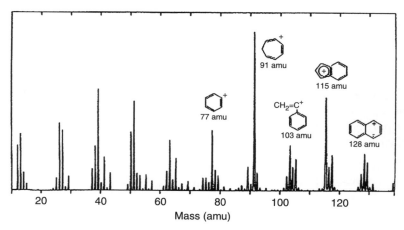

Figure 9.21 Positive ion SIMS spectrum of polystyrene.

Molecular Surfaces

Table 9.8 Sources of information on polymer characterization.

Technique	Reference
X-Ray diffraction	I. H. Hall (ed.), *Structure of Crystalline Polymers*, Elsevier Applied Science, London, 1984. B. D. Cullity, *Elements of X-ray Diffraction*, Addison Wesley, Reading, MA, 1978. F. J. Balta-Celleja and C. G. Vonk, *X-ray Scattering of Synthetic Polymers*, Elsevier, Amsterdam, 1989. L. T. Nguyen, in *New Characterization Techniques for Thin Polymer Films*, ed. H. M. Tong and L. T. Nguyen, Wiley, New York, 1990, p. 57.
Neutron diffraction	S. W. Lovesay, *Theory of Neutron Scattering from Condensed Matter*, Oxford University Press, Oxford, 1984, vol. 1. J. S. Higgins and H. C. Benoit, *Polymers and Neutron Scattering*, Oxford Science Publications, Oxford, 1994. R. A. L. Jones and R. W. Richards, *Polymers at Surfaces and Interfaces*, Cambridge University Press, Cambridge, 1999, p. 94.
Electron diffraction	D. Campbell, R. A. Pethrick and J. R. White, *Polymer Characterization Physical Techniques*, Stanley Thornes, Cheltenham, UK, 2000, ch. 9.
Transmission electron microscopy	D. B. Williams, *Practical Analytical Electron Microscopy in Materials Sciences*, Verlag Chemie International, 1984. E. L. Thomas, in *Structure of Crystalline Polymers*, ed. I. H. Hall, Elsevier Applied Science, London, 1984, ch. 3. D. Campbell, R. A. Pethrick and J. R. White, *Polymer Characterization Physical Techniques*, Stanley Thornes, Cheltenham, UK, 2000, ch. 9.
Scanning electron microscopy	D. B. Williams, *Practical Analytical Electron Microscopy in Materials Sciences*, Verlag Chemie International, 1984. D. Campbell, R. A. Pethrick and J. R. White, *Polymer Characterization Physical Techniques*, Stanley Thornes, Cheltenham, UK, 2000, ch. 10.
Optical microscopy	D. A. Hemsley, *Applied Polymer Light Microscopy*, Elsevier Applied Science, London/New York, 1989. D. Campbell, R. A. Pethrick and J. R. White, *Polymer Characterization Physical Techniques*, Stanley Thornes, Cheltenham, UK, 2000, ch. 11.
X-Ray photoelectron spectroscopy	D. Briggs, in *Electron Spectroscopy: Theory, Techniques and Application*, ed. C. R. Brundle and A. D. Baker, Academic Press, London, 1979, vol. 3. K. Okuno, S. Tomita and A. Ishitani, in *Secondary Ion Mass Spectroscopy SIMS IV*, ed. A. Benninghoven, Springer Series in Chemical Physics, 1984, vol. 36, p. 392. D. Briggs, *Surface Analysis of Polymers by XPS and Static SIMS*, Cambridge University Press, Cambridge, 1998. R. Scruby, *Mater. World*, 2002, **10**(6), 26.
Secondary ion mass spectroscopy	N. J. Chou, in *New Characterization Techniques for Thin Polymer Films*, ed. H. M. Tong and L. T. Nguyen, Wiley, New York, 1990, p. 289. D. Briggs, *Surface Analysis of Polymers by XPS and Static SIMS*, Cambridge University Press, Cambridge, 1998.

Table 9.8 (*continued*)

Technique	Reference
	K. Okuno, S. Tomita and A. Ishitani, in *Secondary Ion Mass Spectroscopy SIMS IV*, ed. A. Benninghoven, Springer Series in Chemical Physics, 1984, vol. 36, p. 392.
Atomic force microscopy	D. H. Reneker, in *New Characterization Techniques for Thin Polymer Films*, ed. H. M. Tong and L. T. Nguyen, Wiley, New York, 1990, p. 327.
Scanning tunnelling microscopy	D. H. Reneker, in *New Characterization Techniques for Thin Polymer Films*, ed. H. M. Tong and L. T. Nguyen, Wiley, New York, 1990, p. 328.

Studies of aromatic-containing polymers, such as polystyrene, indicate that rearrangement of the fragments of the polymer occurs very quickly to form stable ions (Figure 9.21). The most intense peak in the spectrum has a mass of 91 amu and is assigned to the tropyllium cation, $C_7H_7^+$. Other characteristic peaks are at 51 amu ($C_4H_3^+$), 63 amu ($C_5H_3^+$), 65 amu ($C_5H_5^+$), 77 amu ($C_6H_5^+$), 103 amu ($C_8H_7^+$) and 115 amu ($C_9H_7^+$). The important feature to appreciate is that the most obvious fragment from the polymer degradation, $C_6H_5^+$, is not the strongest peak and it is apparent that fragmentation probably involves splitting off of a vinyl-substituted phenyl ion similar to 1034 amu ($C_8H_7^+$) which then rapidly rearranges to form either the stable 91 amu tropyllium cation ($C_7H_7^+$) or by loss of another fragment forms the 65 amu phenyl cation ($C_5H_5^+$). Many papers have been published attempting to sort out the fragmentation patterns for even the simplest polymers. The usual approach that is adopted is to measure the spectrum of the pure component polymers and to use these to carry out subsequent analysis of surface segregation or analysis of blends.

The important difference between XPS and SIMS is the extreme surface sensitivity of SIMS that looks at the topmost layer of the material. It is often found that SIMS studies are very sensitive to contamination and that a monolayer of contaminant may have to be removed before assessment of a surface can be carried out.

The techniques described above are able to provide researchers with the ability to visualize the surface and also to determine quantitatively the atomic composition of both the top layer and the volume close to the surface.

9.8 Fourier Transform Infrared (FTIR) Imaging[40]

The ability to differentiate both spatially and chemically is achievable through coupling FTIR with spatially resolved infrared detectors (FPA).[41] FPA detectors typically have 4096 small detectors arranged in a 64 × 60 grid. The spatial resolution of the system is limited only by diffraction and is of the order of 4–10 μm depending on the wavelength of the band being imaged. By selection of an infrared absorption band specific to a particular polymer, it is possible to create a map of the distribution of that species in the surface. In blends and

Molecular Surfaces

similar systems the morphological features are sufficiently large to allow differentiation using this method, and systems such as polystyrene/low-density polyethylene have been reported.[40]

It is impossible in a single text to cover comprehensively the topic of polymer characterization. A list of useful references is included in Table 9.8.

Recommended Reading

D. Briggs, *Surface Analysis of Polymers by XPS and Static SIMS*, Cambridge University Press, Cambridge, 1998.
R.A.L. Jones and R.W. Richards, *Polymers at Surfaces and Interfaces*, Cambridge University Press, Cambridge, 1999.
J.E. Watts and J. Wolstenholme, *An Introduction to Surface Analysis by XPS and AES*, Wiley, Chichester, UK, 2003.

References

1. J.E. Watts and J. Wolstenholme, *An Introduction to Surface Analysis by XPS and AES*, Wiley, Chichester, UK, 2003.
2. D.R. Randell and W. Neagle, *Surface Analysis Techniques and Applications*, Royal Society of Chemistry, 1990.
3. F. Garbassi, M. Morra and E. Occhiello, *Polymer Surfaces*, John Wiley, Chichester, UK, 1996.
4. S. Wu, *Polymer Interfaces and Adhesion*, Marcel Dekker, New York, 1982.
5. D. Briggs, *Surface Analysis of Polymers by XPS and Static SIMS*, Cambridge University Press, Cambridge, 1998.
6. E.M. McCash, *Surface Chemistry*, Oxford University Press, 2002.
7. G.T. Barnes and I.R. Gentle, *Interfacial Science*, Oxford University Press, 2005.
8. J. Goodwin, *Colloids and Interfaces with Surfactants and Polymers*, Wiley, 2004.
9. D.B. Williams, *Practical Analytical Electron Microscopy in Materials Science*, Verlag Chemie International, 1984.
10. C.M. Chan, *Polymer Surface Modification and Characterization*, Hanser, New York, 1994.
11. B. Cherry, *Polymer Surfaces*, Cambridge University Press, Cambridge, 1981.
12. H.M. Tong and L.T. Nguyem, *New Characterization Techniques for Thin Polymer Films*, SPE Monograph Series, Wiley, New York, 1990.
13. R.A.L. Jones and R.W. Richards, *Polymers at Surfaces and Interfaces*, Cambridge University Press, Cambridge, 1999.
14. I.H. Hall, *Structure of Crystalline Polymers*, Elsevier Applied Science, London, 1984.
15. R.N. Wenzel, *Ind. Eng. Chem.*, 1936, **28**, 988.

16. D. Campbell, R.A. Pethrick and J.R. White, *Polymer Characterization: Physical Techniques*, Stanley Thornes, Cheltenham, UK, 2000, ch. 11.
17. F.S. Baker, J.P. Craven and A.M. Donald, in *Techniques for Polymer Organisation and Morphology Characterisation*, ed. R.A. Pethrick and C. Viney, Wiley, 2003.
18. D.C. Bassett, R.H. Olley and A.S. Vaughan, in *Techniques for Polymer Organisation and Morphology Characterisation*, ed. R.A. Pethrick and C. Viney, Wiley, 2003.
19. D.H. Reneker, *New Characterization Techniques for Thin Polymer Films*, ed. H.M. Tong and L.T. Nguyem, SPE Monograph Series, Wiley, New York, 1990, ch. 12.
20. J.K. Gimzewske, E. Stroll, Schlittler, *Surf. Sci.*, 1987, **181**, 267.
21. T.P. Russel, *Mater. Sci. Rep.*, 1990, **5**, 171.
22. R.K. Thomas, in *Scattering Methods in Polymer Science*, ed. R.W. Richards, Ellis Horwood, London, 1995.
23. M. Born and E. Wolf, *Principles of Optics*, Pergamon Press, Oxford, 1975.
24. W.K. Chu and J.W. Mayer, *Backscattering Spectroscopy*, Academic Press, New York, 1978.
25. R.A.L. Jones and E.J. Kramer, *Phys. Rev. Lett.*, 1989, **62**, 280.
26. D.T. Clark and A. Harrison, *J. Polym. Sci., Polym. Chem.*, 1981, **17**, 957.
27. R.D. Van de Grample, W. Ming, A. Gildenfenning, W.J.H. van Gennip, J. Laven, J.W. Nieantsverdriet, H.H. Brogersma, G. de With and R. van der Linde, *Langmuir*, 2004, **20**, 6344.
28. G. Beamson and D. Briggs, *High Resolution XPS of Organic Polymers*, John Wiley, Chichester, UK, 1992, p. 119.
29. S.G. Gholap, M.V. Badioger and C.S. Gopinath, *J. Phys. Chem. B*, 2005, **109**, 13941.
30. J. Heeg, C. Kramer, M. Wolter, S. Michaelis, W. Plieth and W. J. Fischer, *Appl. Surf. Sci.*, 2001, **180**, 36.
31. M.P. Seah and W.A. Dench, *Surf. Interf. Anal.*, 1970, **1**, 2.
32. D. Briggs, *Encyclopaedia of Polymer Science and Technology*, John Wiley, Chichester, UK, 2002.
33. T.G. Vargo and J.A. Gardella, *J. Vac. Sci. Technol.*, 1989, **A7**, 1733.
34. D. Briggs, *Surf. Interf. Anal.*, 1982, **4**, 151.
35. M.J. Hearn and D. Briggs, *Surf. Interf. Anal.*, 1988, **11**, 198.
36. D. Briggs, *Surf. Interf. Anal.*, 1990, **15**, 734.
37. D. Briggs, *Br. Polym. J.*, 1989, **21**, 3.
38. D. Briggs, in *Encyclopaedia of Polymer Science*, Wiley, New York, 1988, vol. 16, p. 399.
39. D. Briggs, *Surface Analysis of Polymers by XPS and Static SIMS*, Cambridge University Press, Cambridge, 1998.
40. R. Bhargava, S.Q. Wand and J.L. Koenig, *Macromolecules*, 1999, **32**, 2748.
41. R. Scruby, *Mater. World*, 2002, **10**(6), 26.
42. G. Beamson and D. Briggs, *Mol. Phys.*, 1992, **76**, 919.

CHAPTER 10
Polymer Surfaces and Interfaces

10.1 Introduction

In Chapter 9 the techniques which can be used for the study of molecular interfaces were considered and provide us with the tools to ask the question 'What are the factors that make definition of the surface difficult?' In order to help our consideration of the answer it is appropriate to divide the discussion into three types of system: crystalline polymers, amorphous polymers and polymer blends.

10.1.1 Crystalline Polymers

A bulk crystalline polymer, as discussed in Chapter 6, will be constituted from lamellae that are variously aligned and organized in the solid to give a range of microstructures. During the slow crystallization process, the highest melting high molecular weight species will form the first crystallites and the lower molecular weight materials crystallize later. Because of this fractionation process, low molecular weight material either fills the gaps between the lamellae or it segregates to the free surface.

The simplest case would be that of a polymer single crystal grown from solution. Crystals grown in such a manner will have a surface that is predominantly composed of single chains aligned parallel to the surface. The topmost layer of atoms would reflect the surface and since the energy is mostly dominated by short-range interactions the layer will dictate the nature of the surface. If the topmost layer is made up of hydrogen atoms then the surface will have a hydrophobic character. Alternatively, the introduction of oxygen, chlorine, bromine, hydroxyl, carbonyl, *etc.*, groups will make the surface more hydrophilic. This simplified view of the surface is a useful first approximation but is rather inadequate in explaining many of the features observed in real polymer surfaces.

The low molecular weight material segregated at the surface may be more or less disordered than in the ideal single crystal. In the polymer melt there will also exist polymer that has become oxidized, residues of catalysts and processing aids. Often such materials will segregate to an interface. It is therefore not unusual to find that a measurement of the surface energy, contact angle, *etc.*, indicates a value that is unexpected from consideration of the atomic composition of the bulk polymer. Further evidence for the contaminated nature of the surface can be obtained by washing the surface with a suitable solvent. This process will remove the impurities and low molecular weight material and a new value of the contact angle will be observed. As discussed in Chapter 9, surface roughness can also influence the measured surface energy.

However, in general the surface of a crystalline polymer will have physical properties that correlate well with the chemical structure of the bulk polymer, but may reflect the segregation of end groups to the surface.

10.1.2 Amorphous Polymers

The amorphous state is associated with polymers that are unable to crystallize and form a disordered state. The surface of an amorphous polymer can therefore be pictured as an entangled mass of random coil, pseudo-spherical structures (Figure 10.1).

The polymer surface if the random coils were not to be distorted close to the surface would produce a rough surface that is thermodynamically unstable. Forces in the surface will be imbalanced and as a result either the chains in the surface may move closer together to form a smoother surface, densification, or other regions may expand to fill the voids between the spheres, creating a lower density surface. The latter is not usually the case. As in the case of crystalline polymers, impurities—residual catalysts, processing aids—can segregate to the surface and change the surface energy. The ends of the polymer coil will usually have a slightly different surface energy from the rest of the polymer chain and as a consequence they may segregate to the surface.

As an illustration of the tendency for polymer end groups to segregate to the surface we will consider the case of a polystyrene that has fluorine end caps. The polymer was produced by using a dilithium initiator, and the end cap was

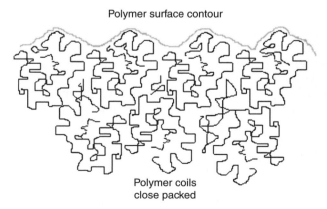

Figure 10.1 Schematic of close-packed polymer coils forming a polymer surface.

$(CH_3)_2Si(CH_2)_2(CF_2)_5CF_3$ end function either on one or both ends.[1] The surface was analysed using a combination of X-ray photoelectron spectroscopy (XPS) and secondary ion mass spectrometry (SIMS) that are both very surface sensitive methods. The very surprising result was found that after a few days' annealing of the samples the surface concentration of terminal fluorine substituted groups that was initially about 85% of the atomic composition of the surface increased to almost 100%. The polymers were all of molar mass at or below the entanglement limit and hence there is little constraint for these materials to attempt to achieve their thermodynamically lowest state that would be the surface segregation of the fluorine groups.

10.1.3 Polymer Blends

Polymer blends may be expected to reflect the balance of forces that control their phase structure in the bulk. If two polymers are compatible in the bulk, they may or may not segregate in the surface depending on the balance of forces. Materials that are able to phase separate will segregate in the surface, and generally speaking the lower surface energy material will move towards the free air surface. This effect was illustrated in Chapter 9 in the case of the surface segregation of the ether soft block in the case of block polyurethane copolymers.

10.2 Theoretical Description of the Surface of a Polymer

10.2.1 Surface Tension of Homopolymers

The precise value of the surface tension is the result of a number of interactions; however, the values for high polymers vary from about 20 mN m^{-1} for highly hydrophobic materials such as polydimethylsiloxane (PDMS) and polyfluorocarbons to values of the order of 45 mN m^{-1} for polar materials such as polyesters and polyamides. The surface tension of water has a value of 80 and hence most materials are only mildly hydrophilic.

As with many other features of molecular interactions, averaged values can be used to describe the behaviour of particular functional groups in a molecule. These group contributions have been estimated for a variety of different functional groups, e.g. oxygen in esters, hydroxyl groups, halogens, nitrogen, sulfur, double and triple bonds, *etc.*,[1,2] and provide a simple method of estimating the expected surface tension for an unknown material. In the estimation of many thermodynamic quantities it is possible to use group contributions to calculate the surface tension. The strength of the interactions can be calculated using the dispersive solubility δ^d which can be calculated from tabulated values of dispersive molar attraction constants F_i according to

$$\delta^d = \sum_i \frac{F_i}{\nu} \tag{10.1}$$

where v is the molar volume of the groups/entities from which the polymer is created.[3] Using this approach the surface tension can be calculated for non-polar polymers using the following empirical relationship:

$$\gamma = 0.2575\delta^2 \rho^{-1/3} \qquad (10.2)$$

where ν is the density in g ml^{-1} and δ is in cal$^{1/2}$ ml$^{-1/2}$. This very simple formula predicts the surface tension to within 5–10% of the measured values for a large number of polymers. In the case of PDMS the group contribution prediction for the backbone is 20.9 mN m^{-1} compared with a value of 20.4 ± 0.07 mN m^{-1} from experiment. One of the interesting problems that can be explored theoretically and is very difficult experimentally is the possible effect of segregation of chain ends into the surface. For non-polar end groups the following expression has been proposed:

$$\gamma = 0.07147 \vartheta^{1/3} (\delta^d)^2 \qquad (10.3)$$

As indicated in previous chapters, low molar mass materials can segregate to surfaces and hence the molar mass dependence of the surface tension is an important factor in considering the performance of a polymer material. Wu[4] has proposed that

$$\gamma^{1/4} = \gamma_\infty^{1/4} - \frac{k}{M} \qquad (10.4)$$

where k is a characteristic constant for that polymer system and Le Grand and Gaines[5] have proposed that

$$\gamma = \gamma_\infty - \frac{k}{M^{2/3}} \qquad (10.5)$$

Both these relationships are theoretically correct and are extremes of the lattice theory, the indices of the molar mass having the value 2/3 for low molar mass and 1 for high molar mass materials.

10.2.2 Theories of Homopolymer Surface Tension

Following the initial approach of Gibbs, the surface tension can be obtained from the equations of state:[5]

$$\gamma = 0.095 P^{*2/3} T^* / \tilde{V}^{2.1} \qquad (10.6)$$

where P, T and V are, respectively, the pressure, temperature and volume. The asterisks indicate these are characteristic parameters of the equation of state and the tilde indicates a parameter reduced by the characteristic parameter. An alternative approach starts from the generated van der Waals square gradient approach in which the effect of the density gradient across the polymer–air interface is assumed to be a gradual transition. The width of this density gradient depends on the models and the type of polymer and can have values between 1 and 100 mm depending on the system. The main change in the density will run over a dimension of less than 1 nm, but the density will only reach its equilibrium value at a distance between 5 and 100 nm. Because the reduced free energy density is dependent on both the reduced temperature and density, it will generally vary through the interface region. The surface tension is related to the excess free energy density $\Delta\tilde{\alpha}$ by

$$\tilde{\gamma} = 2 \int_{-\infty}^{\infty} (\tilde{\kappa}\Delta\tilde{\alpha}|\tilde{\rho}(z)|)^{1/2} \, \partial z \qquad (10.7)$$

where z is the coordinate perpendicular to the surface. Changing the limiting variable produces the equation

$$\tilde{\gamma} = \int_{\tilde{\rho}_{gas}}^{\tilde{\rho}_{liquid}} (\tilde{\kappa}.\Delta\alpha)^{1/2} \, \partial\tilde{\rho} \qquad (10.8)$$

The square gradient coefficient κ is defined by

$$\tilde{\kappa} = -\frac{\partial \tilde{\kappa}_1}{\partial \tilde{\rho}} + \tilde{\kappa}_2 \qquad (10.9)$$

where the first turm on the right-hand side is related to the interaction potential and assumes a value of 0.5 for purely dispersive interactions. The second term arises from non-local entropy effects and is equal to $\tilde{T}/24\tilde{\rho}$. The surface tension is therefore obtained by an integration of the equation of state, $\Delta\tilde{\alpha}(\tilde{\rho}\tilde{T})$. A number of studies have been carried out and by the use of compressible lattice theory it is found to be possible to obtain good agreement between theory and experiment.[1] The reduced surface tension for many different polymers forms a master curve when plotted against reduced temperature which has the following form:

$$\tilde{\gamma} = 0.6109 - 0.06725\tilde{T} - 0.1886\tilde{T}^2 \qquad (10.10)$$

Over small temperature ranges the surface tension varies almost linearly with temperature consistent with the typical linear dependence of density on temperature. The temperature coefficient of surface tension for polymers falls in the range 0.05–0.07 mN m^{-1}.

10.3 Surface Segregation

Most commercial polymer systems are a complex mixture of one or more polymers, plasticizers, antioxidants and processing aids. The surface of such a system will therefore not necessarily be determined by the dominant polymer but will often be influenced by the segregation of low molar mass and low surface tension materials to the air–polymer interface. Surface segregation can occur with or without the formation of a separate phase at the surface. The segregation of low molar mass materials to the surface can produce changes in refractive index and is sometimes referred to as *blooming*. In this case the component i of a miscible mixture adsorbs preferentially to the surface, it produces a change in the surface energy equivalent to $\gamma_i \partial A_i$ where ∂A_i is the surface area occupied by type i molecules and γ_i is their surface tension. Following the usual thermodynamic arguments of Gibbs one obtains

$$\sum_{i=1}^{m} n_i \partial \mu_i = \sum_{i=1}^{m} \gamma_i \partial A_i \qquad (10.11)$$

It also follows that the surface excess obtained using the classical Gibbs adsorption isotherm is given by

$$-\partial \gamma = \sum_i \Gamma_i \partial \mu_i \qquad (10.12)$$

where $\Gamma_i = n_i/A$, in which n_i is the mole fraction of species i in the bulk phase. In practical terms lowering the surface energy decreases the thermodynamic work of adhesion and also in practical terms lowers the energy of adhesion of a material stuck to the polymer surface.

10.4 Binary Polymer Blends

In Chapter 8 the principles of phase segregation in polymer blends were considered. As would be expected surface segregation occurs and usually the lower surface energy component of the mixture will tend to appear at the surface. The surface energy can be directly related to the free energy. The techniques for the study of surfaces are discussed in Chapter 9.

The surface composition and near-surface gradient structure of miscible binary homopolymer blends are determined by a balance between the surface energy driving segregation and the exchange free energy that is associated with the near-surface demixing. At equilibrium, this balance represents a minimum in the overall free energy. If the interactions are sufficiently short range in nature, the surface free energy for a planar interface with a sharp density gradient is given by

$$\gamma = \gamma_s(\phi_b) + \int [\gamma'_s + 2(\kappa \Delta F)^{1/2}] \partial \phi \qquad (10.13)$$

where ϕ_s and ϕ_b refer, respectively, to the surface and bulk volume fractions and γ_s and F are the contributions to the free energy of a unit area of surface and a unit volume of the uniform bulk mixture, respectively. The term $\Delta F = F(\phi) - F(\phi_b^-) - (\phi - \phi_b^-)(\partial F/\partial \phi)_b$ is the free energy cost to exchange composition in the near-surface gradient which eventually balances with the free energy decrease, $\gamma'_s = \partial \gamma_s/\partial \phi$, associated with the lowering of surface energy upon adsorption of the low surface energy polymer at the surface. In the weak segregation limit, the square gradient coefficient κ which is related to the nature of the inter atomic potential takes the form

$$\kappa = \frac{a^2}{36\phi(1-\phi)} \qquad (10.14)$$

the polymers being assumed to be symmetric and with identical statistical segment lengths a. The exchange free energy ΔF can be calculated from the Flory–Huggins form for the free energy of mixing:[7]

$$F(\phi_A) = \left[\frac{\phi_A}{N_A}\right] \ln \phi_A + \left[\frac{(1-\phi_A)}{N_A}\right] \ln(1-\phi_A) + \phi_A(1-\phi_A)\chi \qquad (10.15)$$

where N_A and N_B are the number of lattice units per chain for the two blend components A and B, respectively.[1] The surface composition is obtained by minimizing the surface tension in eqn (10.13) with respect to the surface composition and yields a boundary condition:

$$-\gamma_s = \pm 2(\kappa \Delta F)^{1/2}\bigg|_{\phi=\phi_0} \qquad (10.16)$$

The surface composition gradient can be calculated from the theory, since

$$\frac{\partial \phi}{\partial z} = \pm 2(\kappa \Delta F)^{1/2} \qquad (10.17)$$

which leads to

$$\phi(z) = -\int_{\phi_b}^{\phi_s} \left(\frac{z}{\Delta F}\right)^{1/2} \partial \phi \qquad (10.18)$$

Using experimentally determined binary interaction parameters, the statistical segment length and the surface energy difference between the blend components, the surface composition and the surface concentration profile can be calculated. The profiles obtained closely approximate to an exponential decay:

$$\phi(z) \cong \phi_b + (\phi_s - \phi_b)\exp\left(-\frac{z}{\xi}\right) \qquad (10.19)$$

Equation (10.19) indicates that the surface decay length ξ conveniently characterizes the concentration profile. For a strongly segregated system the decay length is small but becomes large, ~ 10–20 nm, when the concentration fluctuations grow near the critical point. Experimental studies demonstrate that the surface composition scales directly with the surface energy difference between the constituents.[8]

10.5 End Functionalized Polymers

The end functions of many polymers arise as a consequence of the way the reaction has been either initiated or terminated. The end group of a homopolymer will also have a different energy from that of the bulk of the main chain, as discussed in Chapter 7. As a consequence the surface interaction parameter has the form

$$\chi_s = \frac{(\gamma_e - \gamma_r)}{kT} \qquad (10.20)$$

where the subscripts e and r refer to the two components: the end and repeat units, respectively. As a consequence the concentration profiles are as follows:

- The zone next to the surface has an excess of end groups when χ_s is negative.
- The zone next to the surface is depleted of end groups when χ_s is positive.
- The segregation layer is typically about 1 nm.

Studies of fluorosilane-terminated polystyrene (PS-F)[9–11] illustrate this effect (Table 10.1). The data in Table 10.1 indicate that as the molar mass of the end capped PS-F varies so the extent to which end group segregation varies. The highest molar mass polymer has a value approaching that of normal polystyrene. Studies of blends of PS with PS-F indicate that the PS-F is preferentially segregated at the surface. It is quite surprising that such a small component of the bulk can have such an effect on the surface and illustrates the importance of the differential surface energy in controlling the composition of the surface layer.

10.6 Phase Segregation and Enrichment at Surfaces

Perdeuterated poly(methyl methacrylate) is thermodynamically slightly different from poly(methyl methacrylate) This difference in surface energy which is

Table 10.1 Various data for fluorosilane-terminated polystyrene (PS-F).

Polymer	M_n by GPC[a]	Functionality	Surface tension ($mN\ m^{-1}$)	Surface fraction of end groups
Polytetrafluoroethylene			18.2	
5K PS-F	5,300	0.89	19.9	0.87
11K PS-F	10,900	0.78	22.1	0.72
25K PS-F	25,000	0.85	24.1	0.58
148K PS-F	148,000	1.00	29.7	0.19
Polystyrene			32.4	

[a] GPC, gel permeation chromatography.

estimated to be 0.08 mJ m^{-2} is sufficient to allow the perdeuterated polymer to become enriched at the surface. A study of neutron reflectivity and SIMS clearly indicates that surface enrichment is occurring and also demonstrates how the data from these two methods can be combined to allow quantification of the slow diffusion of the species to the surface. Small differences in the thermodynamics of the blend are sufficient to achieve the surface segregation (Figure 10.2).

Atomic force microscopy (AFM) combined with XPS and X-ray reflectivity measurements carried out on deuterated polystyrene and polystyrene containing 1% bromine blends indicate that the surface segregation can be quite marked and depends on the relative composition of the blend (Figure 10.2). The different topographical features arise from the effects of spinodal decomposition of the polymer mixtures during film formation. The initial structures (Figure 10.2a) resemble haystacks and are small islands of the brominated polystyrene. Identification of the nature of the islands is possible through the comparison of the changes in the SIMS spectra as the composition of the mixture is changed. The small haystacks are quite regular and correspond to dimensions that are a little greater than the random coil radius for the polymers. The height of the features is of corresponding dimensions. Increasing the composition leads to a growth in the size of these features as shown in Figure 10.2b. A co-continuous phase structure is observed at a composition of 60% and at higher compositions of the poly(brominated styrene) the surface takes on a cheese-like structure. The depth of the holes has dimensions that are

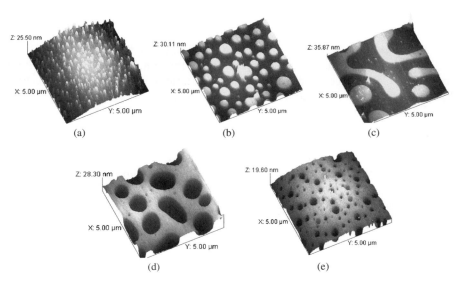

Figure 10.2 AFM images of the surface of polystyrene-d$_8$/poly(1.0% brominated styrene) [P(Br$_{1.0}$S)] blends:[14] (a) 10%, (b) 50%, (c) 60%, (e) 70% and (d) 90%.

once more close to the mean square radius of gyration of the polymer phase indicating that the topography is of the order of the size of a polymer molecule.

10.7 Electrohydrodynamic (EHD) Instabilities in Polymer Films

An interesting feature of the balance of forces which exists at a surface is demonstrated when a thin layer of a polymeric liquid or solution is placed in an electric field.[15–17] The original theory considered the dynamic instability which is created when a dielectric media, polymer liquid, is sandwiched between a conductive liquid and a conductive substrate (Figure 10.3).

It has been shown, however, that provided the gap between the dielectric and another substrate is small, essentially the same EHD instabilities are observed. The equation of motion for the lateral flow of a liquid in a thin film assuming that the fluid is incompressible has the form

$$\frac{\partial}{\partial t}h(x,t) = -\frac{\partial}{\partial x}j(x,t) = C\frac{\partial^2}{dx^2}p[h(x,t)] \qquad (10.21)$$

where x is the lateral coordinate, $h(x, t)$ is the local film thickness and $j(x, t)$ is the lateral liquid flow in the film, integrated along the normal coordinate. The shape of the flow profile and the viscosity of the liquid are absorbed into the positive constant C. The film pressure, $p(x, t)$, can be written as

$$p = \frac{A}{6\pi h^3} + \frac{\varepsilon\varepsilon_0 U^2}{2h^2} - \sigma\frac{\partial^2 h}{\partial x^2} \qquad (10.22)$$

The first term is the disjoining pressure in the film with A the Hamaker constant describing the van der Waals interaction of the film with the surrounding media. The second term represents the electrostatic pressure exerted on the film by an electrostatic potential difference, U, between the conducting media cladding the film, with ε being the dielectric constant of the film material. Finally the third term describes the Laplace pressure in the film with σ donating the interface tension between the film and the upper (liquid) medium.

The liquid will be in thermal motion and there will be a small fluctuation in the thickness in an initially homogeneous film of thickness h_0.

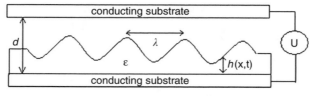

Figure 10.3 Schematic of the EHD experiment.

Equation (10.21) gives

$$-C^{-1}\frac{\partial h}{\partial t} = \left(\frac{A}{2\pi h_0^4} + \frac{\varepsilon\varepsilon_0 U^2}{h_0^3}\right)\frac{\partial^2 h}{\partial x^2} + \sigma\frac{\partial^4 h}{\partial x^4} \qquad (10.23)$$

The overall pressure distribution at the film surface can be written as

$$p = p_0 - \gamma\frac{\partial^2 h}{\partial x^2} + p_{\text{el}}(h) + p_{\text{dis}}(h) \qquad (10.24)$$

where p_0 is atmospheric pressure, the second term stems for the surface tension γ and the fourth term, the disjoining pressure p_{dis}, arises from dispersive van der Waals interactions. The electrostatic pressure for a given electric field in the polymer

$$E_{\text{p}} = \frac{U}{\varepsilon_{\text{p}}d - (\varepsilon_{\text{p}} - 1)h} \qquad (10.25)$$

is given by

$$p_{\text{el}} = -\varepsilon_0\varepsilon_{\text{p}}(\varepsilon_{\text{p}} - 1)E_{\text{p}}^2 \qquad (10.26)$$

For a sufficiently high voltage only the electrostatic interactions need to be considered. In a stability analysis, a small perturbation of the interface with wave number q, the growth rate τ^{-1} and amplitude u is considered: $h(x,t) = h_0 + u\exp[(iqx + t)/\tau]$. The modulation of h gives rise to the lateral pressure gradient inside the film inducing a Poiselle flow j:

$$j = \frac{h^3}{3\eta}\left(-\frac{\partial p}{\partial x}\right) \qquad (10.27)$$

where η is the viscosity of the liquid. A continuity equation enforces mass conservation of the incompressible liquid:

$$\frac{\partial j}{\partial x} + \frac{\partial h}{\partial t} = 0 \qquad (10.28)$$

Combining eqn (10.24), (10.27) and (10.28) a differential equation is obtained that describes the dynamic response of the interface to the perturbation. In a linear approximation a dispersion relation is obtained:

$$\frac{1}{\tau} = \frac{h_0^3}{3\eta}\left(\gamma q^4 + \frac{\partial p_{\text{el}}}{\partial h}q^2\right) \qquad (10.29)$$

Fluctuations are amplified if $\tau > 0$. Since $\partial p_{\text{el}}/\partial h < 0$, all modes with

$$q < q_{\text{c}} = \sqrt{-\frac{1}{\gamma}\frac{\partial p_{\text{el}}}{\partial h}}$$

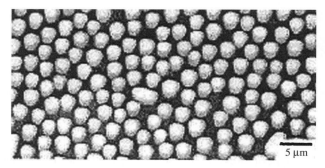

Figure 10.4 AFM image of a 115 nm thick brominated polystyrene film annealed for 1 h at 170 °C. The nearest neighbour distance between the columns is ∼3 μm.

are unstable. With time, the fastest growing fluctuation will eventually dominate, corresponding to the maximum in eqn (10.29):

$$\lambda = 2\pi\sqrt{\frac{\gamma U}{\varepsilon_0 \varepsilon_p (\varepsilon_p - 1)^2} E_p^{-3/2}} \qquad (10.30)$$

It has been shown that application of an electric field causes the polymer to become localized into specific regions and to take on a vertical columnar structure, leaving the regions around the column depleted of polymer (Figure 10.4). This process is similar that which occurs during spinodal decomposition and is responsible for the structures that are illustrated in Figure 10.2.

An extensive discussion of the data on surface segregation can be found elsewhere.[12–14]

The consideration of the polymer surface allows a number of important technological issues to be specifically addressed: contact angle and surface energy data, high porosity, lack of gloss in transparent films, *etc*.

Recommended Reading

R.A.L. Jones and R.W. Richards, *Polymers at Surfaces and Interfaces*, Cambridge University Press, Cambridge, 1999.

J.T. Koberstein, *Encyclopaedia of Polymer Science and Technology*, John Wiley, 2002.

References

1. S. Affrossman, P. Bertrand, M. Hartshorne, T. Kiff, D. Leonard, R.A. Pethrick and R.W. Richards, *Macromolecules*, 1996, **29**, 5432.

2. J.T. Koberstein, *Encyclopaedia of Polymer Science and Technology*, John Wiley, 2002.
3. D.W. Van Krevelen, *Properties of Polymers*, Elsevier, 1980, p. 163.
4. S. Wu, *Polymer Interface and Adhesion*, Marcel Dekker, New York, 1982.
5. D.G. Le Grand and G.L. Gaines, *J. Colloid Interf. Sci.*, 1969, **31**, 162.
6. G. Patterson and A.K. Rastogi, *J. Phys. Chem.*, 1970, **74**, 1076.
7. P.J. Flory, *Principles of Polymer Chemistry*, Cornell University Press, Ithaca, NY, 1971.
8. Q.S. Bhatia D.H. Pan and J.T. Koberstein, *Macromolecules*, 1988, **21**, 353.
9. S. Affrossman, *Macromolecules*, 1994, **27**, 1588.
10. J.F. Elman, *Macromolecules*, 1994, **27**, 5341.
11. M.O.J. Hunt, *Macromolecules*, 1993, **26**, 4854.
12. R.A.L. Jones and R.W. Richards, *Polymers at Surfaces and Interfaces*, Cambridge University Press, Cambridge, 1999.
13. I. Hopkinson, F.T. Kiff, R.W. Richards, S. Affrossman, M. Hartshorne, R.A. Pethrick, H. Munro and J.R.P. Webster, *Macromolecules*, 1995, **28**, 627.
14. S. Affrossman, G. Henn, S.A. O'Neil, R.A. Pethrick and M. Stamm, *Macromolecules*, 1996, **29**, 5010.
15. E. Schaffer, T. Thurn-Albrecht, T.P. Russel and U. Steiner, *Europhys. Lett.*, 2001, **53**(4), 518.
16. M.D. Dickey, E. Collister, A. Raines, P. Tsiartas, T. Holcombe, S.V. Sreenivasan, R.T. Bonnecaze and C. Grant Wilson, *Chem. Mater.*, 2006, **18**, 2043.
17. S. Herminghaus, *Phys. Rev. Lett.*, 1999, **83**(12), 2359.

CHAPTER 11
Colloids and Molecular Organization in Liquids

11.1 Introduction

In many biological systems and some polar polymers, the organization which is observed in the solid is a consequence of processes of pre-assembly occurring in the solution or melt phase prior to solid formation. In Chapter 10 the effects of surfaces on the free energy were considered and in Chapter 3 the possibility of pre-assembly through liquid crystalline phase formation was discussed. An important class of liquid crystalline materials that was not considered in detail in Chapter 3 are *lyotropic* molecules. Molecules such as stearic acid $(CH_3(CH_2)_{16}CO_2OH)$ and palmitic acid $(CH_3(CH_2)_{14}COOH)$ are examples of this type of molecule.[1] The molecular structure is made up of a polar end that is hydrophilic and a non-polar hydrophobic group that usually contains a hydrocarbon chain. The polar group will be soluble in water, whereas the hydrocarbon tail will be insoluble. Attempting to disperse these molecules in water, we find that they exhibit a limited solubility before phase separating. The free energy of the system is favoured by the stearic acid separating out at the water–air interface (Figure 11.1), this being the usual action of a soap.

The segregation of the molecules to the interface is a consequence of the necessity to balance the free energy of the interface. The problem can be considered in terms of a Gibbs surface, which is an arbitrary plane defined in terms of a plane between the two partly miscible liquids. The Gibbs surface is

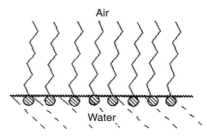

Figure 11.1 Schematic of the organization of stearic acid at the air–water interface.

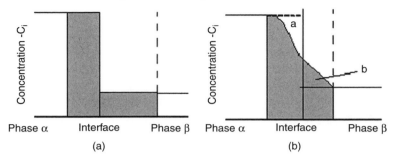

Figure 11.2 Schematic of the surface excess distribution of a surface active species across an interface between two immiscible liquids. (a) Ideal distribution showing ideal shape interface. (b) Distribution across interface. The Gibbs interface is an arbitrary line which in the first case represents the solubility in the two phases α and β and the second (b) is constrained so that the areas are equal.[1]

the interfacial region between the two phases α, β and has a finite thickness implying that there is some solubility of one liquid in the other (Figure 11.2). This is the case for a molecule such as stearic acid that at very low concentrations will be soluble in water but is also soluble in oil.

11.2 Ideal Non-mixing Liquids

The theory considers the distribution of surface active species across an interface between two essentially immiscible liquids (Figure 11.2). The ideal situation is depicted in Figure 11.2a; however, in practice there will be the possibility of a distribution of the surface active molecules across the interface, reflecting that the interface has a finite thickness. The distribution of the surface active component is shown by the shaded area, and in Figure 11.2a there is a clear change in concentration at the Gibbs surface, whereas in the other situations there is distribution of concentration. The generalized theory considers the effect of the surface excess Γ_i on the surface tension γ. The adsorption of a component at the interface will be defined by a decrease in the surface tension γ. We can define the so-called spreading pressure π as

$$(\gamma_0 - \gamma) = \pi \tag{11.1}$$

where γ_0 is the surface tension of the interface in the absence of the surface active component. Using Helmholtz energy instead of Gibbs free energy:

$$\partial A^\sigma = -S^\sigma \partial T - P \partial V^\sigma + \gamma \partial a + \sum_i \mu_i \partial n_i \tag{11.2}$$

where the parameters with superscript σ are the values appropriate for the surface.

Integration of eqn (11.2), keeping the intensive variables constant (*i.e.* T, P, μ_i, γ), yields

$$A^\sigma = -PV^\sigma + \gamma_a \partial a + \sum \mu_i \partial n_i \qquad (11.3)$$

Equation (11.3) can be differentiated to obtain the total derivative:

$$\partial A^\sigma = -P\partial V^\sigma - V^\sigma \partial P + \gamma \partial a + a \partial \gamma + \sum \mu_i \partial n_i + \sum n_i \partial \mu_i \qquad (11.4)$$

Subtraction of eqn (11.4) from (11.2) gives

$$S^\sigma \partial T - V^\sigma \partial P + a \partial \gamma + \sum_i n_i^\sigma \partial \mu_i = 0 \qquad (11.5)$$

If the conditions are isothermal then with T constant and $V^\sigma \approx 0$, we obtain

$$a \partial \gamma + \sum_i n_i^\sigma \partial \mu_i = 0 \qquad (11.6)$$

which rearranges to

$$-a \partial \gamma = \sum_i n_i^\sigma \partial \mu_i \quad \text{and} \quad -\partial \gamma = \sum_i \frac{n_i^\sigma \partial \mu_i}{a} \qquad (11.7)$$

which can be rearranged to have the form

$$-\frac{\partial \gamma}{\partial \mu_i} = \sum_i \frac{n_i^\sigma}{a} \qquad (11.8)$$

which is the number of moles per unit area at the surface and leads to the so-called Gibbs equation:

$$-\frac{\partial \gamma}{\partial \mu_i} = \sum_i \Gamma_i \qquad (11.9)$$

which is the surface excess Γ_i in mol m^{-2}. For a two-component system, *i.e.* a solvent and solute, which is the situation for a soap dispersed in water with an air interface, then

$$-\partial \gamma = \Gamma_1 \partial \mu_1 + \Gamma_2 \partial \mu_2 \qquad (11.10)$$

In this situation it is reasonable to assume that the water cannot cross the interface into the air, but the soap molecule can be located both in the water and air regions. In this case we let Γ_1, the excess area due to component 1, be zero:

$$-\partial \gamma = \Gamma_2 \partial \mu_2 \qquad (11.11)$$

and $\mu_2 = \mu_2^0 + RT \ln(a_2)$, in which a is the activity of the distributed component and μ_2^0 is a constant. Differentiating eqn (11.11) yields

$$\partial \mu_2 = RT \partial \ln(a_2) \tag{11.12}$$

which is referred to as the Gibbs absorption isotherm:

$$-\partial \gamma = \Gamma_2 RT \partial \ln(c_2) \tag{11.13}$$

This equation is often rearranged using $-\partial \gamma = \Gamma \partial \mu$ and $\partial \mu = RT \partial \ln(a)$ to give the surface excess area as

$$\Gamma = -\frac{\partial \gamma}{\partial \mu} = -\frac{1}{RT} \frac{\partial \gamma}{\partial \ln(c)} \text{ or } \Gamma = \frac{-c}{RT} \frac{\partial \gamma}{\partial c} \tag{11.14}$$

The implications of these equations are as follows:

- If $\gamma \downarrow$ as c $\uparrow$, then $\partial \gamma / \partial c$ is negative so Γ is positive and the component is adsorbed at the interface.
- If $\gamma \uparrow$ as $c \uparrow$, then $\partial \gamma / \partial c$ is positive so Γ is negative and depletion of the component occurs at the interface.

This latter situation occurs in the case of some ionic solutions, where as a result of solvation of the species, it would prefer to stay in the bulk rather than at the surface. In the case of a soap, the gain in energy for the hydrophobic tail being transferred across the interface is greater than the solvation energy and segregation tends to occur. In this latter situation, $\pi = (\gamma^0 - \gamma)$ can be expressed as being proportional to the concentration c so that $\pi = kc$ and $k = \pi/c$, so that $\gamma^0 - \gamma = kc$ and $\gamma = \gamma^0 - kc$ leading to $\partial \gamma / \partial c = -k = -\pi/c$, and

$$\Gamma = -\frac{c}{RT} \left(\frac{\partial \gamma}{\partial c} \right) = \frac{-c}{RT} \left(\frac{-\pi}{c} \right) \text{ and } \Gamma = \frac{\pi}{RT} \tag{11.15}$$

The spreading pressure adsorption (mol m^{-2}) is thus $\pi/\Gamma = RT$. If we let $1/\Gamma = \sigma$ (m^2 mol^{-1}), then $\pi \sigma = RT$, *i.e.* the spreading pressure × area occupied by 1 mole = RT. This is the two-dimensional analogue of the ideal gas equation and implies that molecules can move around a surface in the same way that a gas moves around in a three-dimensional solid.

11.3 Minimum Surface Energy Conditions

Although we are primarily concerned with phases which are created within a solution or a solid, looking at an air–liquid interface can help with understanding the conditions that occur in the former conditions. Depending on the surface tension, the vapour–liquid interface will adopt a curvature that reflects the balance of forces that exist in the surface (Figure 11.3).

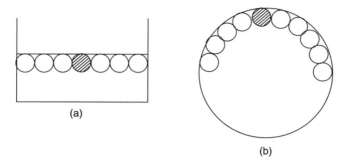

Figure 11.3 Schematic of a plane and curved surface.

In a curved surface (Figure 11.3b), the coordination of any individual molecule will be less than in a planar surface (Figure 11.3a). The coordination is less at a curved surface with the result that there is a greater tendency for the molecules to be able to leave the surface and the effective vapour pressure is higher. Thus using

$$\Delta G = V\mathrm{d}P - S\mathrm{d}T + \sum_i \mu_i \mathrm{d}n_i + \gamma \mathrm{d}A \tag{11.16}$$

we can define $\mu_i = (\mathrm{d}G/\mathrm{d}n_i)_{P,T,n_i a}$ for a flat surface. Thus for a small drop, the addition of a small amount of material $\mathrm{d}n_i$ changes the volume and therefore the area. The addition of $\mathrm{d}n_i$ to a drop of radius r leads to a change in volume $\mathrm{d}V = \sum \overline{V}_i \mathrm{d}n_i = 4\pi r^2 \mathrm{d}r$ where $\overline{V}_i$ ($=\mathrm{d}V/\mathrm{d}n_i$) is the partial molar volume change, and can be expressed for a sphere as

$$\mathrm{d}A = 8\pi r \mathrm{d}r = \frac{2(4\pi r^2 \mathrm{d}r)}{r} = \frac{\sum_i 2\overline{V}_i \mathrm{d}n_i}{r} \tag{11.17}$$

Substitution in eqn (11.16) gives

$$\mathrm{d}G = V\mathrm{d}P - S\mathrm{d}T + \sum_i \mu \mathrm{d}n_i + \frac{\left(\sum_i 2\overline{V}_i \mathrm{d}n_i\right)}{r} \tag{11.18}$$

$$= V\mathrm{d}P - S\mathrm{d}T + \sum_i \left(\frac{2\overline{V}_i \gamma}{r} + \mu_i\right) \mathrm{d}n_i \tag{11.19}$$

We can define for a simple curved surface

$$\mu'_i = \left(\frac{\mathrm{d}G}{\mathrm{d}n_i}\right)_{T,P,n_i} \tag{11.20}$$

so that $\mu'_i = 2\overline{V}_i\gamma/r + \mu_i$ which is equal to the difference in the curved surface and the plane surface, and $\mu'_i - \mu_i = 2\overline{V}_i\gamma/r$. One can compare the equations for the plane and curved surfaces: $\mu_i = \mu_i^0 + RT \ln P_i^0$ for a plane surface and $\mu_i = \mu_i^0 + RT \ln P_i$ for a curved surface. Thus $\mu_i^0 + RT \ln P_i - \mu_i^0 - RT \ln P_i = 2\overline{V_i}\gamma/r$ and $RT(\ln P_i/\ln P_i^0) = 2\overline{V_i}\gamma/r$ which leads to $\ln P_i/\ln P_i^0 = 2\overline{V_i}/RT(r)$. These equations illustrate the observation that droplet and phase separated systems will naturally tend to form a curved surface to minimize the free energy.

11.4 Langmuir Trough[1,2]

In this monograph, we are primarily interested in organization in the solid state; however, since solids are formed from either melts or often solutions it is appropriate to understand the conditions under which pre-assembly of ordered phases is possible in the liquid phase. The pioneering work of Langmuir has helped us gain an insight into the way molecules can organize at liquid–air interfaces and in particular the biologically important water–air interface. The most common method of studying the surface is using a Langmuir balance[3,4] (Figure 11.4).

The trough is usually a constructed from a ceramic and filled to the brim with the substrate (water). The substance to be investigated is carefully added to the trough. In order to define the surface area, a series of movable bars are placed across the surface of the liquid (water). There is usually one bar that is movable and controlled by a screw or pulley. This bar can be moved up and down the trough and allows the area of the surface to be varied. Reduction of the area by movement of the static and movable bars together squashes the molecules trapped in the surface and leads to an increase in the surface pressure. The pressure or force in the surface is easily measured using a piece of filter paper, suspended so as to lie vertically in the surface and its mass is measured using a delicate balance. Typical plots that are obtained are shown in Figure 11.5.

Large molecules, such as sodium stearate, which have no side chains are able to pack into a single monolayer as the pressure is increased. A complete

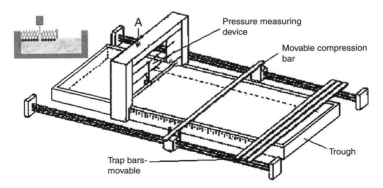

Figure 11.4 Schematic of a Langmuir trough. The marks on the side of the trough indicate distance and hence area of the surface. The inset indicates the compression of the molecules by the moving barrier.[5]

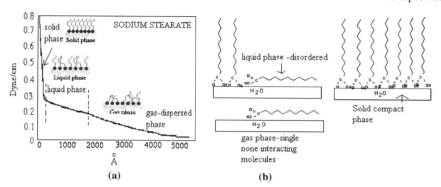

Figure 11.5 (a) Pressure area plot for sodium stearate and (b) schematic diagrams for the various states of compaction of the molecules in the surface. The division of the areas follows the nomenclature of Langmuir.[1]

compressed monolayer will be created at the solid–liquid line and this is essentially a crystalline solid. Below the liquid–gas line, the molecules are more disordered in the surface and at various stages will resemble liquid-like order and, in the extreme, a gas.

11.5 Langmuir–Blodgett Films[3,4]

Over the last twenty years, a vast amount of research has been carried out on ordered films, prepared using a Langmuir trough and a dipping method. The technique involves moving a substrate through the layer that has been compressed to be close to the 'solid' condition. The layers that are picked off from the surface can then be assembled to produce 'idealized' structures that mimic in many cases those found in nature. The process is shown schematically in Figure 11.6. The molecules are first compressed into a monolayer on the liquid surface (Figure 11.6a). A plate, usually a quartz slip, is lifted from below the surface, and moving through the monolayer (Figure 11.6b) picks up a single layer of pre-ordered molecules. If the plate is then pushed back through the surface, a further monolayer is deposited (Figure 11.6c) and so on (Figure 11.6d). Many tens of monolayers have been deposited by this method producing well-organized model structures. The structure of the layers alternates, as the first layer will have the hydrocarbon chains sticking up from the surface and the carbonyl head groups attached to the glass. This hydrophobic surface of the glass slip coated with the monolayer of molecules is then wetted as the plate is returned to the bath by the hydrocarbon chains in the surface monolayer. The molecules in the surface will align with the hydrocarbon chains of the first monolayer changing the surface to being hydrophilic by depositing the second layer in the reverse sense to the first layer. Removal of the plate will once more coat the plate with a further layer aligned in the same sense as the first layer and make the surface hydrophobic. This will then be repeated on

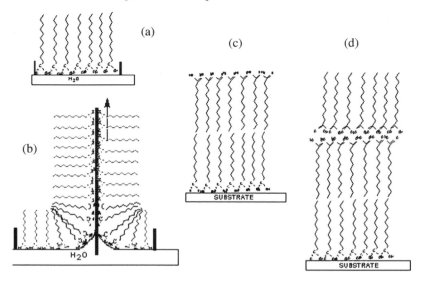

Figure 11.6 Various stages of Langmuir–Blodgett film deposition.

subsequent dipping. The substrates are usually hydrophilic (glass or quartz for optical studies, silver and gold (conductive substrates), silicon or gallium arsenide plates (semiconductive)) or hydrophobic (mica or organically modified glass).

The bilayer structure (two layers organized back to back) is similar to that found in virtually all biological cell structures and forms an effective barrier between two aqueous phases. The layers instead of adopting a planar structure will curve to minimize the energy. The result of the curvature is to form a closed surface with the hydrophilic head groups in contact with the aqueous environment both within and outside the cell. The molecules that can be deposited by this method include many amphiphilic molecules. These molecules usually contain a non-polar hydrocarbon-based chain: alkylbenzene, benzenedodecyl, lauryl, stearic, palmitic oleic, hyrogenated tallows and other fatty acids. The polar head group is usually a sulfate, phosphate, carboxylic acid or amine. Counter ions often will include sodium, potassium or ammonium ions. These molecules can distribute themselves across an interface and change the surface energy.

11.6 Micelle Formation[1,2]

The situation that exists in the trough is rather more complex than this simple picture portrays. In very dilute solution, the molecules may be soluble in the water and hence there is a distribution between the bulk and surface. However, as the concentration is increased the possibility of the surfactant molecules self-aggregating arises. At a critical concentration, the *critical micelle concentration* (CMC), molecular-scale particles are formed in the solution. These molecular-scale particles are called *micelles* and in the simplest form have a spherical shape, with a diameter that is approximately the extended length of the hydrophobic chain. In the case of sodium stearate, there are approximately 50 molecules contained within this spherical micelle structure. Because it has a curved interface with water it will represent a low-energy structure. Because they are of molecular dimensions they will not scatter light and solutions appear clear; however, many physical properties of the solution are changed as a consequence of their presence. The process of micelle formation is illustrated in Figure 11.7. Because the formation of micelles depends on the solubility of the surfactant in the dispersing media, there exists a critical micelle temperature (CMT) below which a micelle can exist and above which it does not exist. There is also another temperature that exists in the surfactant system known as the Krafft point. The Krafft point marks a transition between the dissolution of the soap to form ions and its dissolution to form micelles. The solubility of soap increases sharply with temperature once micelles have appeared by a process that can only be described as self-solubilization that packs more solute into micelles. The Krafft temperature and CMT are closely related.

A micelle will contain a number of molecules, the exact number depending upon the alkyl chain length. As the chain length is increased, so the number of molecules contained in the micelle will proportionally increase:

Number of C atoms in detergent chain, n	Number of molecules in micelle, m
12	33
14	46
16	60
18	78

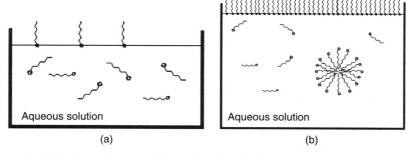

Figure 11.7 Distribution of molecules between the bulk and surface: (a) at a concentration well below the CMC; (b) at a concentration well above the CMC.

Colloidal systems can be divided it two types. *Lyophilic* colloids (solvent loving) are true solutions of a single large molecule or aggregate. They are formed when solute and solvent are brought together and are stable indefinitely. *Lyophobic* (solvent hating) do not form spontaneously and generally need some additional factor to stabilize them. If left long enough the two phases may separate. The general discussion of colloidal systems is beyond the scope of this monograph and the reader is referred to one of the texts listed at the end of this chapter for more detailed discussions on this topic. *Lyophobic* systems can demonstrate the way in which adsorbed molecules on a surface can influence the habit of growth of inorganic crystals and are important in understanding bio-mineralization.

11.7 Stability Energy and Surface Area Considerations in Colloids

The behaviour of the two types of system is reflected in the energy of interaction between the molecules and the dispersing solvent. *Lyophilic* systems are constituted from large molar mass molecules, such as casein, dissolved in a solvent, such as water, to produce a single-phase colloidal solution, milk. The polymer chain is fully solvated by the solvent molecules, all segments are in equilibrium with the surrounding media or solvation shell and the system is thermodynamically and kinetically stable. In *lyophobic* systems the surface tension force acts parallel to the surface pulling inwards to oppose the spreading of the surface. Assuming that the volume is not otherwise constrained, then the smallest surface area for a given volume is a sphere. The total surface energy of the system is equal to the product of the surface tension and the surface area. Suppose we divide a single particle into a series of smaller spheres. For a given mass of the dispersed phase there will be $n = m/4\pi r^3 \rho$ particles, where r is the average radius and ρ is the density. Therefore n is proportional $1/r^3$. The total area of these n particles is $= n \times 4\pi r^2 \propto n \times r^2 \propto r^2/r^3$. Therefore the total surface area is proportional to $1/r$. Therefore as we decrease the radius of the particle we increase the total surface area and hence the total surface energy of the system. Therefore the system becomes less thermodynamically stable in the absence of other factors. Specific surface area A_{sp} is defined as the ratio of the total surface area to the total mass: $A_{sp} = n \times 4\pi r^2 / n \times 4/3\pi r^3 \rho = 3/r\rho$. Although a system may be thermodynamically unstable, many colloids have significant long-term stability (kinetic stability) as a result if a high activation energy for change. Total surface energy for 1 mol of particles of radius 10^{-7} cm is 4 kJ mol^{-1} and should be compared with the activation energy for viscous flow or bond strength, 40–100 kJ mol^{-1}.

11.8 Stability of Charged Colloids

Many colloidal systems and in particular biological systems contain oligomers and macromolecules that can carry a charge. A typical colloidal solution consists of the continuous phase, usually water, a colloidal electrolyte particle and a low molecular weight univalent electrolyte, *i.e.* a simple salt. We designate the colloidal particle as a macroion PX_z, where P is the particle itself carrying z positive charges and is maintained electrically neutral by z counter anions X^-. When such a colloid system is introduced into a vessel separated by a membrane, as in an osmometer, then the small ions M^+, X^- can pass through the membrane but the macro ions (colloid itself) cannot. At equilibrium (Donnan equilibrium):

$$M^+X^- \qquad\qquad M^+X^-P^{z+}$$
$$\beta \text{ phase} \qquad\qquad \alpha \text{ phase}$$
$$(a_{M,\alpha})(a_{X,\alpha}) = (a_{M,\beta})(a_{X,\beta}) \quad \text{Ion activity product}$$

In practice, M or X^- are not evenly distributed but are influenced by the presence of P^{z+}, the colloidal species and the more colloid or macroions present and the higher the charge they carry z, then the greater the unevenness of the distribution of the M^+ ion. In practice, the membrane itself generally becomes polarized because of the macroions on one side and this is similar to the natural membranes in living systems. In dialysis, the Donnan equilibrium allows the removal of ions and traces of impurities. This is a fascinating topic that we do not have space to explore in detail in this monograph.

11.9 Electrical Effects in Colloids

The kinetic effects in lyophilic systems are usually associated with the charge carried by the particle. Proteins, for example, acquire charge through ionization of carboxyl and amino groups to give $-COO^-$ and $-NH_3^+$ ions. The charge on the protein is thus pH dependent because –COOH is a weak acid and $-NH_2$ is a weak base. At low pH the protein will be positively charged and at high pH will be negatively charged. There is an isoelectric point where there will be no charge on the molecule.

With lyophobic systems charge can also readily be acquired. Hydrocarbon droplets in water acquire a negative charge by preferential adsorption of anions, *e.g.* OH^- are less hydrated than the cations H^+ and have greater polarizing power. Even air bubbles can acquire charge and stick to the side of a drinking glass.

11.10 Electrical Double Layer[1,2]

When we have a charged surface in the absence of thermal motion, the charge would be neutralized by adsorption of an equal and opposite number of charged ions (counter ions or gegen ions). In fact, thermal motion prevents formation of such a compact double layer and in practice there is a distribution of counter ions (a screening effect) and a distribution of counter ions in the vicinity of the surface, such that the electrical potential gradually falls to zero at

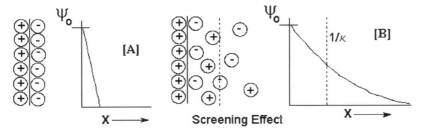

Figure 11.8 Schematic of the ion distribution close to an interface. It is assumed that the particle dispersed in the aqueous phase carries a positive charge and there will be a distribution of negative counter ions in solution. In addition, cations may be present and aid the charge balance.

long distances from the particle surface. This is known as a diffuse double layer and it is possible theoretically to treat the situation quantitatively in a number of ways (Figure 11.8). The first approach is to assume that the ions are not in motion and that they balance out at the interface. This leads to the Gouy–Chapman regime.[5] However, in a real system the ions will move and there is therefore a more diffuse layer close to the interface.

The Gouy–Chapman theory leads to an effective electrical double layer which is represented by the decay of the potential ψ_0 as a function of distance from the interface. The potential is assumed to decay approximately exponentially. The thickness of this layer is characterized by the point at which the potential has dropped to $1/e$ of its initial value and is defined by the double layer thickness $1/\kappa$. The value of κ is defined by $\kappa = \sqrt{8\pi e^2 c_0 z^2 / \varepsilon k T}$, where e is the electron charge, c_0 the bulk electrolyte concentration, z the valence and ε the dielectric constant. The double layer thickness is ~ 10 Å for 10^{-1} M and 100 Å for 10^{-3} M of a univalent ion combination.

A refined version of the potential due to Stern allows for there to be a contact layer of counter ions that does not exchange rapidly; behind this is a more diffuse layer of charge. The Gouy–Chapman[5,6] calculations give too high a charge and this is reduced by the adsorbed counter ions (Figure 11.9).

Original calculations based on point charges gave too high an ionic concentration near the surface. The refined calculations introduce a finite ion size into the calculation that allows for the fact that the ions may have a hydration sheath and allows for specific interactions with counter ions. The theory simply splits the potential function into two parts: the compact contact layer where the potential falls quickly from ψ_0 to ψ_1 and a diffuse layer where the potential drops more slowly to zero.

11.11 Particle (Micelle) Stabilization

The stabilization of a micelle can be achieved by two distinctly different mechanisms depending upon whether the molecules involved are able to

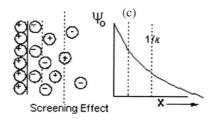

Figure 11.9 Stern potential indicating the contact layer and the diffuse layer profiles.

dissociate into charged species or not. If the stabilizing molecules are able to create ionic species the micelles are able to repel one another and hence there is a barrier that has to be overcome before the size of the particle can be increased to lower the total energy. If the molecules are unable to carry a charge then entropic factors have to provide the stabilization.

11.11.1 Charge Stabilization: Derjaguin–Landau–Verivey–Overbeek (DLVO) Theory

The DLVO theory[7,8] describes the electrostatic stabilization of charged colloidal particles. The charges on the particles may be associated with ionization of specific groups on the molecules, as in the case of proteins or attributed to charges adsorbed onto a particle or molecule. If two particles carry similar charges, then there will be repulsive interactions between them. The double layers described above, model the charge distribution close to such a surface. Unlike charges lead to attractive interactions which will attempt to cause the particles to merge and if the particles are deformable then they will grow to a single-phase large particle and ultimately separate into two phases. The behaviour of charged colloids is a controlled balance of these two effects.

To demonstrate the principles of the theory, it is useful to simplify the problem and initially consider the interaction between parallel plates. If the particle is large, then the parallel plate model is a good approximation to reality. This theory can then be generalized to include spheres by the introduction of a geometrical averaging factor. The so-called Hamaker potentials allow for this generalization of the interaction potentials. Although the spherical case is considered to be more descriptive of colloids, the mathematics are more difficult and the geometrical factors more difficult to understand. The idealized interaction potential is shown in Figure 11.10 for an increasing attractive potential and reflects the effects of change in temperature. The potential is the sum of an attractive term which acts at long range and is being progressively increased and a repulsive term which acts at short range and is essentially constant. The curves simulate the variation of the potential as a function of temperature. A stable colloidal dispersion is only obtained when a minimum exists in the potential as a function of the distance of separation of the particles x. The particles will have kinetic energy and if the average thermal energy kT is

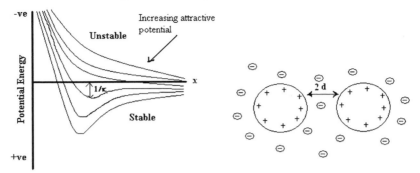

Figure 11.10 Variation of the potential with variation of the strength of the attractive and repulsive contributions to the total potential as a function of particle separation, $2d$.

greater than the stabilization energy depicted as $1/\kappa$, then the particles can move about freely. If the thermal energy is less than kT, then the particles will order at a distance dictated by the location of the minimum energy. The system is able to *self-assemble*.

(i) Repulsive term. The term for two parallel plates held at a distance $2d$ apart and assuming that the double layer has the Gouy–Chapman form:[5,6]

$$V_{\text{rep}} = \left\{ \frac{64 c_0 kT}{\kappa} \right\} \gamma^2 \exp(-2\kappa d) \tag{11.21}$$

where $\gamma = \exp(ze\psi_0/2kT) - 1$.

For the Stern model ψ_0 is replaced by ψ_1, to allow for the effect of contact charges on the shape of the potential.

(ii) Attractive term. Similarly the attractive term has the form

$$V_{\text{att}} = -\frac{A}{48\pi d^2} \tag{11.22}$$

where A is a constant which depends on the density and polarizability of the media. The combination of these two terms gives potential curves of the form shown in Figure 11.10. The balance between the attractive and repulsive forces gives a minimum which is indicative of a stable state. The height of the barrier determines the kinetic stability of the colloid. If $E \gg kT$ then we have kinetic stability; if $E \leq kT$, the system is not stable. The magnitude of κ decreases as the electrolyte concentration decreases and there is a critical value for stability which is determined by $\partial V_{\text{total}}/\partial x = 0$. Hence

$$V_{\text{total}} = V_{\text{repulsive}} + V_{\text{attractive}} = \left\{ 64 \frac{c_0 kT}{\kappa} \right\} \gamma^2 \exp(-2\kappa d) - \frac{A}{48\pi d^2} \tag{11.23}$$

Where V_{total} and $\partial V_{total}/\partial x$ both equal zero, the stabilizing barrier disappears and we get flocculation. Thus

$$dV_{total}/dx = -2kV_{rep} - 2V_{att}/d = 0 \quad (11.24)$$

and

$$\frac{64c_0kT}{\kappa\gamma^2}\exp(-2\kappa d) - \frac{A}{48\pi d^2} = 0 \quad (11.25)$$

For $V_{rep} + V_{att} = 0$ and $\kappa d = 1$, eqn (11.25) becomes

$$\frac{64c_0kT}{\kappa\gamma}\exp(-2) = \left(\frac{A}{48\pi}\right)\kappa^2 \quad (11.26)$$

However, $\kappa = (8\pi c_0 e^2 z^6/\varepsilon kT)^{1/2}$ and therefore $c_0 = 107\varepsilon^2\ k^5\ T^5\gamma^4/A^2 e^6 z^6$. Inserting reasonable values for a colloidal system in water we obtain $c_0 = 8 \times 10^{-22}(\gamma^4/A^2 z^6)$ mmol l^{-1} at high potentials $\gamma \to 1$. The theory predicts flocculation concentration of different electrolytes containing mono-, di- and trivalent counter ions to be in the ratio $1:1/2^6:1/3^6$ or $100:1.6:0.13$.[8] This theory, in a more generalized form, can be used to describe the behaviour of proteins and other macromolecular species. The theory highlights the extreme sensitivity of the system to the charge stabilization and how change in the charge distribution can destabilize the system (salting out).

11.11.2 Steric or Entropic Stabilization?[9]

Many colloids are dispersed in organic solvents, e.g. carbon black in printing ink, TiO_2 in paints. Many polymer systems, because of the length of the chains involved, form micelle structures as a consequence of *entropic* stabilization. In organic solvents, charging effects are much less pronounced and may be almost absent in relatively low dielectric constant solvents. The polymeric nature of the molecules allows them to create disorder around the particle to which they are attached and act as protective lyophilic colloids. Polymer molecules that can be used in this way have a structure which allows them to be distributed across the interface and will usually have polar and non-polar block segments. In an aqueous system, the polar element will usually be water soluble, whereas the non-polar element will be water insoluble. This difference in solubility will naturally create an interface and the flexibility of the chains dispersed in the water produces the required entropy for entropic stabilization. The approach of a pair of particles and hence their ability to flocculate is inhibited by the extending disordered chains and the overall effect is that of steric stabilization (Figure 11.11). Polymer molecules are sometimes designed to be selectively absorbed onto a specific surface of another polymer particle and 'wave around' in the solvent. The waving chain will often have a branched chain structure to inhibit crystallization. This type of stabilizer has the advantage of being less

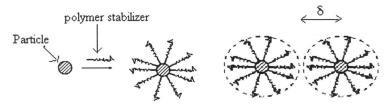

Figure 11.11 Schematic of absorption of a polymer chain on a particle surface and the resultant stabilization. The parameter δ designates the thickness of the disordered layer.

sensitive to pH, electrolyte and temperature changes, and is a vital part of modern non-drip paint systems.

The detailed theoretical modelling of the stabilization process is quite complex; however, a simple model provides the qualitative behaviour of these systems. When two particles approach one another to a distance that is less than 2δ, where δ is the thickness of the adsorbed layer, then the degree of stabilization can be defined in terms of the energy change which occurs upon interaction of the adsorbed layers. The energy term will be a combination of an entropy and enthalpy term: $\Delta G = \Delta H - T\Delta S$. If ΔG is negative, flocculation will occur; if ΔG is positive, then the system will be stabilized.

11.11.3 Entropic Theory

This assumes that absorbed polymer layers are essentially impenetrable and as the layers approach they become compressed. This compression process will involve a change in configurational entropy, *i.e.* ΔS is negative. The change in the enthalpic interaction is usually neglected so that $\Delta G = -T\Delta S$, ΔG is thus positive and flocculation occurs. The simplest way of considering the polymer is to visualize it as a rod hinged at the surface and of length l (Figure 11.12). If the entropy of an unconstrained polymer chain is described by $S_\infty = k \ln W_\infty$ and that of an constrained system by $S_c = k \ln (W_c)$, then the change in the entropy is $\Delta S = S_c - S_\infty = k \ln (W_c/W_\infty)$.

The number of configurations is proportional to the length of the chain. In the case of the unconstrained molecule this has a value l and in the constrained situation a value h. Then the interaction potential V_r is given by

$$V_r = kT \ln\left(\frac{h}{l}\right) = kT\left(1 - \frac{h}{l}\right) \text{ if } \frac{h}{l} \ll 1$$

If there are n adsorbed molecules in the interface then the total repulsive energy per unit surface area is

$$V_r = N_s kT \theta_\infty \left(1 - \frac{h}{l}\right)$$

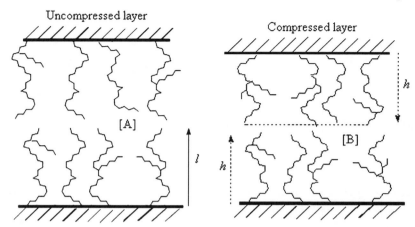

Figure 11.12 Schematic of uncompressed and compressed layers.

where θ_∞ is the fraction of surface covered when the molecules are totally expanded. The theory has been further developed to account for spherical particles of flexible molecules and the effects of the solvent on the configuration. The results are essentially similar except that a shape factor has to be included to allow for the averaging of the potential.

11.12 Phase Behaviour of Micelle Systems[10]

As discussed in Chapter 8, a phase separated system has the tendency to form ordered structures. Self-assembly and molecular organization depend on a number of competing factors. Typically there will be differences in the strength of the intramolecular interactions, the flexibility of the chains and the intermolecular forces. The relative magnitude of the attractive hydrophobic forces between hydrophobic tails, the repulsive electrostatic forces between the charged head groups and the head group hydration effects all influence aggregate architecture and stability. From the concepts outlined above in relation to minimization of the surface energy, it is apparent that the curvature of the structure formed is an important criterion for determining the shape. The curvature can be expressed in terms of a dimensionless parameter know as the 'critical packing parameter' $v/a_0 l_c$, where v is the volume of the hydrocarbon chain, assumed to be fluid and incompressible and expressed in terms of the chain length. The chain length l_c is the critical chain length assumed to be approximately equal to l_{max}, the fully extended chain length. For a molecule containing m chains each with n carbon atoms and having a head group with area a_0 then

$$v = (27.4 + 26.9n)m \tag{11.27}$$

and

$$l_{max} = 1.5 + 1.265n \qquad (11.28)$$

Many surfactants have a molecular structure that involves two long alkane chains being attached to a single polar head group. The double chain structure can significantly influence the packing arrangements. The critical packing parameter can be used as a guide to the aggregate architecture for a given surfactant (Figure 11.13). Typical values and their corresponding aggregate structures are: $v/a_0l_c < 1/3$, spherical micelle; $1/3 < v/a_0l_c < 1/2$, polydispersed cylindrical micelles; $1/2 < v/a_0l_c < 1$, vesicles, oblate micelles or bilayers; $v/a_0l_c > 1$, inverted structures.

As the concentration is increased, the micelles will interact and form a series of regular structures that reflect the detail of the interaction between the species present. This progressive change can be depicted schematically (Figure 11.14). Not all of the phases will be observed, and in the case of micelles, an increase in the temperature will allow the molecules to exceed the stabilization criteria and more open structures can be produced which will eventually flocculate. An idealized surfactant system will usually involve an amphiphilic molecule, a solvent (usually water) and a non-solvent, which may be an oil or other non-polar material. The phase diagram that is obtained is shown in Figure 11.14.

Moving up the water–surfactant axis there is a progressive change in structure. In very dilute solutions, there are isolated surfactant molecules present. Once the CMC has been achieved, micelles will be formed. Increasing the

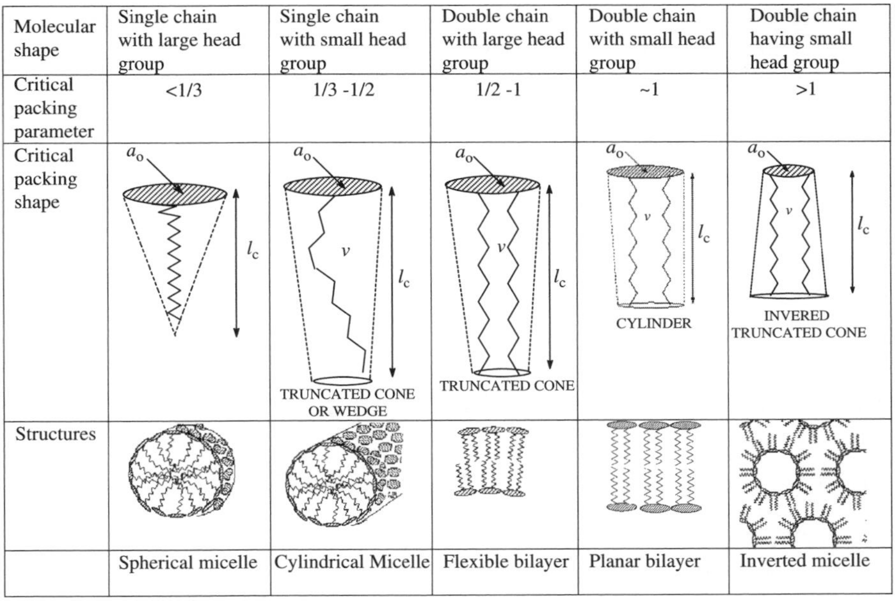

Molecular shape	Single chain with large head group	Single chain with small head group	Double chain with large head group	Double chain with small head group	Double chain having small head group
Critical packing parameter	<1/3	1/3 -1/2	1/2 -1	~1	>1
Critical packing shape	(cone)	(truncated cone or wedge)	(truncated cone)	CYLINDER	INVERTED TRUNCATED CONE
Structures	Spherical micelle	Cylindrical Micelle	Flexible bilayer	Planar bilayer	Inverted micelle

Figure 11.13 Influence of critical packing factors on aggregated structures.

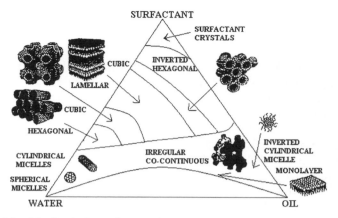

Figure 11.14 Idealized phase diagram for a surfactant–water–oil system.

concentration of surfactant will change the structure of the micelle from a spherical to a cylindrical structure. The first major phase change is between the cylindrical micelles and a hexagonal close packed structure. Subsequent phase changes involve the hexagonal phase changing to a cubic structure and then to a lamellar structure. These are all soft solids and have properties that reflect ever stronger interactions and a progressive change to a hard crystalline state. These intermediate phases will resemble soaps or thick greases. Further increase in concentration causes changes to a cubic phase and then an inverted hexagonal structure and ultimately a crystalline phase is formed if the temperature is below the melt temperature. Moving down the surfactant–oil axis, a simpler set of phase changes is observed which reflect the greater hydrocarbon chain solubility in the oil phase. The phases are therefore the inverse of those in the water–surfactant system and in dilute solution will form inverse micelles. The centre of the phase diagram is the result of the balance of these forces leading to a complex array of phases. Around the centre of the diagram there is a region where the two-phase structures are almost in equilibrium and a co-continuous phase is observed. At low surfactant levels monolayers can be formed at the surface and these will be able to incorporate a small amount of the oil in monolayer. This type of idealized phase diagram is observed for many systems.

11.13 Phase Structures in Polymer Systems

11.13.1 Block Copolymers and Associated Phase Diagrams

Many amphiphilic block copolymers and related oligomeric materials exhibit structure formation at a nanometre scale. By careful selection of the solvent, phase separation can be achieved allowing one block to be dissolved and the other to be immiscible. The head group(s) of a surfactant or the hydrophilic block(s) of a block copolymer will readily dissolve in water and a poor solvent or non-solvent for the other part of the molecule (e.g. the organic tail(s) of the surfactant or the hydrophobic blocks(s) of the block copolymer) will self-segregate or disperse in a hydrocarbon phase. This approach leads to the production of micelles, microemulsions, vesicles, monolayers (on surfaces) and many biologically important systems (e.g. cell membranes). In the case of polymeric materials, the number of polymer chains in one micelle is approximately the same and results in the micelle size distributions being very narrow. For polydisperse block copolymers or surfactants, it is found that to minimize the free energy the aggregates formed are of uniform size. The features that are created at the surface of a blend (Chapter 10) are all of similar size and support this contention. The size of the micellar core is determined by the association number and volume of other solvent-phobic species incorporated into the core. Larger micelles can be produced if the core is filled with species that are solvent-phobic. For triblock copolymers with two end blocks in a poor or non-solvent, supramolecular formation with open structures that tend to obey an open-association mechanism can occur[11-13] (Figure 11.15).

The solubility of block copolymers decreases with increasing temperature and above a certain temperature, phase separation occurs, the 'cloud point' is observed.[13] At higher polymer concentrations, the entanglement of the polymer chains in solution leads to the formation of homogeneous, immobile, gel-like structures. The first gel-like structure is usually a cubic structure (body-centred cubic (bcc) or face-centred cubic (fcc)), formed by an ordered packing of spherical micelles. At even higher polymer concentrations, the arrangement of the hydrophilic and hydrophobic regions leads to the formation of hexagonal

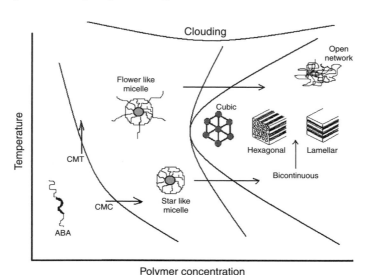

Figure 11.15 Schematic phase diagram for the phase behaviour of amphiphilic block copolymers represented by ABA-type triblock copolymers in aqueous solution.

or lamellar structures. Bicontinuous cubic structure could also be observed between the hexagonal and lamellar regions.[14–16] For the flower-like micelles formed by a block copolymer in a selective solvent for the middle block, an open network without any ordered structure is generally obtained at high polymer concentrations. Copolymer–water–oil ternary phase diagrams can be much more complicated even at a constant temperature.[17–23]

Under these conditions, broad distributions of the micellar size and mass can be observed. Furthermore, as a function of external conditions such as temperature or solvent composition, micellar shapes other than spherical can be observed, e.g. prolate[24–27] or oblate[28] ellipsoids of revolution.

11.13.2 Pluronics

One of the most extensively investigated nonionic block copolymer–solvent systems are the Pluronics. This generic name describes a series of triblock copolymers containing hydrophilic A blocks and hydrophobic B blocks or groups, with A and B being polyoxyethylene, polyoxypropylene and polyoxybutylene. The blocks are all polyethers, but the different structure of the alkyl component varies the degree of hydrophilicity of the block, polyoxyethylene being the most hydrophilic and polyoxybutylene being the least hydrophilic to water. The phase behaviour is usually studied with the addition of a non-polar organic solvent, such as xylene. The phase structure has been studied extensively using small-angle X-ray scattering.

The phase behaviour of Pluronics in the presence of water and xylene has been extensively studied.[21–23] The ternary phase diagram of Pluronic P84 (E19P44E19)/water/p-xylene studied by Alexandridis et al.[23] at room temperature is reproduced in Figure 11.16. The various phases identified are similar to those for the idealized surfactant system (Figure 11.14). At least nine different phases have been identified: normal micellar solution (oil-in-water), cubic, hexagonal, bicontinuous cubic, reversed micellar solution (water-in-oil), cubic, hexagonal, bicontinuous cubic and lamellar phases. These nine phase classes represent all known possible phases in such a ternary system. In general, block copolymer systems can offer various different nanoscale templates such as three-dimensional (3D) cubic packings of spherical units, two-dimensional (2D) hexagonal packings of cylindrical units, one-dimensional (1D) periodic lamellar systems and 3D bicontinuous cubic morphologies. Considering that the total chain length, the block length ratio, the chain architecture and the chemical composition of block copolymers are all adjustable, the templates can be easily tuned to fit specific requirements. Sometimes, micelles with differing morphologies (e.g. ellipsoids or rods) can also be achieved by choosing a suitable combination of block length and solvent quality. Therefore, it is not surprising that block copolymers have been much applied in nanofabrication processes.

The changes in the phase structure can have dramatic effects on the physical properties, and some of the phases are essentially soft solids rather than being liquids. The very rich structures of polymer matrices and the subsequent formation of nanostructured materials often require an extensive range of

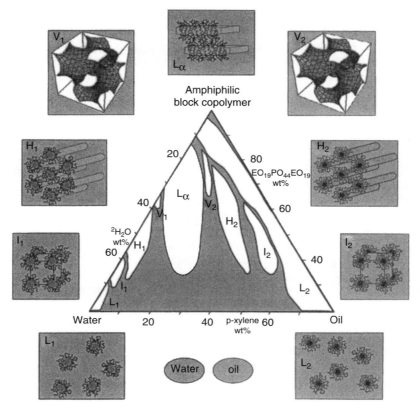

Figure 11.16 Ternary phase diagram of Pluronic P84 (E19P44E19)/water/*p*-xylene at room temperature. Nine different phases can be identified: normal (oil-in-water) micellar solution L1, cubic I1, hexagonal H1, bicontinuous cubic V1, reversed (water-in-oil) micellar solution L2, cubic I2, hexagonal H2, bicontinuous cubic V2 and lamellar phases.[23]

physical techniques to characterize the complex structures over a range of length. It is apparent that with charged polymer systems and also those capable of entropic stabilization, pre-assembly can occur in the fluid phase prior to forming a solid, and this is an important mechanism for the creation of structure in the solid state. Biological systems use this mechanism to great advantage and create complex structures which are capable of executing at a molecular level a specific physical function. This pre-assembly process is able to create a number of the higher order structures that are discussed in Chapter 12.

Recommended Reading

A.W. Adamson and A.P. Gast, *Physical Chemistry of Surfaces*, Wiley Interscience, New York, 1997.

S.Friberg, *Lyotropic Liquid Crystals*, Advances in Chemistry Series 152, American Chemical Society, Washington, DC, 1976.

R.J. Hunter, *Foundations of Colloid Science*, Oxford University Press, 2002.

S. Ross and I.D. Morrison, *Colloidal Systems and Interfaces*, John Wiley, New York, 1988.

References

1. S. Ross and I.D. Morrison, *Colloidal Systems and Interfaces*, John Wiley, New York, 1988.
2. R.J. Hunter, *Foundations of Colloid Science*, Oxford University Press, 2002.
3. A.W. Adamson and A.P. Gast, *Physical Chemistry of Surfaces*, Wiley Interscience, New York, 1997.
4. D.A. Cadenhead, *Ind. Eng Chem.*, 1969, **61**(4), 22.
5. G. Gouy, *J. Phys. Theor. Appl.*, 1910, **9**(4), 457.
6. D.L.A. Chapman, *Philos. Mag.*, 1913, **25**(6), 475.
7. B.V. Derjaguin and L. Landau, *Acta Physicochem. USSR*, 1941, **14**, 733.
8. H. Koelmans and J. Overbeck, *Discuss. Faraday Soc.*, 1954, **18**, 52.
9. D.H. Napper, *Polymeric Stabilization of Colloidal Dispersions*, Academic Press, London, 1983.
10. R.M. Pashley and M.E. Karaman, *Applied Colloid and Surface Chemistry*, John Wiley, Chichester, UK, 2004, p. 74.
11. T. Liu, Z.K. Zhou, C.H. Wu, B. Chu, D.K. Schneider and V.M. Nace, *J. Phys. Chem. B*, 1997, **101**(43), 8808.
12. T. Liu, L.-Z. Liu and B. Chu, in *Block Copolymers: Self-Assembly and Applications*, ed. P. Alexandridis and B. Lindman, Elsevier, Amsterdam, 2000.
13. T. Liu, V.M. Nace and B. Chu, *J. Phys. Chem. B*, 1997, **101**(41), 8074.
14. Z. Tuzar and P. Kratochvil, in *Surface and Colloid Science*, ed. E. Matijevic, Plenum Press, New York, 1993, vol. 15.
15. K. Kon-no, in *Surface and Colloid Science*, ed. E. Matijevic, Plenum Press, New York, 1993, vol. 15.
16. B. Chu and Z. Zhou, in *Nonionic Surfactants*, ed. V.M. Nace, Marcel Dekker, New York, 1996.
17. K. Mortensen, W. Brown and E. Jorgensen, *Macromolecules*, 1994, **27**(20), 5654.
18. G. Wanka, H. Hoffmann and W. Ulbricht, *Colloid Polym. Sci.*, 1990, **268**(2), 101.
19. K. Mortensen, W. Brown and B. Norden, *Phys. Rev. Lett.*, 1992, **68**(15), 2343.
20. P. Alexandridis and K. Andersson, *J. Phys. Chem. B*, 1997, **101**(41), 8103.
21. P. Holmqvist, P. Alexandridis and B. Lindman, *Macromolecules*, 1997, **30**(22), 6788.
22. P. Alexandridis, U. Olsson and B. Lindman, *Langmuir*, 1996, **12**(6), 1419.

23. P. Alexandridis, U. Olsson and B. Lindman, *Langmuir*, 1998, **14**(10), 2627.
24. H. Utiyama, K. Takenaka, M. Mizumori, M. Fukuda, Y. Tsunashi and M. Kurata, *Macromolecules*, 1974, **7**(4), 515.
25. M. Antonietti, S. Heinz, M. Schmidt and C. Rosenauer, *Macromolecules*, 1994, **27**(12), 3276.
26. V. Castelletto, R. Itri and L. Q. Amaral, *J. Chem. Phys.*, 1997, **107**(2), 638.
27. M. Leaver, V. Rajagopalan, O. Ulf and K. Mortensen, *Phys. Chem. Chem. Phys.*, 2000, **2**(13), 2951.
28. G. W. Wu, Z. K. Zhou and B. Chu, *J. Polym. Sci., Part B: Polym. Phys.*, 1993, **31**(13), 2035.

CHAPTER 12
Molecular Organization and Higher Order Structures

12.1 Introduction

In the previous chapters, the organization of polymers and small molecules has been studied at macroscopic, microscopic and molecular level. In many of the systems the structures observed scale relatively simply from the molecular to the macroscopic structure. In the case of a system such as polyethylene, the type of organization observed depends on the scale of the structure being examined. At the nanometre level folding of the chains produces lamellar structures. Because these lamellae are nucleated at the same instant at various points in the crystallizing volume the growing nuclei and/or clusters of lamellae interact and spherulitic structures are observed. Between these spherulites are grain boundaries and these are the dominant feature at the micrometre level. At macroscopic level all these features are not observed and the material appears smooth and homogeneous. Understanding the macroscopic behaviour requires a knowledge as to how these different levels of morphology interact. In biological systems, nature self-assembles complex structures from relatively simple components. Nature recognizes the way in which molecular order scales from molecular to macrostructure, acknowledges the advantages of the occurrence of composite structure in a material and demonstrates the way in which nature builds in rigidity, flexibility and ductility into polymeric materials. It is impossible to cover all natural materials in a book of this size and we will look at only three examples of such systems to illustrate the parallels between natural and synthetic materials.

12.2 Hair

The structure of hair can be considered schematically as in Figure 12.1.

The primary core molecular structure is represented in terms of the α-helix of keratin, which is a major constituent of hair. Keratins are the most abundant proteins and are the main component of the horny layer of epidermis and the

Molecular Organization and Higher Order Structures

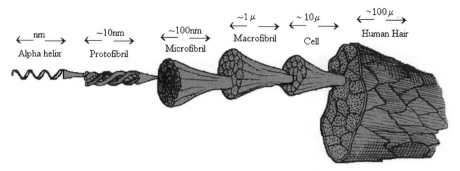

Figure 12.1 Schematic of human hair showing level of morphology.

appendages such as hair, nails, scales, quills, horns and feathers. Keratinous tissues provide strong mechanical support and a chemical barrier. The hard α-keratinous tissue occurs in quills and hair and forms well-ordered and oriented supramolecular architectures. Hair contains profibrils which are made up of keratin combined with other proteins. The protofibrils in turn cluster to form a microfibril that in turn forms part of a macrofibril that are the main part of the cell structure. Hair is constituted from a series of cells. The platelets on the surface of the hair are hydrophobic, whereas the core material is hydrophilic. The principal difference between various forms of hair lies in the nature and cysteine/cystine content of these proteins that are part of the cell walls and are incorporated in the microfibrils. The protein plays the role of a *habit* modifier and controls the organization of the keratin from the initially formed amorphous phase to an organized crystalline structure.

Keratin   Cysteine   Cystine

The chemical structure is in fact much more complex than implied by this simple description, but it illustrates the hierarchy of levels of organization that are present within the hair.[1-5] Pigmentation and the detailed balance of various proteins impact on the colour, structure and texture. Micro X-ray analysis of hair has shown the existence of two major polymorph intermediate filament (IF) architectures. Just outside of the follicle, the filaments are characterized by a diameter of 100 Å and have a low core density. As the hair grows, lateral aggregation of the filaments occurs into a more compact network of filaments and there is a contraction of their diameter to $\sim$75 Å and associated long-range longitudinal ordering of the microfibril. The architecture of the IF in the upper zones is specific to hard α-keratin, whilst other architectures are found in the lower zone.[1] Examination of the structure at high resolution indicates that two

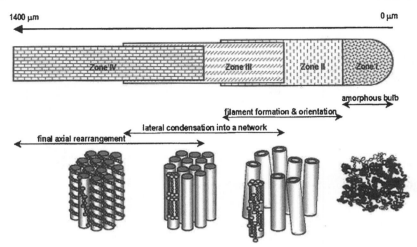

Figure 12.2 Schematic of the changes in morphology as a hair grows.

~450 Å long keratin chains are assembled into a dimeric molecule with a rod-like character and these form the α-helical coils. These dimeric molecules are assembled, both longitudinally and laterally, forming long cylinder-shaped filaments with an axial well-defined periodic ordering. The number of chains across a filament is ~32 chains. These polymorphic changes give hair its strength and the process can be represented schematically as in Figure 12.2.

Water can plasticize hair by interaction with the core of the cell but loses the water when dried. The outer structure contains cystine that has a disulfide linkage and these crosslinks are important in determining the overall organization of the chains within the hair structure and control the detailed morphology that can be adopted. Differences between hair types are reflected in different distributions of the keratin chains, the nature and distribution of the proteins and the type and distribution of the pigment. The proteins, in this case, are playing an important role in the control of the formation of the crystal structure within the core fibres.

12.3 Structure in Cellulose Fibres[2–6]

A classic example of the creation of macrostructure from molecular organization is cellulose-based fibre materials. Cellulose is the dominant polysaccharide in plant cell walls and is often touted as being the most abundant biopolymer on earth. A basic cellulose unit, known as the elementary fibril, contains thirty-six 1,4-β-D-linked polyanhydroglucopyranose chains[7] (Figure 12.3a), and may eventually be coated with non-cellulosic polysaccharides to form the cell wall microfibril. These microfibrils are then crosslinked by hemicelluloses/pectin matrixes during cell growth. The cellulose molecule is constrained to adopt

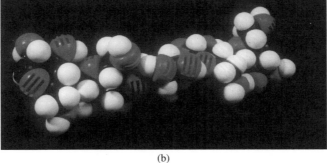

Figure 12.3 Composition and conformation of the cellulose molecule.

certain conformations by various intra- and intermolecular hydrogen bondings which are created when the polymer chains are brought close to one another (Figure 12.3b).

Close inspection of the relative orientation of the rings indicates that the pendant hydroxyl groups can interact to give a lower energy structure if the bridge ether is slightly twisted. As a consequence the backbone of the cellulose chain forms a helical structure with a pitch of 12 rings per rotation. In this respect there are similarities with the keratin helical structure in hair. As with the case of hair, these microfibrils are then themselves organized to form the overall fibre structure (Figure 12.4). Transmission electron microscopy (TEM) of the cell wall indicates that microfibrils have a diameter of between 2 and 10 nm. In native *Valonia* cellulose crystal the packing appears to be hexagonal with dimensions of 2–3 nm.

To understand the physical properties of cellulose, it is appropriate to consider organization at three different levels.[10]

1. *The molecular level.* Cellulose as a consequence of the favoured hydrogen bonding interactions adopts an extended rod-like form which aids packing into crystalline structures. Molecular modeling[4] has allowed the development of a detailed knowledge of the conformational preferences of polymeric systems and realistic models of the chain backbone structure to be created. The local structure of the chain is assumed to involve the β-D-glucose existing in the pyranose form and the ring adopts a $^4C^1$ chair formation, which is the lowest energy conformation for β-D-glucopyranose. The presence of intramolecular

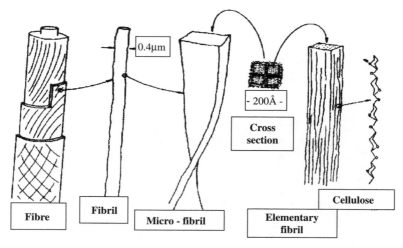

Figure 12.4 Morphology of a cellulose fibre.[8–10]

hydrogen bonding between the hydroxyl group in the 3 position and the ring ether oxygen in the 5 position imposes a twist on the C(1)–O(1) and the C(4)–O(4) glucosidic bonds which results in the chain adopting a helical conformation. Varying how the rigid chains pack together leads to the possibility of polymorphic forms being obtained.[8,9] The study of these structures has been the subject of active research for many years and reflects the fact that nature can control the way in which the chains organize into the larger structures that ultimately generate the variety of physical properties that make these materials useful. For instance, cotton fibres in the bud form are known to exist in a disordered structure and rapidly form the cellulose I structure when the bud opens. Cellulose I structure is a thermodynamically a metastable crystalline structure, which on dissolution in a suitable solvent is not re-formed when the polymer is reprecipitated.

2. *The supermolecular level.* The individual chains are aggregated into polymorphs that form microcrystallites and fibrils. The packing of the cellulose molecules in these hydrogen bonded systems is very dense and none of the hydroxyls in the interior of the crystal are easily accessible. The ability to exchange hydrogen has been explored by investigation of the change in the infrared spectra when cellulose is suspended in deuterium oxide. Native cellulose exhibits the cellulose I structure.[7,8] The mercerized or regenerated cellulose exhibits different structure designated cellulose II. The latter is considered the more thermodynamically stable of the two forms and the nature of the hydroxyl group interactions is different in the two forms. The average hydrogen bond length in the case of cellulose I is 2.72×10^{-10} m which is markedly shorter than the value of 2.80×10^{-10} m in cellulose II. The more dense structure and greater involvement of hydrogen bonding are reflected in the

lower susceptibility to reaction of the regenerated cellulose with the cellulose II structure. Besides the cellulose I and cellulose II polymorphic lattice structures, there are cellulose III and cellulose IV crystal modifications. The cellulose III structure is formed when the reaction product of native cellulose fibres with liquid ammonia is decomposed and closely resembles the cellulose II structure. The cellulose IV structure is obtained by treatment of regenerated cellulose fibres in a hot bath under conditions of the fibres being stretched. The degree of order inside and around these fibrils and their perfection of orientation with respect to the fibre axis are all variables that require to be characterized. A variety of different methods have been used to examine the 'crystalline/amorphous' ratio and the accessibility of the crystalline regions. Among the methods that have been used are the initial loss of matter in hydrolysis, heterogeneous oxidation, deuterium exchange studies, formic acid esterification, hydrogen evolution from the reaction of metallic sodium with cellulose, water and nitrogen adsorption, density measurements, X-ray diffraction and infrared spectroscopy to name just a few. The microstructure of the cell walls is readily visualized using atomic force microscopy (AFM) (Figure 12.5).

The space between the cellulose fibres is filled with hemicellulose, the form of the structure and composition of which varies from plant to plant:

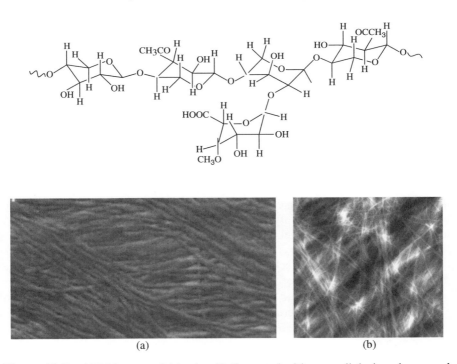

Figure 12.5 AFM images of (a) microfibrils coated with non-cellulosic polymer and (b) the primary wall structure showing microfibrils aligned parallel to one another and the macrofibrils.[2]

and *lignins* which are complex aromatic-containing hydrocarbon polymers:

Lignin

Both the hemicellulose and lignin structures are representative indicating that unlike cellulose they are unable to form the well-ordered and ultimately crystalline structure which gives these materials their useful properties.

3. *The morphological level*. The fibrillar crystalline strands are in turn organized into the 'cross-morphology' of the fibre, the existence of native cell wall layers of the skin–core structure in synthetic fibres, the presence of interfibrillar interstices and voids, *etc*. In particular, the primary wall and outer layer of the secondary wall (S1) in which the fibrils are laid down in a criss-cross fashioned helical manner strongly affect the swelling behaviour and the physical and chemical properties. The increase in tensile strength of cotton fibres on wetting and liquid ammonia treatments could be traced to the hindrance of lateral swelling and the swelling compression exerted on the fibrils in the inner secondary wall layers.

The scale of the organization can be visualized; the fibre structures require to be examined at different levels of magnification[5–8] (Figure 12.6). The physical and chemical properties of the cellulose fibre are mainly affected by:

- The constitution of the cellulose molecule: the presence of the 1,4-glycoside linkages between glucose units, the three reactive hydroxyls and their involvement in intra- and intermolecular hydrogen bonding which leads to the twisted helical structure.
- The length of the cellulose molecules, especially its relation to the length of the elementary crystals or fused fibrillar aggregations. Recent studies favour the idea that the cellulosic material is dispersed in a matrix of non-cellulosic material. The ordered cellulosic fibres are organized in a pseudo-liquid crystalline form.[2–6] This type of model is consistent with many physical properties of these materials (Figure 12.6b).

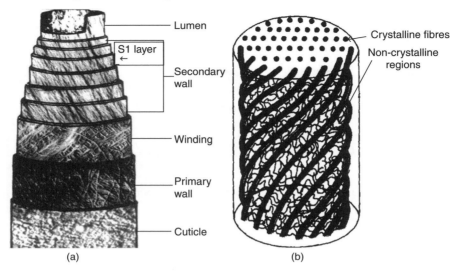

Figure 12.6 (a) Schematic of the various levels of morphology to be found in a cotton linters fibre. (b) Model of fibre structure based on rigid fibres in an amorphous matrix.[4]

- The criss-cross dimensions in the well-ordered fibrillar elements.
- The accessible surface of the fibrils or their aggregations in relation to the fibrillar dimensions.
- The accessible regions interlinking the crystallites in the elementary fibrils.
- The presence of interfibrillar interstices, voids and capillaries and their relationship to the degree of orientation of the fibrillar elements with respect to the fibre axis.

In the case of cotton and other cellulose-based systems, the structure is influenced by the outer layer around the fibre (cuticle) that is composed mainly of hemicellulose. The role of these hemicellulose polymers is critical in the determination of the macroscopic physical properties of the fibre. The winding wall plays an important role in retaining the rigidity of the structure and the layered criss-cross structure will immediately be recognized by engineers as an important element of composite design, allowing extensibility, flexibility and load-bearing properties to be achieved in the same structure. The lumen is a hollow section that runs the length of the fibre and imparts high impact properties as well as allowing the rapid loss of moisture which occurs when the linters emerges from the cotton bud.

The detailed texture of the fibre has been a point of some recent discussion.[14] The cell wall textures are known as axial, transverse, crossed helicoidal and random (Figure 12.7).

Other types of texture are derived from these basic types by the successive deposition of different textures as shown in Figure 12.6. In elongating plants,

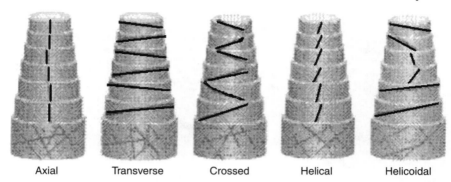

Figure 12.7 Schematic of the typical fibre orientations found in cell walls. Bold lines indicate the fibre alignment.

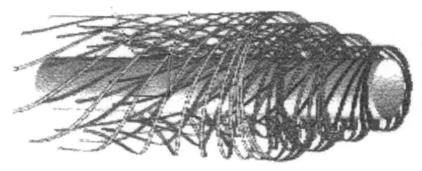

Figure 12.8 Visualization of the geometry development for long fibres.[14]

the cortical microtubules as well as the cellulose microfibrils that are being deposited are both transverse to the elongation direction of the cells. Scientists faced with helicoidal cell walls in which microtubes and the cellulose microfibrils did not match realized that the helicoidal walls resemble cholesteric liquid crystals that are known to self-assemble spontaneously. However, this type of equilibrium assembly can only take place when there is sufficient bulk material present. It has been proposed[14] that since cellulose microfibrils are long slender structures the angle that the fibres adopt relative to the axis is dictated by the number of tapes (N), their width (w) and the cell diameter (D). The angle is defined by

$$\sin \alpha = \frac{Nw}{\pi D} \tag{12.1}$$

The number of tapes is dictated by the number of active synthetic sites in the membrane. In an *E. hyemale* root hair membrane this distance between the tapes is ~ 150 nm. This model allows the generation of a number of different textures observed in plants. A visualization of the fibre forming around the central core is shown in Figure 12.8.

12.4 Natural Silks[14]

Natural silk is the archetypal supramolecular assembly of polymer fibres.[14] Silk is produced by *Bombyx mori* silk worms and spiders and is generated from protein molecules that are able to aggregate into a strong fibre. Spiders produce several different types of silk, covering a wide range of properties and applications. The golden orb weaver (*Nephilu cluvipes*) secretes from the major ampullate glands a dragline, which forms the radial threads of the web as well as providing a 'safety rope' that the spider can rely on if it falls. The fibres have a high tensile strength and modulus and strain to failure. In contrast, the capture thread (viscid silk) produced by the flagelliform glands is highly compliant and able to absorb the kinetic energy of captured flying insects. Other silks spun by the spider are used for swathing prey and are tough and resistant to chemical or photochemical degradation.

12.4.1 Silk Fibre Chemistry

Each form of silk is slightly different. In the case of *Nephilu clavipes*, DNA analysis suggests the existence of two similar proteins of molar mass 320 kDa in the silk produced by the spider. Spidroin 1 is characterized by short runs (5–7 residues) of polyalanine separated by approximately five repeat units of a Gly-Gly-X motif, where X is principally one of the large side-chain amino acids Gln, Tyr and Leu. Spidroin 2 has a slightly longer run of polyalanine (6–10 residues) and also contains several proline-containing pentapeptides that include Gly-Tyr-Gly-Pro-Gly, Gly-Pro-Gly-Gly-Tyr and Gly-Pro-Gly-Gln-Gln. The proline content of spidroin 2 precludes its forming regular, crystallizable conformations. Spidroins 1 and 2, respectively, are 719 and 628 residues long, each including a 'tail' that is slightly more hydrophilic. These C-terminal regions contain approximately 60 residues and no repeating motif is apparent. The complex and imperfectly repeating monomer sequence has several significant consequences for fibre processing, microstructural assembly and product properties.

12.4.2 Silk Fibre Processing

The DNA of silk fibre is synthesized in aqueous solution in the spinneret and at this stage is amorphous. It is forced out through an orifice in a process that is rather akin to polymer extrusion into air. The spider using its legs to draw silk through the spinnerets may augment this spinning process. Within the spinnerets, the 30 wt% aqueous solution of the protein is a high-viscosity fluid. The spinning process involves formation of a lower viscosity, shear-sensitive liquid crystalline phase as protein concentration is increased by evaporation of water.

Figure 12.9 Model of liquid crystalline domain in silk secretion.[14]

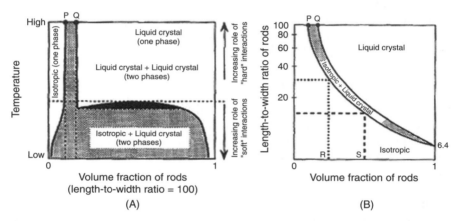

Figure 12.10 Liquid crystal phase formation by rods (length : diameter = 108 : 1): prediction of a simple lattice model.[14]

The lowering of the viscosity is associated with the formation of supramolecular rod-like structures created by aggregation of the solubilized protein chains. Figure 12.9 shows a simple model of isotropic fibroin coils, dissolved in water and arranged into orientationally ordered supramolecular rods.

This simple model accommodates the existence of rod-like structures needed for liquid crystal formation, without recourse to anisotropy of individual molecules. The secretions are biphasic (mixed isotropic/liquid crystalline) over a narrow concentration range implying that they have the thermodynamic characteristics typical of an athermal liquid crystalline solution (Figure 12.10).

The molecules must have a rigid rod structure that can align and have a rod axial ratio with an upper limit of around 30, since the biphasic regime starts at a concentration greater than approximately 30 wt%. The concentration of protein in the extruded fibre is ~24%, typical of natural silk secretion prior to drying, and the rod axial ratio has a lower limit greater than the minimum value needed to sustain liquid crystallinity, ~6.4. The theoretical prediction of the phase diagram for an assembly of rods (Figure 12.10a) shows the athermal behaviour when the solvent–rod interactions cause the biphasic region of the phase diagram to form a 'chimney' that spans only a narrow range of concentrations. The concentration range covered by the 'chimney' depends on the length-to-diameter ratio of the rods (Figure 12.10b). The concentration at R is 24 vol.%, corresponding to the value typical of silk secretion stored in the gland, suggesting a maximum rod length-to-diameter ratio of approximately 30. The concentration at S is a 50 vol.% solution and is associated with the maximum predicted from birefringence and occurs with a rod length-to-diameter ratio of ~15, in agreement with experiment.

12.4.3 Silk Fibre Microstructures

The strong, stiff, tough fibre microstructure is created by processing the liquid crystalline fluids that are denatured as a consequence of the shear process. The fibre microstructure is composed of ordered regions of varying size and degree of internal perfection. The packing of these entities can be described in terms interaction between sequences that consist of crystallizable polyalanine and are disrupted by the GGX motifs. The effect of shear is to create a metastable, β-strand conformation. The resultant structure is composed of ordered regions ~20 Å across of pure polyalanine that are distributed throughout the matrix, and characterized by a uniform intersheet spacing. At the other end of the size spectrum, there are imperfectly ordered regions of mixed (polyalanine and GGX) composition that are a few hundred nanometres across; internally, their composition varies on the approximate scale of the dominant motif lengths, a few nanometres, which in turn leads to variations in the local intersheet spacing and the creation of regions of disorder and order defined by the number of nearest-neighbour matches. A liquid crystalline globular phase provides a processable intermediate between water-soluble polymer and insoluble fibre. The reinforcement and matrix are both elements in a complex hierarchical microstructure that extends over length scales ranging from molecular to macroscopic. Collectively, this is responsible for the unique combination of mechanical properties exhibited by spider silk fibres.

The above discussion of silk fibres produced by spiders due to Viney,[15] indicates how liquid crystalline order can be a precursor to the formation of the more ordered crystalline structures that become load bearing. We can see in nature combinations of the processes that are sometimes seen to occur individually in synthetic polymer systems.

A general discussion of the influence of chain conformation in bipolymers on their ability to create specific structures has been presented elsewhere[16,17] and indicates how local interactions lead to the creation of rod-like, helical and folded structures which are of crucial importance in nature achieving the structural feature that it requires for particular applications.

In summary, nature manages to combine the use of chemical complexity with kinetics and thermodynamics to produce materials that are fit for specific purposes. Many of the principles that control the formation of structure and morphology in synthetic materials are demonstrated in a more diverse sense in naturally occurring materials. However, the same physical principles operate and are controlling the creation of these various structures. It is impossible in this short monograph to do justice to the wide variety of structures found in nature but hopefully the reader will recognize behind this plethora of morphologies some of the principal building blocks discussed in this book. Good hunting.

References

1. M. Er Rafik, F. Briki, M. Burghammer and J. Doucet, *J. Struct. Biol.*, 2006, **154**, 79.
2. S.-Y. Ding and M.E. Himmel, *J. Agric. Food Chem.*, 2006, **54**, 597.
3. C. Ververis, K. Georghiou, N. Christodoulakis, P. Santas and R. Santas, *Ind. Crops Products*, 2004, **19**, 245.
4. A.K. Bledzki and J. Gassan, *Prog. Polym. Sci.*, 1999, **24**, 221.
5. J. Einfeldt, D. Meissner and A. Kwasniewski, *Prog. Polym Sci.*, 2001, **26**, 1419.
6. P. Zugenmeir, *Prog. Polym. Sci.*, 2001, **26**, 1341.
7. G.M. Brown and H.A. Ley, *Science*, 1965, **147**, 1038.
8. S.E. Doyle, A.M. Gibbons and R.A. Pethrick, in *Wood and Cellulose*, ed. J.F. Kennedy, G.O. Phillips, D.J. Wedlock and P.A. Williams, Ellis Harwood, Colchester, UK, 1987, ch. 7, p. 71.
9. C. Viney and A.H. Windle, in *Wood and Cellulose*, ed. J.F. Kennedy, G.O. Phillips, D.J. Wedlock and P.A. Williams, Ellis Harwood, Colchester, UK, 1987, ch. 8, p. 77.
10. K.C. Ellis and J.O. Warwicker, *J. Polym. Sci.*, 1962, **56**, 339.
11. J. Blackwell and R.H. Marchessault, in *Cellulose and Cellulose Derivatives*, ed. N.M. Bikales and L. Segal, Wiley Interscience, New York, 1971, part IV, p. 151.
12. R.A. Pethrick, S. Doyle, R.K. Harris, K.J. Packer and F. Heatley, *Polymer*, 1986, **27**, 19.
13. D.J. Crofton, S. Doyle and R.A. Pethrick, in *Wood and Cellulose*, ed. J.F. Kennedy, G.O. Phillips, D.J. Wedlock and P.A. Williams, Ellis Harwood, Colchester, UK, 1987, p. 237.
14. A.M. Emons and B.M. Mulder, *Trends Polym. Sci.*, 2000, **5**, 35.
15. C. Viney, *Supermolec. Sci.*, 1997, **4**, 75.
16. A. Tonelli and M. Srinivasarao, *Polymers from the Inside Out*, Wiley Interscience, Chichester, UK, 2001.
17. H. Zhu, J. Ji and J. Shen, *Biomacromolecules*, 2004, **5**, 1933.

Subject Index

ABA triblock copolymer 303
absorbed polymer layers 299–300
AB-type block copolymer 227
acetal, crossed polars 125
acoustic velocity 104
activation enthalpies 103, 193
adamantane 100
adsorbed layers 298–9
AFM *see* atomic force microscopy
ageing 181, 198–200
air–liquid interface 235, 284–5, 287–9
air–solid interface 235
alignment *see* molecular alignment
alkanes 8–10, 103–5
n-alkanoic acids 55, 57
alkylbiphenyl esters 58–9
alkyl chain branching 68
alpha-helix 308–10
alpha relaxation process 188, 192, 193, 194
amine-cured epoxy resins 220–2, 223
amorphous structure 12, 107, 147, 149, 179–205, 272–3
AMU *see* atomic mass units
angular momentum conservation law 76, 77
angular resolution 262
anisotropic axis 53
anisotropic electronic polarizability 71
annealing phenomena 123–4
antiplasticization 195
applications 82, 230–1

Argand diagrams 182–3
Arrhenius behaviour 27, 187, 188, 203
assessment methods 238–40
atactic form 11–12, 13–14, 109
atomic force microscopy (AFM) 44, 244–7, 268, 279, 282, 313
atomic mass units (AMU) 265, 266, 268
atomic polarization 184
atomic ratios 263
ATR (attenuated total reflection) 247
attachment 29, 32, 34–5, 142, 143
    Lauritzen–Hoffman theory 160–2
    Sadler–Gilmer theory 170
attenuated total reflection (ATR) 247
attenuation length 260
Auger process 254, 255
average energy 3
Avrami equation 156–9, 176
Avrami exponent (n) 158, 159
axial cellulose fibre structure 315–16
axialite 120

backbone structure 82–3, 89–91
B/A ratio 114
BCF *see* screw dislocations
bend 80
benzaldehyde 42
benzoate esters 69
benzoic acid 42–3
benzophenone-3,3′,4,4′-tetracarboxylic anhydride 88
benzophenone 16, 48

beta process 188
BFDH (Bravais–Friedel–Donnay–Harker approach) 32–3
bicycloctane 100
bilayers 290–1
binary polymer blends 215–16, 217–19, 276–7
binding energy (eV) 254, 257, 258
biological systems 290–1, 303, 308–20
birefringence 130–1
bischloroformate-derived polyurethanes 86
blends 142–3, 173, 195, 207–31, 273, 276–7
blind attachment 170
Bloch waves 47
block copolymers 14, 217, 224–9, 263, 302–4
blood-handling materials 263–4
blooming phenomenon 276
Boltzmann distribution 7, 27
Bombyx mori (silk worm) 317
bonds 8, 9, 33–4, 184
Borrmann effect 47
Bragg diffraction condition 46, 47
branched chain structures 224
Bravais–Friedel–Donnay–Harker (BFDH) approach 32–3
Bravais structures 18, 19
brominated polystyrene thin film 282
8-bromo-2,2′-dimethyl-6-nitro-1′-phenyl-(2H-[1]benzospiropyran-2,2′-indoline) 201
n-butane 8, 9
butyl methacrylates 197

Cahn–Hilliard diffusion equation 218–19
calcite 44
calorimetric methods 136–7
camphene 100
carbon-13 NMR 193
carbon tetrabromide 99, 100, 101
carboxy-terminated butadiene acrylonitrile (CTBN) 221–2

CED (cohesive energy density) 194
cellulose fibres 310–17
cellulose I/II structure 312, 313
central rigid core length 61
chains
 axis 108
 end groups 272–3, 274, 275–6, 278
 end location 150
 entanglement 141, 148
 folding 112–13, 115–16, 142, 145–9
 mixing 104
 packing 113
 rigidity 55–7
 sequence structure effects 58
characterization, surfaces 238
charged colloids 293–4, 296–8
chemical shifts 255, 256
chemical structure effects 196–8
chiral centres 67
chiral smectic materials 55
2-chloro-2-nitropropane 101, 102
cholesteric liquid crystals 55, 56, 67
classifications 12–14, 18, 82–3
Clausius–Mossotti equation 184
clearing point 74
cloud point 303
CMC (critical micelle concentration) 292, 301
CMT (critical micelle temperature) 292
cohesive energy density (CED) 194
coils 272
collision polarization 185
colloids 284–305
colour 82
comonomers 148, 195–6
compatibilizers 208
compatible blends 207
compressible lattice theory 275
compression theory 299–300
computer prediction 40, 41, 78–9
conformational states 8–10, 11–13
conservation laws 75–7
contact angle 234, 235, 237, 238–40
contamination 17, 42–3, 271–2
cooling rate, glassy solids 180

Subject Index

copolymers 154, 207, 217
cotton fibres 312, 314–15
Coulombic interactions 3–4, 280–1, 293–5
creep behaviour 199, 200
criss-cross structure 315
critical condition 215–16
critical micelle concentration (CMC) 292, 301
critical micelle temperature (CMT) 292
critical molar mass (M_{crit}) 172
critical packing factor 300–2
critical phase separation temperature 208, 213
critical surface tension 239, 240
crossed cellulose fibre structure 314, 315–16
crossed polars 125
cross polarized microscopy 53, 79–80
crystal growth 141–76
  see also nucleation
  attachment energy 34–5
  attachment methods 32
  Bravais–Friedel–Donnay–Harker approach 32–3
  chain end location 150
  chain folding 112, 113, 115, 116, 142, 145–9
  crystal habit 24
  crystallization kinetics 150–3
  dislocations 37–40
  equilibrium melting temperature 153–6
  general Avrami equation 156–9
  heterogeneous 30–1
  homogeneous 25–30
  Ising model surface roughening 35–6
  lamellae stacks 148–50
  macrosteps 42–3
  metastable phases 171–2
  microstructural examination 46–9
  Miller's indices 20
  minimum energy conditions 143–6
  molecular fractionation 172–5
  nucleation 142–3

orientation-induced crystallization 175–6
periodic bond chains 33–4
pharmaceutical chemicals 17
relative rates 40
small molecular systems 16–49
stages 29–30
theories 159–71
thermodynamics in melt 141–2
crystalline/amorphous ratio 313
crystalline polymers 11–12, 107–39, 271–2
crystalline solid formation criteria 13–14
crystallization 21–3, 150–3, 154
  see also crystal growth; nucleation
crystallizing units see stems
crystallography 108–10, 114, 115
crystal morphology see morphology
CTBN see carboxy-terminated butadiene acrylonitrile
cubic crystals 18, 19
curved surfaces 288–9, 300–2
4-cyano-4'-*n*-pentylbiphenyl 60
4-cyano-4''-*n*-pentyl-*p*-terphenyl 60
cyano biphenyl molecules 58
cyclohexadiene rings 63
cyclohexane rings 63
cysteine 309
cystine 309, 310

dark brushes 79–80
data analysis 43–4
DDS see 4,4'-diaminodiphenylsulfone
deconvolution 255
defects in crystal growth 37, 79–81, 111, 169
de Gennes reptation theory 161
degree of crystallinity 135–8
degrees of freedom 99
densities, polymers 111
density methods 136–7
depth profiling 262–4
Derjaguin–Landau–Verivey–Overbeek (DLVO) theory 296–8

deuterated polystyrene 279
DGEBA *see* diglycidyl ether of bisphenol A
4,4'-diaminodiphenylsulfone (DDS) 223
diazomolecules 57–8
diblock copolymers 226
DIC *see* differential interference contrast
2,2-dichloropropane 102
dicyclohexylphthallate 196
dielectric permittivity 185–6
dielectric relaxation spectroscopy (DRS) 183–9
diesel fuel 43
differential interference contrast (DIC) 132
differential scanning calorimetry (DSC)
  alkane mixtures 104
  carbon tetrabromide 101
  crystal growth kinetics 159
  degree of crystallinity determination 136
  glass–rubber transition 181
  polyethylene crystals thickening process 124
diffuse double layer 295
diffusion equation 218
diglycidyl ether of bisphenol A (DGEBA) 220, 223, 224
diluents 195
dimers 3, 21, 90
dipolar effects 71, 101–2, 183–9
director (n) 53, 72, 79–81
disinclinations 53, 79–81
dislocations 37–40, 79–81
disordered *see* amorphous
distribution function $Z(p)$ 201–2
DLVO (Derjaguin–Landau–Verivey–Overbeek) theory 296–8
DNA analysis 317
dodecane 5–8, 141
Doi–Edwards model 176
domain interface 228–9
Donnan equilibrium 294

droplet example 234–8
DRS *see* dielectric relaxation spectroscopy
DSC *see* differential scanning calorimetry
dyes 200–2, 226
dynamic mechanical thermal analysis 181–3
dynamic modulus measurement 181, 182

EHD *see* electrohydrodynamic...
elastic behaviour 75–7
elastic electron scattering 244
electrical colloid effects 294
electrical double layer 294–5
electrohydrodynamic (EHD) instabilities 280–2
electronic polarization 184
electron mean free path 258–62
electron microscopy (EM)
  liquid crystal systems 81
  molecular surfaces 242–4, 245
  phase separation 221, 222, 223
  polymer spherulites 119
  SBS copolymers 225, 226
  technique 128–9, 133
  X-ray emissions 244
electron recoil 252–3
electron scattering 243–4
electron spectroscopy for chemical analysis *see* X-ray photoelectron spectroscopy
electrostatic interactions *see* Coulombic interactions
elongation 78–9, 181
empirical nucleation description 27–9
emulsifiers 208
enantiomeric mesophase 83
end functionalized polymers 278
end groups 272–3, 274, 275–6
energy conservation laws 76, 77
enrichment 278–80
entanglement 141, 148
enthalpy 110, 111, 211–12, 237

Subject Index 325

entropic model/theory 172, 299–300
entropic stabilization 298–9
entropy 172, 211–12, 237, 298–300
environmental scanning electron microscopy (ESEM) 242, 245
epoxy resins 220–4
equation of state theories 209–10
equilibrium melting temperature 153–6
equilibrium shape 143–4
escape depths 260–2
ESEM (environmental scanning electron microscopy) 242, 245
esters 58–9, 188
etching samples 133–4
eV *see* binding energy
excluded volume effect 141
experimental techniques 124–33, 159–70

face-centered cubic (fcc) arrangement 234
factor α 35
fcc *see* face-centered cubic...
ferroelectric liquid crystals 91
films *see* thin films
flat surfaces 32
flexible chains 57, 90
flexible spacer nature/length 89
flocculation 298, 299
Flory–Huggins equation 277
Flory–Huggins interaction parameter 211, 228
fluorescence 247
fluorine 255
fluorine end caps 272–3
fluorosilane-terminated polystyrene (PS-F) 278
folded structures 112, 113, 115, 116, 142, 145–9
force field 71
form birefringence 131
forward recoil spectroscopy 252–3
Fourier transform infrared (FTIR) imaging 268–9
fractal structures 20

fractal surfaces 121, 122
fractionation process 149, 172–5, 271
fragility 202–4
Frank impurity mechanism 45
freedom *see* degrees of freedom
free energy barrier (ΔG^*) 152
free energy (ΔG) 151
  see also Gibbs free energy
free energy difference (ΔG) 25–6
free volume 190–2, 194, 195, 198, 199–202
freezing points, paraffin series 8
Fresnel reflectivity/reflectance 250–1
FTIR *see* Fourier transform infrared
functional groups 273–4

G *see* Gibbs free energy
gauche state
  n-alkanes 8–10
  dodecane molecule 5, 6, 7
  polyethylene 108
  polymer crystal growth 141–2
  polymer solidification 107
Gay–Berne (GB) potential 78
general Avrami equation 156–9, 176
geometry of crystal growth 158
GGX motifs 317, 319
Gibbs absorption isotherm 287
Gibbs equation 286
Gibbs free energy (G)
  binary polymer blend surfaces 276–7
  end group segregation to surface 276
  homogeneous crystal growth 25–6
  homopolymer surface tension 274–5
  Lauritzen–Hoffman theory 161
  phase separation 208–9, 213
  surfaces 233–8
Gibbs interface 284–5
Gibbs–Thomson relationship 27
glass–rubber transition (T_g) 179–205
  chemical structure effects 196–8
  comonomer incorporations 195–6
  dielectric relaxation spectroscopy 183–9
  fragility 202–4

Kauzmann paradox 198
molar mass effects 194–5
phenomenology 180–90
physical ageing 181, 198–200
physical characteristics 194–8
plasticization effect 195
pressure dependence 198
process 192–4
SBS block copolymer 226
temperatures 197
theories 204–5
globular molecules 99, 100, 101, 102
glucose, quenching behaviour 198–9
Gouy–Chapman regime 295, 297
growth of crystals 141–76
  see also habit; lamellae; nucleation
  chain end location 150
  chain folding 112, 113, 115, 116, 142, 145–9
  equilibrium melting temperature 153–6
  general Avrami equation 156–9
  kinetics 150–3
  Lauritzen–Hoffman theory 160–9
  minimum energy conditions 143–6
  rate equations 169
  regimes 164–9
  Sadler–Gilmer theory 169–70
  thermodynamics in the melt 141–2

habit 17, 23–5, 113–16, 309
hair 308–10
Hamaker constant 237, 280, 296
handedness, screw dislocations 38
Hartman–Perdok model 40, 41, 45
heats of fusion 110, 111
helical structures 109, 110, 111, 308–11, 315–16
Helmholtz energy 285
hemicellulose 313, 314, 315
heterogeneous crystal growth 30–1
hexagonal crystals 18, 19
hexamethylethane 99, 100, 236
higher energy states 7, 107
higher order structures 308–20

high-impact polystyrene (HIPS) 219–21, 222
high molar mass samples 120
HIPS (high-impact polystyrene) 219–21, 222
history 2, 52
hkl values 20–1
homogenous crystal growth 25–30
homologous series 7–8, 58–9
homopolymers 273–5
H-shaped dimers 90
hydrodynamic flow 75
hydrogen bonding 312

ideal gas equation 287
ideal non-mixing liquids 285–7
imbalance of forces, surfaces 233–4
immiscible blends 208
immiscible liquids 285–7
impact properties 219–20
impurities 17, 42–3, 271–2
indolinobenzospirans 200
inelastic electron scattering 243–4, 260
inelastic mean free path 260
instability condition 209
interaction potentials 237
interfaces
  see also surfaces
  block copolymer phases 228–9
  Gibbs free energy 233–8
  liquid droplet example 234–8
  molecular surfaces 233–69
  types 233
interference microscopy 133
intermediate hair filaments 309
internal rotation 196–8
inverse square law 237
investigation techniques 107–39
ion beam analysis 252–3
ionic solids 2–5
ions 293–5
Ising model 35–6
isotactic structure 11–12, 13–14, 109, 110
isothermal volume contraction 181
isotropic–mesophase temperature 84

Subject Index

J_n (mass nucleation rate) 27–8

Kauzmann paradox 198, 204
KDP *see* potassium dihydrogen phosphate
kebab structures 121, 122
keratin, hair 308–10
kinetics 21–3, 150–3, 218–19
kinked faces 32
Krafft point 292

lamellae 115–17
   chain ends location 150
   distortions 148, 149
   melt-crystallized 117–18
   polymer crystal growth 143, 144
   spherulite formation 118–23
   stacks 148–50
   supercooling effect on thickness 145
   surface area 146
Lang diffraction technique 46, 48
Langmuir–Blodgett films 290–1
Langmuir trough 289–90
lattices 3, 17, 18, 19, 217
Laue X-ray diffraction 46, 48
Lauritzen–Hoffman (LH) theory 160–9
layer polymers 88
LCST (lower critical solution temperature) 213
Legrande polynomials 72, 73
length scale 128
Lennard–Jones potential 78–9
Leslie–Erickson theory of hydrodynamic flow 75–7
LH (Lauritzen–Hoffman theory) 160–9
light microscopy techniques 53, 79–80, 127–33, 241–2, 267
lignins 314
like sign dislocations 38–9
linear cascade regime 264
linear dimers 90, 91
linear expansion coefficient 180–1
linear momentum conservation law 75–6, 77
linear polyethylene 150, 169, 173

linking groups 62–4
liquid–air interface 235, 284–5, 287–9
liquid crystal phase 52–92
   applications 82
   common molecular features 59–66
   computer simulations 78–9
   definition 52
   molecular structure effects 55–9
   silk 318
   structure visualization 92
liquid droplet example 234–8
liquids
   micelle formation 291–3
   molecular organization 284–305
   phase structures 302–5
   surface tension values 235
liquid–solid interface 235, 236
loss factor 183
lower critical solution temperature (LCST) 213
lowest energy structures 2–3, 5–7, 11
low molar mass material 8–10, 149, 155
lyophilic colloids 293, 294
lyophobic systems 293, 294
lyotropic molecules 53, 54, 284

macrosteps 42–3
macrostructures, biological 308–20
Madelung constant 4
Maier–Saupe theory 73–4
main chain crystalline polymers 84–8
mass conservation law 75
mass nucleation rate (J_n) 27–8
M_{crit} *see* critical molar mass
MCS (modulation contrast system) 132
MDI (methylenediphenyldiisocyanate) 229
mean field theory 5, 71, 72, 73
mechanical model development 72
melting points 84, 111, 145, 146
melting process 99–100, 143
melts 117–23, 141–2
membranes 290–1, 303
mesogenic phase 59–60

mesogens, definition 53
mesomorphic *see* mesophase
mesophase 53, 78, 83–4, 85, 86–7
metallocene polymers 148, 173–5
metastable limit (ΔC_{max}) 28
metastable phase 22, 23, 171–2
methane 103
methylenediphenyldiisocyanate (MDI) 229
4-methylpentenene-1 111
α-methylstyrene-alkane copolymer system 193
Metropolis technique 78
micelles 291–3, 295–6, 300–2
microfibrils 122, 309–10, 311, 312
microscopy
  see also electron microscopy; *individual microscopy techniques*
  optical 53, 79–80, 125–34, 241–2, 267
microstructures 46–9, 319–20
microtomes 126–7
Miller's indices 20
minimum energy conditions 143–6, 287–9
miscibility 207, 209–12, 215–17
  see also phase separation
modelling 204–5, 209–10, 264–5
modulation contrast system (MCS) 132
modulus 181–3
molar mass effects 83, 194–5, 215–17
molar mass segregation/fractionation 149, 172–5, 271
molecular organization
  cellulose fibres 311–12
  higher order structures 308–20
  liquid crystal computer simulations 78–9
  liquid crystal mechanical model development 71–5
  liquids 284–305
  nematic liquid crystals 53, 80
  rotation 8–11, 102–3, 196–8
  smectic liquid crystals 54–5
molecular solids 1–14
molecular surfaces *see* surfaces

monoclinic crystals 18, 19
monolayers 123, 164, 290–1
monomers 91, 197
Monte Carlo approach 78, 151, 170
morphology
  annealing phenomena 123–4
  block copolymers 225
  cellulose fibre 312, 314–15
  chemical structural polymer type classification 13
  crystal lamellae 115–17
  crystalline polymers 107–39
  crystal structures 23–5, 40, 41
  experimental techniques 124–33
  hair 309–10
  melt grown crystals 117–23
  molecular surfaces 233
  polymer spherulites 118–23
  prediction 24–5
  single polymer crystals 112–15
  snow crystals 20
  solid polymers 1
motional freedom *see* degrees of freedom
moving element (T_g process) 192–4
multilayer reflectivity 251–2
multiple nucleation 165–6, 167

n see Avrami exponent; director
N *see* nematic liquid crystal phase
naphthalene 41
2,6-naphthalene-based polymers 88
natural silks 317–20
nature *see* biological systems
nearly miscible blends 213–14
nematic liquid crystal phase (N)
  characteristics 6, 7, 53
  distributions/order 72–5
  elastic behaviour modelling 75–7
  linking group influence on thermal stability 62–4
  molecular arrangements 80
  molecular features 60–2
nematic to isotropic transitions *see* transitions

Subject Index

Nephilu cluvipes 317
neutron diffraction analysis 247–52, 267
n-nonadecane 104
non-planar geometries 116–17
non-polar end groups 274
non-wetting droplets 235–6
norbornylene 99, 100
n-pentane 10
nuclear magnetic resonance (NMR) 193
nucleation 31–2, 36–41
  Avrami equation 156–8
  crystal growth rates 31–2
  crystallization 22, 25
  empirical description 27–9
  Lauritzen–Hoffman regime II/III growth 165–7
  polymer crystal growth 142–3, 152
  site sources 36–41
  spherulite formation 119–20

octadecane 103
opposing sign dislocations 39
optical display applications 82
optical microscopy 53, 79–80, 125–34, 241–2, 267
order parameters 71, 72–5
orientation birefringence 130
orientation-induced crystallization 175–6
orthorhombic structure 18, 19, 108
ortho spin positronium (o-Ps) 189–91
oscillation 181, 182, 184
osmium tetroxide staining 221–2, 226
Ostwald ripening 22
overlayer method 260

packing 113, 300–2
paired ions 2–3
palmitic acid 284
PALS (positron annihilation lifetime spectroscopy) 189–90
paracetamol 17
paraffin homologous series 7–8

parallel plate model 296
para spin positronium (p-Ps) 189
partially miscible blends 207
PBC (periodic bond chains) 33–4
PDMS (polydimethylsiloxane) 13
pendant groups 65–6, 68–9
Pendellösung effect 47, 48
periodic bond chains (PBC) 33–4
permittivity 101, 102, 185–6
persistence length 169
PES *see* poly(ether) sulfone; potential energy surface
pharmaceutical chemicals 17
phase contrast microscopy 129
phases
  behaviour 300–2, 303–5
  diagrams 213–14, 215–17, 227, 302–4
  inversion 220
  liquid polymer systems 302–5
  segregation at surfaces 278–80
  separation 142–3, 172–5, 207–31
  smectic liquid crystals 54–5, 56
m-phenoxy ether 190, 191
phonon spectra 135
physical properties 2–11, 110, 111, 181, 194–200
pick off annihilation 189
pinning attachment 170
pivalic acid 99, 100
planes of crystal growth 20–1
plane surfaces 288–9
plant cell walls 310–17
plastic crystals 99–105
plasticization effect 195
platelet formation 43
Pluronics 304–5
PMMA *see* poly(methyl methacrylate)
point group 19
Poiselle flow 281
poisoned growth 44
polarization colours 131
polarized light microscopy 53, 79–80, 118, 119, 120, 129–30
trans-poly(1,4-butadiene) 172

poly(4-methylpent-1-ene) 117
polyalanine 319
1,4-β-D-linked polyanhydroglucopyranose chains 310, 311
polydimethylsiloxane (PDMS) 13
poly(ester anhydrides) 87
poly(esterimide)s 87
poly(ether sulfone) (PES) 223
poly(ether-urethane) block copolymers, XPS 263
polyethylene
  chemical shift 256
  crystal growth 112–13, 114, 148, 154, 169
  crystallography 108
  growth rate equations 169
  low MW highly linear metallocene-synthesized 123
  melt-crystallized lamellae 117–18
  minimum energy conditions 144–5
  non-planar geometries 117
  nucleation in melt 143
  Sadler–Gilmer theory 169
  shape vs. growth conditions 114
  SIMS 265, 266
  single crystal growth 112–13
  spherulites 119
  structure 2
  thickening process 123–4
poly(ethylene oxide) 110, 114
polyimides 85–6
polyisobutylene 265, 266
polymer dispersed liquid crystals 82
polymeric liquid crystalline materials 82–91
polymerization, history 2
polymer network stabilized liquid crystal phase 91
polymer–polymer miscibility 209–11
polymer spherulites 118–23, 125–33, 308
poly(methylene) 10–11
poly(methyl methacrylate) (PMMA) 179, 188–9, 196, 278–9

polymorphism 108, 109–10, 142
polyoxymethylene (POM) 110, 114, 115, 117, 156
polypropylene 11–12, 109–10, 265, 266
polysaccharides 310–17
polystyrene 219–20
  chemical shift 256
  fluorine end caps 272–3
  SIMS spectrum 266
  spherulites 119
  structure 179
  XPS shake up satellites 255, 258, 259
polystyrene-*block*-polybutadiene-*block*-polystyrene (SBS) block copolymer 224–9
polystyrene-*block*-polyisoprene-*block*-polystyrene 224–5
polystyrene–poly(methyl methacrylate) (PS-b-PMMA) 230
polysubstituted aromatics 102, 103
polytetrafluoroethylene (PTFE) 156, 184, 256
polythiophene 259
poly(trimethyl terephthallate) 154
polyurethanes 86, 229–31, 263
poly(vinyl acetate) 199
poly(vinyl alcohol) (PVA) 255
poly(vinyl bromide) 256
poly(vinyl chloride) (PVC) 200, 256, 262
polyvinylidene fluoride 258
positron annihilation lifetime spectroscopy (PALS) 189–90
potassium dihydrogen phosphate (KDP) 45
potential energy 102
potential energy surface (PES) 204
p-Ps *see* para spin positronium
prediction, crystal morphology 24–5
pressure 199, 217
primary nucleation 22, 152
primitive lattice cells 18
propane 103
propylene–ethylene copolymer 121
proteins 294

Subject Index 331

PS-b-PMMA (polystyrene–poly(methyl methacrylate)) 230
pseudo-liquid crystals 314
PS-F *see* fluorosilane-terminated polystyrene
PVC *see* poly(vinyl chloride)
pyramidal shape 112, 113
pyromellitic acid dianhydride 256
pyromellitic dianhydride 88
pyromellitic diimide 257

quenching behaviour 198–9
quills 309

radius of gyration 147
Raman scattering 104–5, 135
random copolymers 14
random lamellar structures 120
real polymer molecules 11–13
reciprocal lattices 21
reflectance 247
reflectivity 247–52
refractive index 125, 248–50
regimes of growth 164–9
regular chain folding 149
relative crystal growth rates 40
relaxation behaviour 184, 193, 202–4
reptation theory of de Gennes 161
restricted chain mobility 141
rigid backbones 89–91
rigid block size 57–8
rigid rod-like polymers 84, 85, 86
RIG (rough interface growth) mechanism 40
ring closure process 201–2
ring structures 57–69
RISM *see* Rotational Isomeric States Model
rod-like molecules 67, 84, 85, 86
rod-like silk structures 318–19
'rooftop' structure 229, 230
rotation 8, 9, 10–11, 102–3, 196–8
Rotational Isomeric States Model (RISM) 11, 141, 147
rotator phase 172

rotor–plastic behaviour 101–5
roughening temperature (T_r) 169
rough interface growth (RIG) mechanisms 40
roughness 35–6, 239–40, 241
Rouse modes 175
rubber elasticity 175, 180, 225
rubber toughened epoxy resins 220–2
Rutherford backscattering 252–3

Sadler–Gilmer theory 169–71
SALS (small-angle light scattering) 126
sample preparation for EM 133–4
sampling depths, XPS 263
SANS (small-angle neutron scattering) 147
saturated solutions 22–3
SAXS (small-angle X-ray scattering) 135
SBS (polystyrene-*block*-polybutadiene-*block*-polystyrene) 224–9
scanning electron microscopy (SEM) 121, 122, 242, 243–4, 267
scanning tunnelling microscopy (STM) 267
scattering 21, 243–4, 252–3
Scherrer shape factor 135
Schiff bases 63, 69
screening effect 294–5, 296
screw dislocations 37–8, 39–40
secondary ion mass spectrometry (SIMS) 264–8
secondary nucleation 22, 152
sectorization 116
segregation 149, 172–5, 271–6, 278–80
self-assembling systems 297, 316
SEM *see* scanning electron microscopy
sensor applications 230–1
seven crystal systems 18
shake up satellites 255, 258
shape of crystal *see* habit
sheet structure 3
shift factor 192
shish-kebab structures 121, 122
side chain liquid crystalline polymers 89, 91

silks 317–20
SIMS (secondary ion mass spectrometry) 264–8
single crystals 5–8, 16, 99, 100, 112–15
single knock off regime 264
size of rigid block 57–8
small-angle light scattering (SALS) 126
small-angle neutron scattering (SANS) 147
small-angle X-ray scattering (SAXS) 135
small molecular systems 16–49
smectic liquid crystals 6, 7, 54–5, 56, 67–71, 110
snow crystals 20
soaps 284, 287, 292
sodium chloride 2–5, 17
sodium stearate 289–90, 292
solid–air interface 235
solidification process 107
solid–liquid interface 235, 236
solid state 11–13
solubility curve, typical 22, 23
solution-grown crystals 112, 115, 271
solutions 22–3, 29–30
solvent etching 133–4
space group 19
specific free energy of melting 143
specimen preparation 127
spectroscopy 183–90, 247, 252–8, 262–4, 267
specular reflection 247, 248
spherical *see* globular
spherulites 118–23, 125–33, 308
spiders 317–19
spidroin 1/2 317
spinning process, spiders 317–18
spinodal condition 214, 215–16
spinodal decomposition 209, 218–19
spirals 38–9
splays 80, 120
spreading of drops on surfaces 239, 240
spreading films 235
sputtering 264
stability of colloids 293–4, 296–8
stains 127, 133–4, 221

static method 181
statistical methods 71, 72
stearic acid 284
steepness index 204
stems 146, 160–2
step growth 32, 37–8, 43–4
steric hindrance 197–8
steric stabilization 298–9
Stern model 296, 297
STM (scanning tunnelling microscopy) 267
strain birefringence 130–1
stress tensors 75–7
structure–property relationships 1–14
   classification of polymers 12–14
   ionic solids 2–5
   liquid crystal phase formation 55–9
   low molar mass hydrocarbons 8–10
   molecular solids 5–8
   physical basis 2–11
   polymers 1–14
   poly(methylene) chains 10–11
   real polymer molecules in solid state 11–13
   sodium chloride solid crystal 2–5
styrene–butadiene–styrene block copolymers 131
styrene–polybutadiene–polystyrene 220
substituent changes 59
succinonitrile 100
supercooling (ΔT) 151–2, 153, 163
super-folding model 148
supersaturated solutions 22–3, 41
surfaces 233–69
   see also interfaces
   amorphous polymers 272–3
   area 289–90, 293
   assessment methods 238–40
   binary polymer blends 276–7
   characterization 238
   crystalline polymers 271–2
   electrohydrodynamic instabilities in polymer films 280–2
   end functionalized polymers 278
   energy 5, 145, 146, 231, 233–8

enrichment 278–80
FTIR imaging 268–9
ion beam analysis 252–3
neutron diffraction analysis 247–52
phase segregation 278–80
polymer blends 273
roughness 35–6, 239–40, 241
spectroscopy 247
theoretical description 273–5
vacuum techniques 253–68
visualization 240–4
X-ray analysis 247–52
surface tension 234–8, 273–5
surfactant–water–oil system 302
'Swiss cheese' structure 123
switchboard model 147
symmetry operations 18–19
syndiotactic polypropylene 11–12

tactic polypropylene 11–12
take-off angle, XPS 263
TDI (toluenediisocyanate) 229
TEM *see* transmission electron microscopy
temperature dependence
  dodecane *gauche/trans* states 7
  liquid crystal colour 82
  liquid crystalline phase modelling 71, 73, 74
  main chain crystalline polymers 84
  polymer crystal growth 151
terminal groups 64–5, 66, 67–8
o-terphenyl 180
tertiary nucleation 152
TETA (triethylenetetra-amine) 220, 222
tetragonal crystals 18, 19
T_g *see* glass–rubber transition
thermal expansion behaviour 180–1
thermal stability 61, 62–4
thermodynamics 21–3, 141–2, 155, 208–12
thermoplastic toughened epoxy resins 222–3
thermotropic liquid crystals 53, 54
thickening process 123–4

thickness 145, 162, 228–9
thin films 280–2, 290–1
Thompson–Gibbs equation 145, 146, 154, 155, 163
tight folds 149
toluenediisocyanate (TDI) 229
torsional angles 8–10
T_r *see* roughening temperature
trans-crystalline polyethylene 108, 123
transitions
  see also glass–rubber transition
  alkyl chain branching effects 68
  linking group effects 64
  nematic liquid crystal molecular features 60–2
  pendant group effects 66
  terminal substituent effects 66
translation 100, 102
transmission electron microscopy (TEM) 112, 221, 222, 242, 243
trans polymer solidification 107
trans state 5, 6, 7, 8–10
transverse cellulose fibre structure 315–16
triblock copolymers 303–5
1,1,1-trichloroethane 102
triclinic crystals 18, 19
triethylenetetra-amine (TETA) 220, 222
trigonal crystals 18
trimer structure 3
T-shaped dimers 90, 91
twist 67, 80
two-dimensional nucleation 36–7

UCST *see* upper critical solution temperature
unit lattice cells 17
universal curve 261
upper critical solution temperature (UCST) 213, 217
urea 229
urethane 229

vacuum techniques 253–68
van der Waals interactions 280–1

VFT *see* Vogel–Fulcher–Tammann
virtual mesophase 83
Vogel–Fulcher–Tammann (VFT) equation 203
volume variation 180–1

water–air interface 284–5
wave front analogy 156–8
weight fraction 225, 227
wetting droplets 235
white beam X-ray topography *see* Laue X-ray diffraction
Williams–Landau–Ferry (WLF) equation 153, 161, 165, 190–2, 200

Wulff plot 40

XPS *see* X-ray photoelectron spectroscopy
X-ray photoelectron spectroscopy (XPS) 253–8, 262–4, 267
X-rays 247–52, 267
 see also crystallography
 diffraction 46–8, 134–5, 150
 emissions 244
 scattering 104, 109, 137–8
xylene solutions 112, 114

Young's equation 234–8